Green Energy and Technology

Climate change, environmental impact and the limited natural resources urge scientific research and novel technical solutions. The monograph series Green Energy and Technology serves as a publishing platform for scientific and technological approaches to “green”—i.e. environmentally friendly and sustainable—technologies. While a focus lies on energy and power supply, it also covers "green" solutions in industrial engineering and engineering design. Green Energy and Technology addresses researchers, advanced students, technical consultants as well as decision makers in industries and politics. Hence, the level of presentation spans from instructional to highly technical.

Indexed in Scopus.

Indexed in Ei Compendex.

Abdul Sattar Jatoi • Nabisab Mujawar Mubarak
Editors

Bio-Electrochemical Conversion of Waste to Fuel

Springer

Editors
Abdul Sattar Jatoi
Department of Chemical Engineering
Dawood University of Engineering and Technology
Karachi, Pakistan

Nabisab Mujawar Mubarak
Department of Chemical Engineering
College of Engineering, Imam Mohammad Ibn Saud Islamic University (IMSIU)
Riyadh, Saudi Arabia

ISSN 1865-3529 ISSN 1865-3537 (electronic)
Green Energy and Technology
ISBN 978-981-95-4309-0 ISBN 978-981-95-4310-6 (eBook)
https://doi.org/10.1007/978-981-95-4310-6

This Springer imprint is published by the registered company Springer Nature Singapore Pte Ltd.
The registered company address is: 152 Beach Road, #21-01/04 Gateway East, Singapore 189721, Singapore

In the name of Allah, the Most Gracious and the Most Merciful. First of all, I would like to express my heartfelt gratitude and appreciation to Allah SWT for the permission, guidance, wisdom, and blessings that I have received over the years, culminating in my reaching this important destination in my life, where I have accomplished my goal. Finally, I would like to present my most heartfelt and warmest appreciation to the great parents and parents-in-law (may Allah SWT bless and reward them).
Dr. Abdul Sattar Jatoi

In the name of Allah, the Most Gracious and the Most Merciful. First of all, I would like to express my heartfelt gratitude and appreciation to Allah SWT for the permission, guidance, wisdom, and blessings that I have received over the years, culminating in my reaching this important destination in my life, where I have accomplished my goal. Finally, I would like to extend my most heartfelt and warmest appreciation to my parents and parents-in-

law (may Allah SWT bless and reward them), as well as my brothers and sisters, who always encouraged and supported me during the completion of this book. Special and heartfelt gratitude to my dearest wife, Muna, and our children, Fayyad, Fawwaz, and Faleeha, for their unwavering encouragement, endless sacrifices, patience, understanding, ideas, and inspiration throughout the process of completing the book smoothly and on time.

Dr. Nabisab Mujawar Mubarak

Preface

Over the past two decades, extensive research has been conducted to enhance the performance of microbial electrochemical systems (MES) in wastewater treatment and product generation, as well as their industrial application scale. In expanding these techniques, several economic issues related to the feasibility of the process have been examined. This economic feasibility must be evaluated in terms of environmental sustainability and efficiency, that is why these techniques are considered in the first place. Recent advancements in nanomaterials and nanocomposites for bioelectrochemical systems open up new research areas for the development of fuel cells. The performance of the cathode and anode materials used in fuel cells is responsible for converting chemical energy into electrical energy through a redox reaction. In recent years, various studies aimed at improving the efficiency and production of fuel cells have recognized the importance of electrode materials. The progress made in microbial fuel cell (MFC) electrode materials is discussed in this book. The primary objective of this application is to investigate the enhancement of energy production by modifying anode materials to facilitate interaction with microorganisms.

The arrangement of the book's contents provides researchers working in or affiliated with environmental, wastewater management, biological, and composite nanomaterials applications with a one-stop solution. It also discusses the functionality of nanomaterial-based electrodes for microbial electrochemical systems. This book offers recent advancements and progress in applying nanomaterials and nanocomposites-based materials for bioelectrochemical systems. Additionally, it offers various properties, pollutant removal efficiency, application as a biosensor, and numerous other features related to the bioelectrochemical system. We are delighted to have received contributions from esteemed authors worldwide, and we are deeply grateful for their support. We thank them all for their interest and quick submission. The enthusiasm and efforts of the authors in the subject of bioelectrochemical systems, particularly in the context of using bioelectrochemical systems, which we believe will be required reading for the growing number of researchers in this field. Each chapter has been thoroughly examined through several phases, including quality and plagiarism checks. Those chapters that passed the initial evaluation were peer-reviewed for professional comments by two independent peer reviewers. Only chapters that got good feedback were forwarded to the writers for editing. We got 24 abstract submissions. However, only 17 were selected for full-paper submission.

We appreciate the writers' honesty in refining their chapters and their constructive responses to comments and suggestions, and we thank them for their patience and understanding. We would like to thank all competent peer reviewers for taking the time to examine the work and provide helpful recommendations and feedback. Their support allowed the book to meet international requirements. Finally, we editors would like to express our heartfelt appreciation to Elsevier's editorial staff (Rafael G. Trombaco), who guided us through every stage of the book's development, beginning with the proposal, creating ELSA, updating as needed, producing proofs, creating a professional cover page, and finally publishing. Without their help, this book would not have reached this point.

Karachi, Pakistan — Abdul Sattar Jatoi
Riyadh, Saudi Arabia — Nabisab Mujawar Mubarak

Bio-Electrochemical Conversion of Waste Into Fuel

In the face of mounting environmental challenges and increasing energy demands, *Bio-Electrochemical Conversion of Waste into Fuel* presents a timely and groundbreaking exploration of innovative, sustainable solutions. This authoritative volume delves into the rapidly evolving field of bio-electrochemical systems (BES), where waste materials are transformed into valuable biofuels through the synergistic integration of microbiology, electrochemistry, and environmental engineering.

Written by leading experts in the field, the book provides a comprehensive overview of microbial fuel cells (MFCs), microbial electrolysis cells (MECs), and microbial electrosynthesis (MES). It examines how genetically engineered microbes and novel electrode materials are pushing the boundaries of waste-to-energy technologies. From wastewater treatment to industrial biomass valorization, each chapter offers in-depth insights into the science, engineering, and practical applications of BES for energy recovery and environmental sustainability.

Whether you are a researcher, engineer, student, or policymaker, this book equips you with the knowledge and tools needed to understand and implement next-generation biotechnologies for a cleaner, greener future.

Key Features:

- In-depth analysis of microbial and electrochemical interactions
- Case studies on real-world applications and pilot-scale systems
- Emerging trends in synthetic biology and electrode innovation
- Future perspectives on circular bioeconomy and climate resilience

Unlock the potential of waste as a renewable resource—discover the transformative power of bio-electrochemical systems.

Acknowledgments

I thank Engr. Prof. Dr. Samreen Hussain, Vice Chancellor of Dawood University of Engineering and Technology, and colleagues in the Department of Chemical Engineering for their support and encouragement.

Dr. Abdual Sattar

I want to express our heartfelt gratitude to his excellency Prof. Ahmed S. Alaameri, President of Imam Mohammad Ibn Saud Islamic University (IMSIU) and the higher management of Imam Mohammad Ibn Saud Islamic University (IMSIU), as well as my colleagues in the Department of Chemical Engineering, for their continuous support and encouragement. My special thanks go to all my co-editors and authors for their valuable contributions.

Dr. Nabisab Mujawar Mubarak

Contents

About the Editors

Abdul Sattar Jatoi is an Assistant Professor in the Department of Chemical Engineering at Dawood University of Engineering and Technology, Karachi, Pakistan. He has over 13 years of experience in academics and research. Dr. Jatoi has expertise in food bioprocesses and their applications in various food industries. His research interests include microbial electrochemical systems for environmental and energy applications. He has published over 140 papers, including 30 book chapters and one book. He has two books. He serves as a reviewer for more than 50 scientific journals and has presented 70 works in various international conference proceedings.

Address: Department of Chemical Engineering, Dawood University of Engineering and Technology, Karachi, Pakistan

Email address: abdul.sattar@duet.edu.pk

Research Interests: Bio-processing industries, microbial electrochemical system, emerging pollutant treatment via different approaches, sustainable nanomaterials development, wastewater treatment, green-water treatment technology, separation and purification techniques, environmental analytical techniques

Nabisab Mujawar Mubarak is an Associate Professor in the Department of Chemical Engineering, College of Engineering, Imam Mohammad Ibn Saud Islamic University (IMSIU), Riyadh, Saudi Arabia. He serves as a scientific reviewer for numerous chemical engineering and nanotechnology journals. In the field of research, Dr. Mubarak has published over 470 journal articles and 40 conference proceedings, authored 90 book chapters, and has an h-index of 91. His areas of interest include the synthesis of carbon nanomaterials, the production of magnetic biochar using microwaves, and wastewater treatment using advanced materials. He is the recipient of the Curtin Malaysia Most Productive Research Award, the Outstanding Faculty of Chemical Engineering Award, the Best Scientific Research Award (London), and the Exceptional Scientist Award in publication and citation by i-Proclaim, Malaysia. Since 2020, he has been listed among the world's top 2% of most influential scientists in the fields of chemistry and energy. The *List of the Top 2% of Scientists in the World*, compiled and published by Stanford University, is based on international scientific publications, the number of scientific citations for research, and participation in the review and editing of scientific research. Dr. Mubarak is a Fellow of the Institution of Engineers Australia, a Chartered Professional Engineer (CPEng) of the Institution of Engineers Australia, and a Chartered Chemical Engineer of the Institution of Chemical Engineers (IChemE), UK. He has published 11 books and is a co-editor for ongoing Elsevier-edited books: (1) *Nanomaterials for Carbon Capture and Conversion Technique*, (2) *Advanced Nanomaterials and Nanocomposites for Bioelectrochemical Systems*, (3) *Water Treatment Using Engineered Carbon Nanotubes*, (4) *Hybrid Nanomaterials for Sustainable Applications: Case Studies and Applications*, (5) *Sustainable Nanotechnology for Environmental Remediation*, (6) *Nanotechnology for Biomedical Applications*, (7) *Nanotechnology for Electronic Applications*, (8) *Application of Bio-Additives for the Food Industry*, (9) *Contemporary Nanomaterials in Material Engineering Applications*, (10) *Fundamentals of Biomaterials: A Supplementary Textbook*, and (11) *Emerging Water Pollutants: Concerns and Remediation Technologies*.

Address: Department of Chemical Engineering, College of Engineering, Imam Mohammad Ibn Saud Islamic University (IMSIU), Riyadh 11432, Saudi Arabia

E-mail: MNSMujawar@imamu.edu.sa; mubararak.yaseen@gmail.com

Research Interests: Advanced carbon nanomaterials synthesis via microwave technology, MXene synthesis and its application in wastewater treatment and energy storage, graphene/CNT buckypaper for strain sensor application, biofuels, magnetic buckypaper, immobilization of enzymes, protein purification, magnetic biochar production using a microwave, and wastewater treatment using advanced materials.

Introduction to Bioelectrochemical Conversion

Arshad Iqbal Jarwar, Muhammad Azam Usto, Bazil Owais, Majid Iqbal, Faizan Ahmeed, Rafi-uz-zaman Brohi, Abdul Karim Shah, and Abdul Sattar Jatoi

Abstract Bioelectrochemical conversion is an emerging interdisciplinary field that integrates microbiology, electrochemistry, and environmental engineering to harness the metabolic capabilities of microorganisms for energy and resource recovery. At the heart of this technology lie bioelectrochemical systems (BES), such as microbial fuel cells (MFCs) and microbial electrolysis cells (MECs), which utilize electroactive microbes to catalyze oxidation–reduction reactions at electrode surfaces. These systems offer a promising approach for converting organic waste, wastewater, and carbon dioxide into valuable products, such as electricity, hydrogen, methane, and biofuels, under mild operating conditions. The versatility and sustainability of bioelectrochemical conversion make it a compelling alternative to conventional energy and waste treatment technologies. This introductory overview highlights the fundamental principles, key components, and potential applications of bioelectrochemical systems, laying the groundwork for their broader adoption in clean energy generation, waste valorization, and environmental remediation.

Keywords Bioelectrochemical system · Overview · Microbial fuel cell · Microbial electrolysis cell

A. I. Jarwar (✉) · M. A. Usto · B. Owais · A. K. Shah · A. S. Jatoi
Chemical Engineering Department, Dawood University of Engineering and Technology, Karachi, Pakistan
e-mail: arshad.iqbal@duet.edu.pk

M. Iqbal
Federal Urdu University of Science and Technology Karachi, Karachi, Pakistan

F. Ahmeed
NED University of Engineering and Technology, Karachi, Karachi, Pakistan

Rafi-uz-zaman Brohi
Environmental Protection Agency Government of Sindh, Karachi, Pakistan

A. Sattar Jatoi, N. M. Mubarak (eds.), *Bio-Electrochemical Conversion of Waste to Fuel*, Green Energy and Technology,
https://doi.org/10.1007/978-981-95-4310-6_1

1 Introduction

New technologies are being developed in this sector to manage waste sustainably and generate new energy sources from its nutrients. The bioelectrochemical system (BES) platform effectively transforms biodegradable materials into usable fuels. BES uses the metabolic processes of bacteria to promote chemical changes and electron transfer reactions. Bioelectrochemical systems (BES) and bioelectrocatalysis are utilized to produce a range of green energy forms, chemicals, and minerals. Additionally, the growth of waste-to-energy technologies, pollution bioremediation, and low-carbon energy has raised interest in developing microbial fuel cells (MFC) and bioelectrochemical fuel cells. The most sophisticated type of BES is the MFC, which uses microorganisms to produce energy from chemical sources. Bhat et al. defined biomass as an energy source from plants, animals, and garbage. Over the years, numerous authors have explored the possibility of applying contemporary techniques to produce clean energy from plant and animal waste.

Innovative solutions have emerged in response to the growing demand for sustainable waste management and renewable energy. One of the most promising methods for converting trash into useful fuels is the bioelectrochemical system (BES). By utilizing the metabolic activity of microorganisms, BES offers an electron transfer system that propel chemical changes, minerals chemicals and gain power are produced via bioelectrical catalytic processes and bioelectrochemical system (BES) microbial fuel cell (MFC) and bioelectrical fuel cell are examples of BES that have gained attention due to the increased interest in waste polarization low carbon energy and the sustainable bioremediation of environmental pollutants the most sophisticated beings that can guarantee the bathrobe transformation of chemical energy into electrical energy is the MFC [1].

According to Bhat et al., biomass is a carbon-neutral energy source that may be produced from waste plants and animals [2]. Using cutting-edge technology, several writers have explored the potential benefits of incorporating plant and animal waste into clean, renewable, and sustainable energy sources over the years [3].

The increasing need for clean and renewable energy sources, driven by the rapid depletion of fossil fuels, has spurred these innovations [4]. Furthermore, mounting concerns about global warming and climate change have reaffirmed the need to create a low-carbon economy [5]. To address these global issues, researchers are now investigating renewable energy technologies (RET), such as solar, wind, and biomass technologies [6]. Gasifiers, pyrolysis bioreactors, digesters, fuel cells, and other biological systems can transform biomaterials and biowaste into bioenergy, biofuels, and biochemical products [7]. The increasing need for sustainable waste management has led to the development of creative solutions for renewable energy. The bioelectrical chemical system beds are one of the most promising techniques for converting waste into usable fuels; beds provide electron transfer mechanism that drives chemical transformations by harnessing in our nursing the metabolic activity of microorganisms, minerals, chemicals, and gain powers are produced using biological system (BES) and bioautocatalytic processes the growing interest in waste mineralization low carbon energy and sustainable bioremediation of environmental pollutants has drawn attention to PES such as microbial fuel cell MFC and biological fuel cells BFC the MSC is the most advanced BES that can ensure microbial conversion of chemical energy into electrical energy. The principles of bioelectric chemical systems, together with their types, difficulties, and advancements, are thoroughly covered in this chapter.

1.1 Fundamentals of Bioelectrochemical Systems (BESs)

Microorganisms must be able to initiate electrochemical processes to convert organic and inorganic waste into energy-rich products, such as electricity, hydrogen, methane, and other biofuels. This is the foundation of bioelectrical bioelectrochemical conversion. Bioelectric electrochemical systems (BESs) are the main force behind this process. They utilize the metabolic activity of electroactive microorganisms to accelerate oxidation–reduction reactions at the anode of these systems. Organic substrates, such as industrial effluents, agricultural waste, and wastewater pollutants, act as electron donors. Microorganisms oxidize these substrates, releasing protons and electrons [8]. The returns are moved to the anode, either directly weird conductive nanowires or cytochromes or indirectly via soluble redox mediators. These electrons then travel to the cathode via an external circuit. The power reduction processes include the proton reduction to hydrogen gas in microbial electrolysis cells (MEC) or the oxygen reduction to water in microbial fuel cells. This control of electron flow is produced by electricity or fuels, offering a sustainable and clean alternative to energy generation based on fossil fuels.

Furthermore, bees are crucial to modern waste and energy management programs due to their stability, which enables them to be utilized in environmental applications such as wastewater treatment and bioremediation [1]. Apart from microbial interactions, reactor design, operating temperatures, electrode materials, and system scalability are other critical factors that affect the effectiveness of bioelectrochemical conversion [9]. Examples of electrode materials that are crucial for boosting electron transfer rates and microbial adhesion—two factors that enhance system performance—include carbon-based or metal-modified surfaces. Reactor designs, whether single-, dual-, or hybrid, affect ion mobility and electrochemical efficiency. Furthermore, pH, temperature, and substrate availability affect microbial activity and reaction kinetics, necessitating careful system optimization. Recent advances in synthetic biology, biofilm engineering, and nanotechnology have greatly expanded the potential of BESs by enabling the creation of extremely lucrative and useful bioelectrochemical platforms [10].

1.1.1 Basics of Bioelectrochemical System Mechanisms

The basic idea behind bioelectrochemical systems (BESs) is extracellular electron transfer (EET), in which electroactive microbes facilitate electron transfer between an anode and a cathode through redox reactions that produce energy or facilitate chemical synthesis. A two-chamber arrangement divided by a proton exchange membrane, as illustrated in Fig. 1, can be used to comprehend the fundamental operation of (BESs) protons and neutrons are produced as metabolic byproducts when bacteria such as Shewanella and geobacter oxidize organic substrates close to the anode the electrons are moved to the anode by either indirect electron transfer (IET) which employs soluble redox mediators like flavins and quinones or direct electron transfer Det which uses conductive nanowires or cytochromes [11]. These electrons then proceed to the cathode via an external circuit, where they participate in reduction processes, such as reducing hydrogen ions to hydrogen gas or oxygen to water, thereby maintaining charge neutrality. In the meantime, protons cross the membrane to the cathode chamber. The fundamental processes that enable the generation of hydrogen in electrolysis cells, microbial electrochemical cells (MECs), and bioelectricity in microbial fuel cells are highlighted by this electrochemical interaction [12]. Several factors, including the type of microbial population, the properties of the Jordan material, reactor design, and operating parameters such as pH, temperature, and substrate concentration, influence the performance of a bioelectrochemical system. Recent advances in BES research have primarily focused on enhancing electron transfer mechanisms and system topologies to achieve greater energy efficiency and broader applicability. They enhance BES design by using non-structural electrodes, which significantly boost the effectiveness of microbial adhesion in electron transport [13]. Combining graphene-based anodes with bioengineered microbial strains has increased power densities and fuel generation by accelerating the electron transfer rate. Yes, these are promising technologies for sustainable development, as they have been modified for various applications, such as bioremediation of wastewater and biosensing, through the integration of

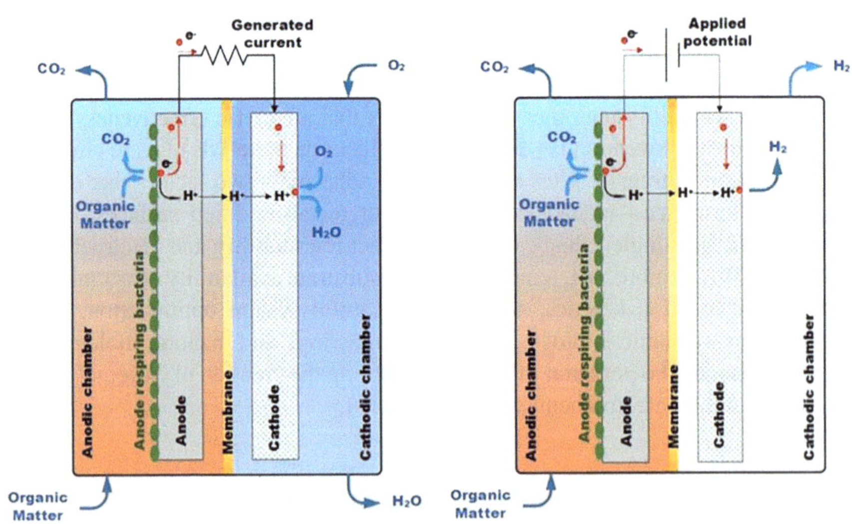

Fig. 1 Schematic representation of electricity production (left) and hydrogen production (right) through MFC and MEC, respectively [1]

microbial metabolism and abiotic catalysts. Hybrid bees open new possibilities for chemical synthesis and carbon capture [14]. All things considered, electrochemical processes present a feasible path toward generating clean energy.

1.1.2 Types of Bioelectrochemical Systems

Microbial electron transport is involved in specialized technologies known as bioelectrochemical systems (BES), which encompass energy production, wastewater treatment, desalination, and biochemical synthesis. The BESs studied most include microbial fuel cells (MFC), microbial electrolysis cells (MEC), microbial desalination cells (MDC), and microbial electrosynthesis, the most well-known BES technology. Microbial fuel cells MFC use electroactive bacteria to turn organic waste into electrical power. At the anode of MFCs, microorganisms decompose organic materials, producing electrons that are then sent via an external circuit to the cathode, where oxygen reduction occurs. This method, which simultaneously cleans wastewater and produces renewable electricity, has drawn interest as a sustainable alternative to traditional energy sources [15]. To boost power output and efficiency, recent developments in MFCs have focused on enhancing microbial activity, strengthening electrode materials, and increasing system scalability [13].

Microbial Electrolysis Cell (MEC)

Because at least one of their electrode processes involves electrochemical interactions with microbes, BESs can be regarded as electrochemical systems. Because of their ability to transfer electrons from a biodegradable substrate to a solid electrode, certain microorganisms—also referred to as anode-respiring bacteria (ARBs)—are typically required for the anodic process. Figure 2 shows a schematic representation of the MFC and MEC operating principles. The biological anode reactions in MECs and MFCs share several commonalities. One of the main byproducts of ARB metabolism, electrons, is transferred to the anode together with protons and CO2 before moving to the cathode through an external link. When the BES is utilized as an MFC, oxygen (or another oxidant) is present at the cathode, and electrons flow freely through an electrical load, where energy is recovered. A power source is required to initiate the reaction at the electrodes because no oxidant is allowed on the cathode in an MEC [16].

The main advantage of MECs is that they produce hydrogen from low-energy waste sources more efficiently than conventional water electrolysis. MECs are, therefore, a promising technology for sustainable hydrogen production, wastewater treatment, and energy recovery. However, challenges remain to be addressed, including the need for scalable systems, efficient electron transport techniques, and reasonably priced electrode materials. Advances in electrode design, microbial community optimization, and reactor architecture have significantly enhanced MEC performance, increasing its viability for industrial applications [17]. To illustrate the

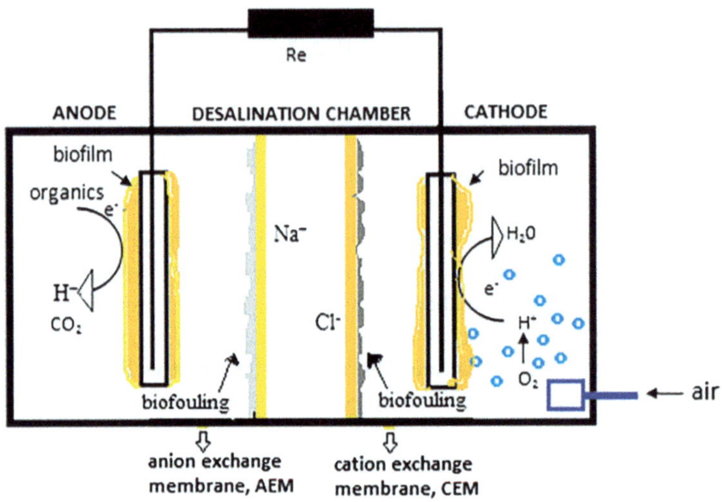

Fig. 2 Microbial desalination cell [2]

creation of hydrogen gas and electron transport, MEC drawings typically show a two-chamber arrangement with a membrane dividing the anode and cathode.

Microbial Desalination Cell (MDC)

Microbial desalination cells (MDCs), a novel use of BES technology, combine desalination and microbial activity to produce fresh water and power simultaneously. Although they work similarly to MFCs, MDCs have additional ion-exchange membranes that aid in removing salts from brackish or salted water [18]. Electroactive microorganisms generate an electrical potential that drives ions across membranes by oxidizing organic molecules at the anode and producing electrons that flow to the cathode. Cations move to the cathode chamber, and anions move to the anode chamber to reduce salinity in the central desalination compartment.

MDCs appeal to areas with freshwater constraints because they provide the twin benefits of desalination and wastewater treatment. They are an environmentally favorable alternative to conventional desalination techniques, such as reverse osmosis, because they use less energy. MDC diagrams usually show electron flow, salt ion movement across ion exchange membranes, and the simultaneous production of bioelectricity and desalination.

Microbial Electrosynthesis (MES)

Microbial electrosynthesis (MES), a novel biocatalytic electrochemical system (BES) application, converts carbon dioxide into beneficial organic molecules using electroactive microorganisms as biocatalysts. MES devices use an externally

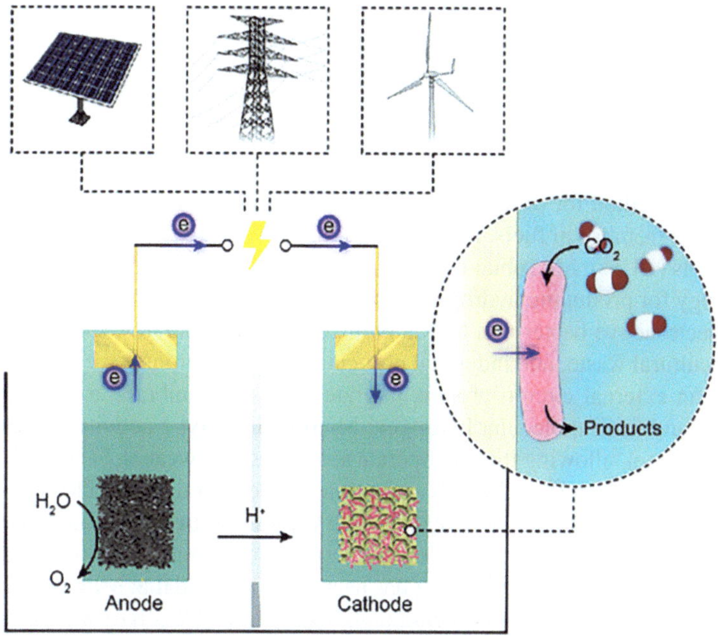

Fig. 3 Microbial electrosynthesis. Schematic representation of a microbial electrosynthesis cell powered with electricity [3]

supplied voltage to initiate microbial metabolic pathways at the cathode, where CO_2 is converted into compounds such as ethanol, acetate, or methane [19]. In addition to producing biological precursors and biofuels, this method tackles climate change and offers a workable way to absorb and store carbon. The primary benefit of MES is that it provides a sustainable alternative to traditional chemical synthesis methods by utilizing renewable electricity to drive biosynthesis. However, poor microbial growth rates, limited electron transfer efficiency, and the high cost of cathode materials are significant obstacles. To increase MES viability, advancements in electrode engineering, microbial strain selection, and process optimisation are necessary [20]. Figure 3 of MES systems often depicts the reduction of CO_2 at the cathode, microbial electron uptake, and the production of organic compounds as end products.

1.2 Applications of BESs in Waste-to-Fuel Conversion

Bioelectrochemical systems, or BESs, provide a novel approach to converting organic waste into beneficial biofuels while addressing environmental concerns. These systems utilize electroactive microorganisms to drive redox processes, enabling the sustainable production of biofuels, including ethanol, hydrogen, and methane. When incorporated into waste management systems, they reduce greenhouse gas emissions and improve resource recovery.

1.2.1 Biohydrogen Production

A practical method for producing hydrogen sustainably, as an alternative to hydrogen derived from fossil fuels, is the creation of biohydrogen using bioelectrochemical systems (BESs). Microbial electrolysis cells (MECs) are the most efficient BES technology for producing hydrogen. Protons and electrons are released at the anode when electroactive bacteria in MECs oxidize organic substrates, such as wastewater, agricultural waste, and industrial effluents. While electrons travel to the cathode through an external circuit, protons migrate across a membrane to combine with electrons and form molecular hydrogen. Microbial catalytic activity makes MECs energy-efficient, allowing them to operate at a voltage between 0.2 and 0.8 V, which is significantly lower than that of conventional water electrolysis. The technique is a dual-purpose technology, as it enables both wastewater treatment and reuse simultaneously. Shewanella and Geobacter are two important microbial species that may transfer electrons outside cells [4]. It has been demonstrated that MECs can increase overall energy recovery by producing high hydrogen yields (up to 90% of the theoretical maximum) under specific conditions [21]. Figure 4 provides a detailed schematic of MEC-based hydrogen synthesis, including electrode reactions, microbial interactions, and proton and electron transport.

1.2.2 Biogas Production

By combining electrochemical reactions with microbial metabolism, bioelectrochemical systems (BESs) have emerged as a promising method for enhancing biogas production. The microbial electrolysis cell (MEC), one of the best BES technologies for producing biogas, releases protons and electrons when organic waste is broken down by electroactive bacteria at the anode. In addition to producing hydrogen, MECs can also produce methane (CH_4) when a methanogenic biocathode is added. To increase biogas's total production and purity, electro-methanogenesis utilizes an applied voltage to facilitate the direct microbial conversion of CO_2 and protons into methane [22]. Furthermore, by accelerating electron transfer between microbial populations, BESs enhance anaerobic digestion (AD) and increase methane production. According to studies, methane production can be increased by 30%–50% when BESs are integrated with anaerobic digesters (AD-BES systems) compared to conventional anaerobic digesters (AD systems) [23]. The method effectively reduces organic pollutants in wastewater while recovering bioenergy, making it a sustainable waste management solution. Moreover, direct electron transfer while facilitating efficient methane production requires the Methanosaeta and Geobacter species [20]. A schematic representation of

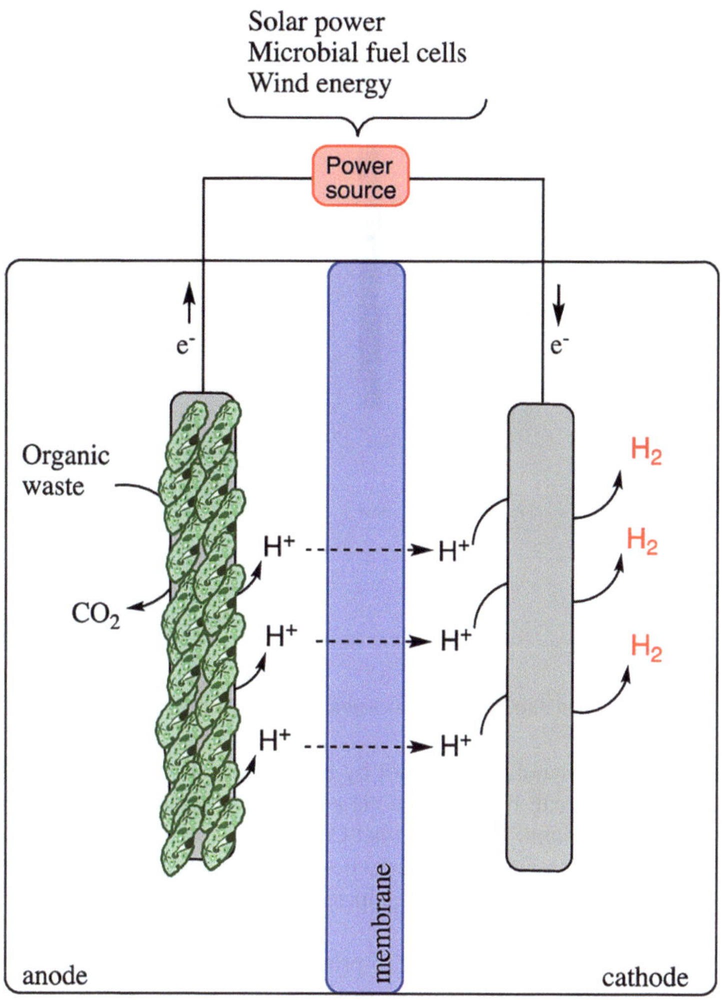

Fig. 4 Microbial electrolysis cell for H_2 production [4]

BES-driven biogas generation, illustrating microbial interactions and electron transport, is presented in Fig. 5.

1.2.3 Liquid Fuels

An efficient method for turning organic which into liquid biofuels like ethanol butanol and acetate is to use bioelectrochemical system BES which helps produce sustainable energy; one important BES technology is microbial electrosynthesis, which allows electroactive microorganisms to directly convert CO_2 and organic waste into liquid fuels microbes near the (MES) cathode convert CO_2 into useful compounds

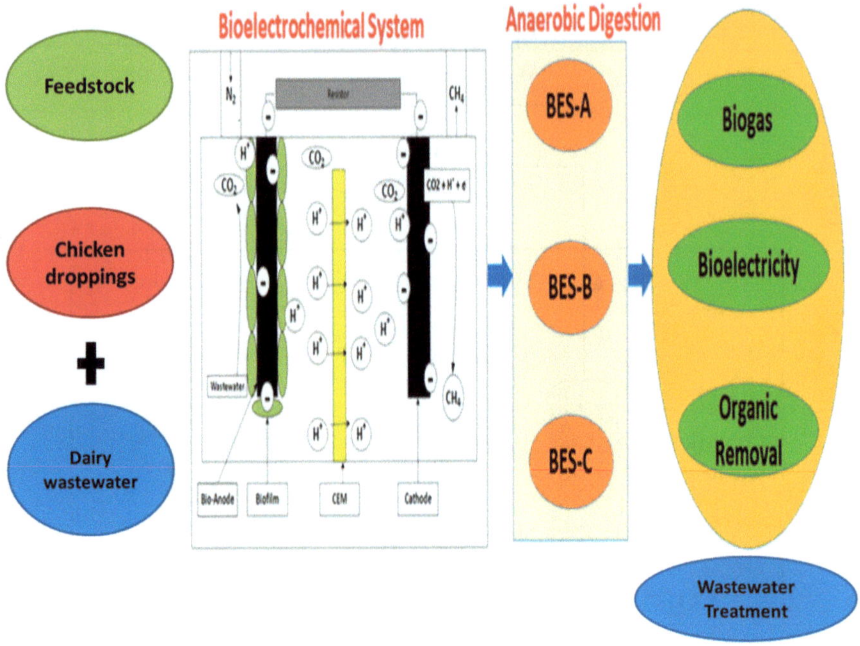

Fig. 5 Biogas production through bioelectrochemical system from waste [5]

including acetate, ethanol, and butanol by using neutrons from an external power source [19]. Biofuel can be produced via additional microbial fermentation using acetate as a starting point. Microbial Fuel Cells (MFCs) and Microbial Electrolysis Cells (MECs) can also aid in the production of liquid fuel by providing electrons that drive fermentation processes and enhance the synthesis of bioethanol from agricultural waste [24].

This process is a sustainable way to produce biofuels, as it not only generates biofuels but also facilitates wastewater treatment and carbon capture. Figure 6 illustrates a schematic of the BES-driven waste-to-liquid fuel conversion process, along with the pathways for fuel production and microbial electron transfer. This image illustrates the microbial electrosynthesis (MES) process in a bioelectrochemical system (BES) for converting trash and CO_2 into beneficial biofuels and chemicals. Using electrons (e^-) from an external power source, microbes in the anaerobic cathode chamber efficiently convert CO_2 into hydrogen (H_2), formate ($HCOO^-$), and other fuels. Conversely, the aerobic compartment contributes less than 10% of fuel generation and encourages biomass growth. During the process, nitrate (NO_2^-) and ammonia (NH_3) are also decreased. MES is a potentially useful technology for waste-to-fuel conversion and carbon capture in the production of sustainable energy [25].

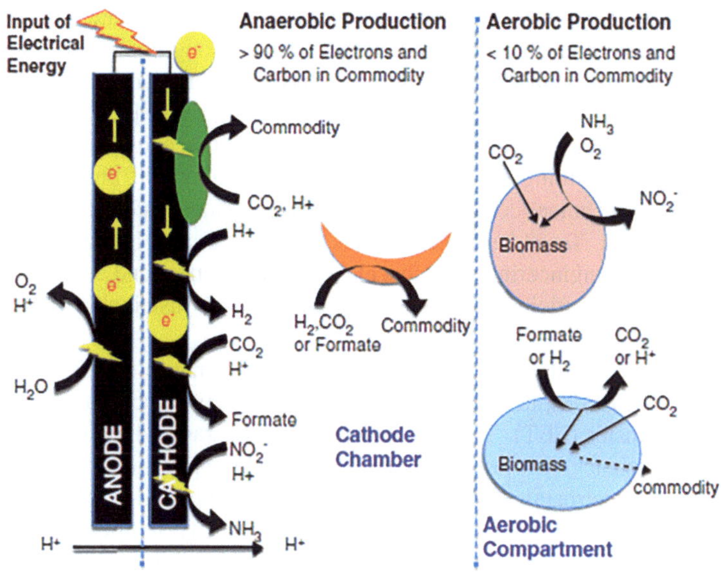

Fig. 6 Microbial electrosynthesis to liquid fuels [6]

1.3 Challenges in Bioelectrochemical Waste Conversion

Converting bioelectrochemical waste presents numerous challenges, particularly in terms of cost and microbiological effectiveness. Electroactive microorganisms with extracellular electron transfer (EET) capacities are essential to bioelectrochemical systems (BESs). However, slow electron transfer rates, biofilm instability, and competition with non-electrogenic microorganisms limit the system's effectiveness. Because undesirable microbial growth can outcompete electroactive bacteria, thereby reducing the production of fuel and bioelectricity, the diversity of microbial communities in waste environments makes the process more challenging. Long-term stability is also challenging because environmental variables, such as pH variations, substrate availability, and electrode surface conditions, impact microbial function [26].

The high cost of materials and system scalability further restricts general use. Membrane separators, electrode materials, and catalysts are the main contributors to the overall cost. Although platinum-based catalysts enhance performance, their high cost has led to research into less costly alternatives, such as carbon-based electrodes and metal oxides [13]. Nevertheless, these substitutes typically have lower longevity and effectiveness. Additionally, maintaining BESs in practical applications still entails significant operational and technical costs, especially for large-scale wastewater treatment or biofuel generation. To address these issues, research is also being conducted on microbial genetic engineering, low-cost electrode fabrication, and enhanced system designs to increase efficiency and financial viability for sustainable waste-to-fuel conversion [24]. The high cost of materials and system scalability further restricts general use. Membrane separators, electrode materials, and catalysts are the main contributors to the overall cost. Although platinum-based catalysts enhance performance, their high cost has led to research into less costly alternatives, such as carbon-based electrodes and metal oxides [13].

Nevertheless, these substitutes typically have lower longevity and effectiveness. Additionally, maintaining BESs in practical applications still entails significant operational and technical costs, especially for large-scale wastewater treatment or biofuel generation. To address these issues, research is also being conducted on microbial genetic engineering, low-cost electrode fabrication, and enhanced system designs to increase efficiency and financial viability for sustainable waste-to-fuel conversion [24].

Microbial constraints significantly impact the performance of bioelectrochemical systems (BESs) despite the requirement of electroactive bacteria for extracellular electron transfer (EET). The slow electron transfer rate, which reduces energy conversion efficiency, is a primary issue. Large-scale applications are challenging because many electroactive bacteria, such as Shewanella and Geobacter, require specific environmental parameters, including an ideal pH, temperature, and substrate availability. Biofilm detachment, microbial competition, and the presence of non-electrogenic bacteria can sometimes impair system performance, although biofilm development and electrode stability are also crucial for long-term operation [27]. BESs are also restricted in adapting to different waste streams due to their low microbial diversity. While genetic engineering and electroactive bacterial enrichment are promising methods, further research is required to improve their robustness and effectiveness [13]. These microbial hurdles must be overcome to advance BES technology toward practical applications for converting waste into fuel. One of the primary problems with Bioelectrochemical systems (BES) is scalability. Even while small-scale BES has demonstrated promising outcomes in environmental remediation, wastewater treatment, and energy generation, scaling up these systems remains challenging. The difficulties in scaling up are preserving efficient microbial activity, ensuring steady feeding and electron distribution, and minimizing losses due to system inefficiencies. Furthermore, the size makes it more challenging to regulate electrode materials, biofilm formation, and microbial populations, which usually leads to worse performance or higher operating costs.

1.4 Innovation and Future Direction

Bioelectrochemical systems (BES) have shown great promise for several applications, including wastewater treatment, environmental remediation, and bioenergy production. Recent developments have focused on enhancing the efficiency and utility of BES by refining electrode materials, optimizing microbial fuel cell (MFC) architecture, and modifying the microbial populations involved. Advanced electrode materials, including metal nanoparticles, conductive polymers, and carbon-based composites, have been developed to enhance electron transfer and improve system performance. The enhanced surface area, conductivity, and stability of these materials are critical to the long-term operation of BES. Developing innovative biocompatible materials and coatings that enhance microbial adherence and mitigate biofouling may enhance the efficiency of microbial electrolysis and fuel cells. Enhancing energy output and enabling large-scale deployment are two more goals of system design innovations, such as creating modular and scalable systems. Furthermore, research is being done on synthetic biology methods to develop microorganisms with enhanced electron transport and biocatalytic properties. This would enable a higher current density and improved control over energy or waste.

Future BES research endeavors will address current concerns, including long-term performance, cost-effectiveness, and system scalability. The creation of hybrid BES, which combines bioelectrochemical processes with renewable energy sources such as solar or wind, to ensure a more reliable and sustainable energy supply is a significant area of research. This integration would lessen the intermittent character of renewable energy, increasing the attractiveness of BES systems for large-scale applications. Furthermore, bioelectrochemical technology has considerable room to grow in the fields of carbon capture and bioremediation. Future research may focus on developing specialized microbial communities capable of managing various contaminants, including heavy metals and pharmaceuticals. The use of BES is to reduce carbon dioxide and transform it into valuable molecules, such as alcohols or methane, is another fascinating field that could offer practical solutions to the climate crisis. Lastly, enhancements in material costs, system integration, and operational efficiency are required to make BES more economically feasible. The full potential of bioelectrochemical systems as a sustainable solution for waste-to-energy conversion, environmental remediation, and clean energy generation would require combining these developments, such as microbial process improvement and BES technological scalability. The application of BES to lower carbon dioxide emissions and transform it into useful substances, such as alcohol or methane, is another fascinating field that may offer fresh approaches to combating climate change. Ultimately, higher material costs, system integration, and improved operational efficiency will be necessary to enhance the economic viability of BES. The full potential of bioelectrochemical systems as a sustainable solution for waste-to-energy conversion, environmental remediation, and clean energy generation would require combining these developments, such as microbial process optimization and BES technology scalability.

1.5 Conclusion

Bioelectrochemical systems that convert waste into fuels can efficiently and sustainably address waste management issues and energy demands. By utilizing the metabolic activity of microbes to transform organic waste into useful biofuels, such as hydrogen, methane, and ethanol, bioelectrochemical systems (BES), including microbial fuel cells (MFCs) and microbial electrolysis cells (MECs), offer innovative solutions. By processing wastewater and reducing greenhouse gas emissions, these systems help clean up the environment and provide an alternative energy source. Recent developments in BES technologies have enhanced the scalability and efficiency of these processes for practical applications by utilizing improved electrode materials, modified microbial populations, and refined system designs. Integrating BES technology with renewable energy sources, such as solar and wind, further enhances their sustainability by reducing reliance on fossil fuels and improving energy output consistency. Numerous issues persist despite these developments, including the need for increased system scalability, cost-effectiveness, and reliability. To overcome these obstacles, future studies should focus on developing more effective conversion procedures, exploring new waste feedstocks, and optimizing microbial activity to enhance the overall process efficiency. Realizing the full potential of waste-to-fuel conversion also requires further development in hybrid systems, such as integrating biorefinery systems (BES) with other renewable technologies and synthetic biology to create high-performance microorganisms. These technologies have the potential to drastically change how we produce energy, reduce environmental harm, and handle waste as they develop and become more economically feasible. This would greatly aid a more sustainable and circular economy. To create the future bioelectrochemical waste-to-fuel systems, bioengineering, environmental science, and energy technology must continue their interdisciplinary cooperation.

References

1. Escapa A, Mateos R, Martínez EJ, Blanes J (2016) Microbial electrolysis cells: an emerging technology for wastewater treatment and energy recovery. From laboratory to pilot plant and beyond. Renew Sust Energ Rev 55:942–956. https://doi.org/10.1016/j.rser.2015.11.029
2. Zhang H, Wen Q, An Z, Chen Z, Jun N (2016) Analysis of long-term performance and microbial community structure in bio-cathode microbial desalination cells. Environ Sci Pollut Res 23. https://doi.org/10.1007/s11356-015-5794-7
3. Fang X, Kalathil S, Reisner E (2020) Semi-biological approaches to solar-to-chemical conversion. Chem Soc Rev 49. https://doi.org/10.1039/C9CS00496C
4. Logan BE, Call D, Cheng S, Hamelers HVM, Sleutels THJA, Jeremiasse AW et al (2008) Microbial electrolysis cells for high yield hydrogen gas production from organic matter. Environ Sci Technol 42(23):8630–8640. https://doi.org/10.1021/es801553z
5. Fazal T, Ur Rehman MS, Mushtaq A, Hafeez A, Javed F, Aslam M et al (2019) Simultaneous production of bioelectricity and biogas from chicken droppings and dairy industry wastewater employing bioelectrochemical system. Fuel 256:115902
6. Lovley DR, Nevin KP (2013) Electrobiocommodities: powering microbial production of fuels and commodity chemicals from carbon dioxide with electricity. Curr Opin Biotechnol 24(3):385–390. https://doi.org/10.1016/j.copbio.2013.02.012
7. Kataya G, Cornu D, Bechelany M, Hijazi A, Issa M (2023) Biomass waste conversion technologies and its application for sustainable environmental development—a review. Agronomy 13(11):2833
8. Shemfe M, Ng KS, Sadhukhan J (2017) Bioelectrochemical systems for biofuel (electricity, hydrogen, and methane) and valuable chemical production

9. Mahmoud RH, Gomaa OM (2022) Bio-electrochemical frameworks governing microbial fuel cell performance: technical bottlenecks and proposed solutions. RSC Adv 12(10):5749–5764
10. Li F, Tang R, Zhang B, Qiao C, Yu H, Liu Q et al (2023) Systematic full-cycle engineering microbial biofilms to boost electricity production in Shewanella oneidensis. Research (Washington, DC) 6:0081
11. Lovley DR (2006) Microbial fuel cells: novel microbial physiologies and engineering approaches. Curr Opin Biotechnol 17(3):327–332
12. Schröder U (2007) Anodic electron transfer mechanisms in microbial fuel cells and their energy efficiency. Phys Chem Chem Phys PCCP 9:2619–2629
13. Wang H, Ren ZJ (2013) A comprehensive review of microbial electrochemical systems as a platform technology. Biotechnol Adv 31(8):1796–1807
14. Harnisch F, Schröder U (2010) From MFC to MXC: chemical and biological cathodes and their potential for microbial bioelectrochemical systems. Chem Soc Rev 39(11):4433–4448
15. Logan BE, Regan JM (2006) Microbial fuel cells—challenges and applications. Environ Sci Technol 40(17):5172–5180
16. Rabaey K, Angenent L, Schroder U, Keller J (2010) Bioelectrochemical systems: from extracellular electron transfer to biotechnological applicaiton
17. Escapa A, Mateos R, Martínez E, Blanes J (2016) Microbial electrolysis cells: an emerging technology for wastewater treatment and energy recovery. From laboratory to pilot plant and beyond. Renew Sustain Energy Rev 55:942–956
18. Mehanna M, Saito T, Yan J, Hickner M, Cao X, Huang X et al (2010) Using microbial desalination cells to reduce water salinity prior to reverse osmosis. Energy Environ Sci 3(8):1114–1120
19. Nevin KP, Woodard TL, Franks AE, Summers ZM, Lovley DR (2010) Microbial electrosynthesis: feeding microbes electricity to convert carbon dioxide and water to multicarbon extracellular organic compounds. MBio 1(2). https://doi.org/10.1128/mbio.00103-10
20. Rabaey K, Rozendal RA (2010) Microbial electrosynthesis—revisiting the electrical route for microbial production. Nat Rev Microbiol 8(10):706–716
21. Escapa A, Lobato A, Garcia D, Morán A (2013) Hydrogen production and COD elimination rate in a continuous microbial electrolysis cell: the influence of hydraulic retention time and applied voltage. Environ Prog Sustain Energy 32(2):263–268
22. Villano M, Aulenta F, Ciucci C, Ferri T, Giuliano A, Majone M (2010) Bioelectrochemical reduction of CO_2 to CH_4 via direct and indirect extracellular electron transfer by a hydrogenophilic methanogenic culture. Bioresour Technol 101(9):3085–3090
23. Xie Q, Lu Y, Tang L, Zeng G, Yang Z, Fan C et al (2021) The mechanism and application of bidirectional extracellular electron transport in the field of energy and environment. Crit Rev Environ Sci Technol 51(17):1924–1969
24. Deepak Pant DP, Bogaert G, Diels L, Vanbroekhoven K (2012) A comparative assessment of bioelectrochemical systems and enzymatic fuel cells. In: Microbial biotechnology: energy and environment. CABI, Wallingford, pp 39–57
25. Ibrahim I, Salehmin MNI, Balachandran K, Hil Me MF, Loh KS, Abu Bakar MH et al (2023) Role of microbial electrosynthesis system in CO_2 capture and conversion: a recent advancement toward cathode development. Front Microbiol 14
26. Logan BE, Rabaey K (2012) Conversion of wastes into bioelectricity and chemicals by using microbial electrochemical technologies. Science 337(6095):686–690
27. Pinck S, Ostormujof LM, Teychené S, Erable B (2020) Microfluidic microbial bioelectrochemical systems: an integrated investigation platform for a more fundamental understanding of electroactive bacterial biofilms. Microorganisms 8(11):1841

Types of Organic Waste for Bioelectrochemical Systems

Hala Bakr El-Nassan

Abstract Bioelectrochemical systems (BESs) are considered a new and efficient technology for waste management and bioenergy production. BESs have dual actions to reduce environmental pollution and provide electricity from readily available, sustainable, and cost-effective sources. Microbial fuel cells (MFCs) can efficiently convert chemical energy obtained from organic waste into electric energy. Microbial electrolysis cells (MECs) convert energy into biofuels. Moreover, BESs contribute to the recovery of metals, water, and nutrients. The organic waste that serves as a substrate in fuel cells can be in the form of solid masses or wastewater. The decomposition of solid organic wastes is more difficult than that of organic wastewater. This chapter provides a brief overview of the various types of organic waste used in BESs.

Keywords Bioelectrochemical systems · Microbial fuel cells · Microbial electrolysis cells · Organic waste · Waste treatment

1 Introduction

BES is a new, evolving technology that can generate electricity from organic waste. In BES, the microorganisms can generate or accept electrons and act as catalysts for oxidation or reduction. This also reduces the load of organic contamination in organic wastes. Therefore, BES is considered a highly effective strategy for waste management [1]. The BESs generate electricity, treat various types of waste, recover metals from solid waste, and remediate water [2–5]. BES can be used efficiently to treat organic pollutants, such as azo dyes, polycyclic aromatic hydrocarbons,

H. B. El-Nassan (✉)
Pharmaceutical Organic Chemistry Department, Faculty of Pharmacy, Cairo University, Cairo, Egypt
e-mail: hala.el-nassan@pharma.cu.edu.eg

A. Sattar Jatoi, N. M. Mubarak (eds.), *Bio-Electrochemical Conversion of Waste to Fuel*, Green Energy and Technology,
https://doi.org/10.1007/978-981-95-4310-6_2

organochlorine compounds, and pesticides, which resist biodegradation under conventional conditions [1, 6]. The BES consists of an anode and a cathode placed in two separate chambers and connected via an external circuit. Electron donors, such as sulfides, are oxidized at the anode, while electron acceptors, like nitrate or oxygen, are reduced at the cathode. There are two main types of BES. The microbial fuel cell (MFC) is used to generate electricity from organic waste, while the microbial electrolysis cell (MEC) utilizes an electric current to produce methane or hydrogen from waste [7]. In MFC, the anodic chamber contains the waste and the anaerobic microorganisms. On the contrary, the cathodic chamber contains acidic water or phosphate-buffered saline (PBS). It is kept in aerobic conditions, as the oxygen is combined with protons and electrons to form water molecules. The electricity generated is collected through the external circuit. MFC can generate electricity, recover some types of metals, and provide water for irrigation [7].

In the MEC, an external voltage is supplied to reduce protons, generating hydrogen and/or methane gas. The production of fuel gases and electricity from organic wastes can also reduce the emission of greenhouse gases [7].

The microorganisms used in BES cells are either exoelectrogenic or microorganisms that oxidize organic materials [7]. Exoelectrogenic microorganisms, such as *Geobacter sulfurreducens*, *Shewanella oneidensis*, and *Lactococcus lactis*, are primarily utilized in microbial fuel cells (MFCs). This microorganism can transfer the electrons generated inside its cells by oxidizing organic matter to the electron acceptors outside their cells (either electrodes or metal ions). These microorganisms can form biofilm on the surface of the electrodes, resulting in a significant enhancement of current production [7]. The organic-matter-oxidizing organisms degrade organic matter into simple organic molecules that exoelectrogens can use. Examples of this type include *Clostridium*, *Bacteroides*, and Proteobacteria strains [7]. The bacteria can also be used as a biocathode, thus eliminating the need for expensive metal cathodes and reducing the overall production cost of the reactor [8, 9]. Several factors influence the performance and output of the BES, including temperature, pH, the type and particle size of the organic waste, cell design, electrode types, electrode spacing, and the strain of the microorganism [4, 10, 11]. This chapter focuses on the kinds of organic waste that can be used as substrates in BES. They can be divided into solid organic wastes and wastewater.

2 Solid Organic Wastes

Solid organic wastes are used to generate electricity or valuable chemicals using biogas energy systems (BESs). In these methods, the microorganism utilizes biodegradable organic matter as a source of nutrients and breaks it down to produce electrons, which are collected as electrical energy. On the other hand, solid waste is decomposed, its volume is reduced, and it is converted into valuable products. The microorganisms also aid in the removal of pollutants and toxic compounds from waste. The BES technology is thus a sustainable and environmentally friendly approach to waste management. The products obtained from BESs can be electricity, biohydrogen, or chemicals like organic acids or alcohols [1, 10]. Thus, BES can be utilized to valorize solid waste and promote energy recovery, thereby reducing the need for landfilling or incineration [12–14]. MFCs typically use water as a solvent to facilitate the transfer of ions. However, solid-phase microbial fuel cells (SMFCs) were recently developed to operate efficiently using organic solid waste. However, SMFCs generate smaller amounts of power output than MFCs, and further research is still needed to increase the efficiency of SMFCs [12].

The organic solid waste can be one of the following types, depending on its source and components:

2.1 *Municipal Solid Waste (MSW)*

The municipal solid waste (MSW) is the domestic and industrial waste produced in towns and cities. It contains a large amount of biodegradable organic matter and is thus ideal for use in BES cells [15, 16]. However, solid waste is difficult for microorganisms to hydrolyze. Therefore, pre-fermentation or mixing with a co-substrate are two useful strategies to enhance the biodegradability of municipal solid waste [15, 17]. Co-digestion with a rapidly biodegradable co-substrate offers several advantages, including improved hydrolysis of organic waste, reduced amounts of harmful substances, adjusted C/N ratios, and increased biogas production. Examples include using dairy wastewater with municipal solid waste [15]. The leachate from municipal solid waste landfills, which arises from the decomposition of municipal solid waste (MSW) in landfills, represents a significant environmental problem that can be addressed using BES [18–23]. The MSW leachate contains organic materials that can serve as a nutrient source for bacteria; however, it also includes a significant amount of heavy metals, ammonia, and other organic and inorganic contaminants that are biotoxins. MSW leachate can be mixed with dairy wastewater to enhance its degradation [15]. Moreover, municipal solid waste landfill leachate has proven to be an efficient substrate for biohydrogen production [3, 16, 24].

The algae-cathode microbial fuel cell (MFC) was utilized to generate electricity and reduce chemical oxygen demand (COD) in landfill leachate (5%–40%) [25]. The study results indicated a 97% COD removal in the anode chamber across all leachate percentages. The removal percentage of nitrogen and phosphorus was increased upon using a low percentage of leachate (less than 10%) [25]. In another study, diluted landfill leachate (15% v/v) was used as substrate in algae-cathode MFCs. The hydraulic retention times (HRTs) were varied, and it was found that an increase in HRTs was accompanied by an increase in nutrient and chemical oxygen demand (COD) removal. The maximum removal efficiencies were 26% for COD, 76.4% for ammonium, and 86.3% for phosphorus at a 60-hour hydraulic retention time (HRT) [26].

Using aged landfill leachate in air-cathode single-chamber MFC-MEC systems enhanced the removal of COD (38.9%) and ammonia (90%) compared to an independent MFC. The electricity generation was also enhanced with the use of the MFC-MEC system. A voltage of 421 mV and a power density of 1330.7 mW/m^3 were achieved [27]. A comparison between using young and old landfill leachate as substrates in MFC revealed a higher power output (96.8 mW/m^2) and enhanced COD removal of 90.0 ± 1.2% when using young leachate (60%) in batch mode. Using 100% young leachate in continuous mode decreased the power output to 75 mW/m^2 and reduced the COD removal percentage to 55.5% [28]. The use of air-cathode MFCs at conductivity (73,588 μS/cm) and high levels of nitrogen concentration (6033 mg/L) for the generation of electricity and the treatment of urban landfill leachate was reported by Puig *et al.* [29]. The results indicated the generation of approximately 344 mW/m^3 and the removal of up to 8.5 kg COD/m^3/day of biodegradable organic matter.

A comparison between the electricity generation from aged landfill leachate using a chemical-cathode microbial fuel cell (CMFC) and an aerobic biocathode microbial fuel cell (ABMFC) was reported in 2017 by Hu *et al.* The results indicated that after 45 days of treatment, the maximum output voltages from CMFC and ABMFC were 699.0 mV and 459.4 mV, respectively. The maximum output powers and internal resistances were also higher with CMFC. The COD removal ratios using CMFC and ABMFC were 56.5% and 64.3%. Likewise, the ammonia nitrogen removal ratios using CMFC and ABMFC were 53.8% and 58.1%, respectively [30].

2.2 *Agricultural Waste*

It consists of crop waste and animal dung and is also perfect for BES systems as it is rich in organic matter. Examples include rice hulls, stalks, bean residue, and ground coffee [31].

2.3 *Lignocellulose Biomass*

Lignocellulose biomass represents one of the most valuable and sustainable sources of organic solid waste. It is obtained from plant and forest biomass and is formed of cellulose, hemicellulose, and lignin. It can be used efficiently as a BES substrate. However, a pretreatment step is needed to decompose the polymeric structure of lignocellulose and facilitate the fermentation by microorganisms in MFC cells [32–34]. For example, the treatment of pinewood flour and hazelnut leaves with sulfuric acid produced a hydrolysate that can efficiently generate electricity (0.43 V at 1000 Ohms external resistance) and hydrogen in single-chamber MFC and MEC, with the removal of more than 88% COD [32].

The combination of thermophilic dark fermentation of cellobiose and the use of MFC resulted in a significant increase in energy and hydrogen production. The effluent from fermentation was used as a substrate in MFC [33]. Alternatively, the hydrolysis of cellulose from corn cobs and sugarcane bagasse to glucose was achieved by a freshwater cyanobacterium (*Oscillatoria annae*). This was followed by electricity generation using three-chamber MFC and a co-culture of *Acetobacter aceti* and *Gluconobacter roseus*. The maximum power output obtained from corn cobs and sugarcane bagasse was 6.73 W/m^3 at 17.28 A/m^3 and 8.78 W/m^3 at 20.95 A/m^3, respectively [34].

2.4 Sewage Sludge

Sewage sludge is the mud-like residue obtained from the treatment of domestic sewage water, and it is often disposed of in rivers and lakes, leading to contamination of freshwater, groundwater, soil, and air. BESs can be used to treat sewage sludge, thereby reducing environmental pollution. Simultaneous energy production and COD removal can be achieved by utilizing sludge, which serves as both a substrate and an inoculum [35–39]. The sewage sludge can also generate biomethane using a double-chamber microbial fuel cell (MFC) [40].

3 Wastewater

Wastewater can be obtained from various sources, which affects its composition. The best performance in electricity generation was obtained using wastewater rich in carbon sources (acids, sugars, alcohols) or nitrogen sources (amino acids and proteins). The primary sources of wastewater are domestic and food wastewater, as well as agricultural, brewery, food and dairy industrial, and textile wastewater. Other sources include petroleum industry wastewater, hospital wastewater, pharmaceutical industry wastewater, and sewage sludge [4, 41–45].

Domestic wastewater is rich in organic matter and nutrients that can be effectively utilized in BES systems [39, 46]. MFCs are more efficient and can save energy by using low- or medium-strength wastewaters with high chemical oxygen demand (COD) content. However, when using high-strength wastes, the MFC system requires either dilution of the feeding solution or a longer hydraulic retention time (HRT) to prevent leakage of organic matter into the cathodic chamber [47].

BESs were efficiently utilized for wastewater treatment, which can also be utilized in irrigation [11, 48]. The high organic matter content in wastewater can block the soil pores due to biofilm formation. Moreover, oxygen availability in the soil is reduced as it is consumed by bacteria growing on the organic matter. The presence of nitrogen, phosphorus, and heavy metals in high concentrations in wastewater is also harmful to the soil and crops. It can accumulate in crops, presenting hazards to human health. The removal of organic matter and electricity generation were efficiently achieved using the MFC system [4, 41, 42, 49]. Scaling up studies is crucial for providing large amounts of water for irrigation; indeed, MFC scale-up and pilot-scale studies have been reported for volumes ranging from 1 to 1000 L [50–55]. However, the cost of MFC is still high, which limits its application in the agricultural sector.

BES's use to remove nitrate and perchlorate from wastewater and groundwater is called bio-electro remediation [56–58]. Moreover, BESs contribute to removing metals like arsenic, cadmium, and chromium from groundwater. Both arsenic and organic matter are oxidized at the anode. The perchlorate and other elements, like vanadium, chromium, and cadmium, are reduced at the cathode [58, 59].

The cattle rumen and rumen fluid contain microorganisms that can ferment plant food and produce sugars. The rumen is obtained from slaughterhouses and can be used in BESs as an inoculum and substrate [60]. Human urine is another waste that can be effectively utilized in MFCs to generate electricity [61–63], remove chemical oxygen demand (COD), and recover essential nutrients, such as nitrogen, phosphorus, and potassium ions, which can be further utilized as fertilizers [64, 65]. Like rumen, human urine can be used in MFC as both an inoculum and substrate, affording a comparable power output to urine with sludge [66]. It can be used undiluted or with inoculation with bioactive bacteria [63].

4 Conclusion and Future Perspectives

BESs represent a sustainable technology that can be applied efficiently to generate electricity or fuels, treat waste, and remove pollutants from wastewater. Thus, BESs can resolve environmental and energy problems and recover resources. Different substrates, including organic solid and liquid waste, can be used in BESs.

Despite the promising results obtained in most research conducted using BES and various substrates, electrodes, and microorganisms, certain limitations may prevent the application of this research on a large scale. The primary challenge is scaling up MFC systems, particularly for wastewater treatment. The cost of materials, particularly electrodes and membranes, is another challenge that hinders the widespread application of BESs. The amount of energy and electricity consumed and produced, especially on a large scale, is not fully studied, and further research is still needed to assess the method's energy efficiency. The use of solid organic waste is limited by its composition, which may contain plastics and polymers that are not biodegradable. Co-digestion, which involves mixing wastewater or other biodegradable solid waste, may be a solution to this problem.

References

1. Pant D, Singh A, Van Bogaert G, Irving Olsen S, Singh Nigam P, Diels L, Vanbroekhoven K (2012) Bioelectrochemical systems (BES) for sustainable energy production and product recovery from organic wastes and industrial wastewaters. RSC Adv 2:1248–1263. https://doi.org/10.1039/c1ra00839k

2. Azizul Moqsud M (2021) Bioelectricity from organic solid waste. In: Strategies of sustainable solid waste management. IntechOpen, London, pp 1–10. https://doi.org/10.5772/intechopen.95297
3. De Gioannis G, Dell'Era A, Muntoni A, Pasquali M, Polettini A, Pomi R, Rossi A, Zonfa T (2023) Bio-electrochemical production of hydrogen and electricity from organic waste: preliminary assessment. Clean Technol Environ Policy 25:269–280. https://doi.org/10.1007/s10098-022-02305-1
4. Sonawane JM, Mahadevan R, Pandey A, Greener J (2022) Recent progress in microbial fuel cells using substrates from diverse sources. Heliyon 8:e12353. https://doi.org/10.1016/j.heliyon.2022.e12353
5. Iskander SM, Brazil B, Novak JT, He Z (2016) Resource recovery from landfill leachate using bioelectrochemical systems: opportunities, challenges, and perspectives. Bioresour Technol 201:347–354. https://doi.org/10.1016/j.biortech.2015.11.051
6. Fernando E, Kyazze G, Ahsan A, Fernando P (2022) Expedited biodegradation of organic pollutants and refractory compounds using bio-electrochemical systems. In: Biodegradation technology of organic and inorganic pollutants. IntechOpen, London, pp 1–17. https://doi.org/10.5772/intechopen.99229
7. Tommasi T, Salvador GP, Quaglio M (2016) New insights in microbial fuel cells: novel solid phase anolyte. Sci Rep 6:29091. https://doi.org/10.1038/srep29091
8. He Z, Angenent LT (2006) Application of bacterial biocathodes in microbial fuel cells. Electroanalysis 18:2009–2015. https://doi.org/10.1002/elan.200603628
9. Wang H, Zhai P, Long X, Ma J, Li Y, Liu B, Xu Z (2023) Research progress on using biological cathodes in microbial fuel cells for the treatment of wastewater containing heavy metals. Front Microbiol 14:1270431. https://doi.org/10.3389/fmicb.2023.1270431
10. Parwate SA, Xue W, Koottatep T, Salam A (2024) Organic waste for bioelectricity generation in microbial fuel cells: effects of feed physicochemical characteristics. PRO 12:1110. https://doi.org/10.3390/pr12061110
11. Madondo NI, Rathilal S, Bakare BF, Tetteh EK (2023) Application of bioelectrochemical systems and anaerobic additives in wastewater treatment: a conceptual review. Int J Mol Sci 24:4753. https://doi.org/10.3390/ijms24054753
12. Budihardjo MA, Syafrudin AJ, Effendi S, Hidayat C, Purnawan AID, Lantasi FI, Muhammad BSR (2021) Waste valorization using solid-phase microbial fuel cells (SMFCs): recent trends and status. J Environ Manag 277:111417. https://doi.org/10.1016/j.jenvman.2020.111417
13. Chandrasekhar K, Kumar AN, Raj T, Kumar G, Kim SH (2021) Bioelectrochemical system-mediated waste valorization. Syst Microbiol Biomanuf 1:432–443. https://doi.org/10.1007/s43393-021-00039-7
14. Maddirala S, Bhadra S, Mozumder MSI, Garlapati VK, Sevda S (2024) Advancements in bio-electrochemical systems for solid organic waste valorization: a comprehensive review. PRO 12:805. https://doi.org/10.3390/pr12040805
15. Bolognesi S, Cecconet D, Callegari A, Capodaglio AG (2021) Bioelectrochemical treatment of municipal solid waste landfill mature leachate and dairy wastewater as co-substrates. Environ Sci Pollut Res 28:24639–24649. https://doi.org/10.1007/s11356-020-10167-7
16. Florio C, Nastro RA, Flagiello F, Minutillo M, Pirozzi D, Pasquale V, Ausiello A, Toscano G, Jannelli E, Dumontet S (2019) Biohydrogen production from solid phase-microbial fuel cell spent substrate: a preliminary study. J Clean Prod 227:506–511. https://doi.org/10.1016/j.jclepro.2019.03.316
17. Mahmoud M, Parameswaran P, Torres CI, Rittmann BE (2014) Fermentation pre-treatment of landfill leachate for enhanced electron recovery in a microbial electrolysis cell. Bioresour Technol 151:151–158. https://doi.org/10.1016/j.biortech.2013.10.053
18. Huang L, Li X, Cai T, Huang M (2018) Electrochemical performance and community structure in three microbial fuel cells treating landfill leachate. Process Saf Environ Prot 113:378–387. https://doi.org/10.1016/j.psep.2017.11.008

19. Zhang G, Jiao Y, Lee DJ (2015) Transformation of dissolved organic matters in landfill leachate-bioelectrochemical system. Bioresour Technol 191:350–354. https://doi.org/10.1016/j.biortech.2015.05.082
20. Li S, Chen G (2018) Effects of evolving quality of landfill leachate on microbial fuel cell performance. Waste Manag Res 36:59–67. https://doi.org/10.1177/0734242x17739969
21. Zhang G, Jiao Y, Lee DJ (2015) A lab-scale anoxic/oxic-bioelectrochemical reactor for leachate treatments. Bioresour Technol 186:97–105. https://doi.org/10.1016/j.biortech.2015.03.022
22. Vázquez-Larios AL, Solorza-Feria O, Poggi-Varaldo HM, De Guadalupe González-Huerta R, Ponce-Noyola MT, Ríos-Leal E, Rinderknecht-Seijas N (2014) Bioelectricity production from municipal leachate in a microbial fuel cell: effect of two cathodic catalysts. Int J Hydrog Energy 5:16667–16675. https://doi.org/10.1016/j.ijhydene.2014.05.178
23. Hernández-Flores G, Poggi-Varaldo HM, Romero-Castañón T, Solorza-Feria O, Rinderknecht-Seijas N (2017) Harvesting energy from leachates in microbial fuel cells using an anion exchange membrane. Int J Hydrog Energy 42:30374–30382. https://doi.org/10.1016/j.ijhydene.2017.08.201
24. Choi J, Ahn Y (2015) Enhanced bioelectricity harvesting in microbial fuel cells treating food waste leachate produced from biohydrogen fermentation. Bioresour Technol 183:53–60. https://doi.org/10.1016/j.biortech.2015.01.109
25. Nguyen HTH, Kakarla R, Min B (2017) Algae cathode microbial fuel cells for electricity generation and nutrient removal from landfill leachate wastewater. Int J Hydrog Energy 42:29433–29442. https://doi.org/10.1016/j.ijhydene.2017.10.011
26. Nguyen HTH, Min B (2020) Leachate treatment and electricity generation using an algae-cathode microbial fuel cell with continuous flow through the chambers in series. Sci Total Environ 723:138054. https://doi.org/10.1016/j.scitotenv.2020.138054
27. Feng Q, Xu L, Xu Y, Liu C, Lu Y, Wang H, Wu T, Wang R, Chen Y, Cheng Y (2020) Treatment of aged landfill leachate by a self-sustained microbial fuel cell-microbial electrolysis cell system. Int J Electrochem Sci 15:1022–1033. https://doi.org/10.20964/2020.01.19
28. Hassan M, Wei H, Qiu H, Jaafry SWH, Su Y, Xie B (2018) Power generation and pollutants removal from landfill leachate in microbial fuel cell: variation and influence of anodic microbiomes. Bioresour Technol 247:434–442. https://doi.org/10.1016/j.biortech.2017.09.124
29. Puig S, Serra M, Coma M, Cabré M, Dolors Balaguer M, Colprim J (2011) Microbial fuel cell application in landfill leachate treatment. J Hazard Mater 185:763–767. https://doi.org/10.1016/j.jhazmat.2010.09.086
30. Hu J, Xu L, Jing Q, Xie M, Qing D (2017) Differences in contaminants removal efficiency and electricity production in disposing leachate with chemical-cathode and aerobic bio-cathode MFC. Kem Ind Chem Chem Eng 66:519–524. https://doi.org/10.15255/KUI.2017.013
31. Wang CT, Liao FY, Liu KS (2013) Electrical analysis of compost solid phase microbial fuel cell. Int J Hydrog Energy 38:11124–11130. https://doi.org/10.1016/j.ijhydene.2013.02.120
32. Catal T, Liu H, Fan Y, Bermek H (2019) A clean technology to convert sucrose and lignocellulose in microbial electrochemical cells into electricity and hydrogen. Bioresour Technol Rep 5:331–334. https://doi.org/10.1016/j.biteb.2018.10.002
33. Varanasi JL, Roy S, Pandit S, Das D (2015) Improvement of energy recovery from cellobiose by thermophillic dark fermentative hydrogen production followed by microbial fuel cell. Int J Hydrog Energy 40:8311–8321. https://doi.org/10.1016/j.ijhydene.2015.04.124
34. Krishnaraj RN, Berchmans S, Pal P (2015) The three-compartment microbial fuel cell: a new sustainable approach to bioelectricity generation from lignocellulosic biomass. Cellulose 22:655–662. https://doi.org/10.1007/s10570-014-0463-4
35. Taşkan B, Taşkan E, Hasar H (2020) Electricity generation potential of sewage sludge in sediment microbial fuel cell using Ti–TiO2 electrode. Environ Prog Sustain Energy 39:e13407. https://doi.org/10.1002/ep.13407
36. Rudenko RR, Vasilevich EE, Zhdanova GO, Chizhick KI, Topchiy IA, Stom DI (2018) The use of urban sewage sludge as a substrate in a microbial fuel cell. Int J Eng Technol 7:277–280. https://doi.org/10.14419/ijet.v7i2.23.11931

37. Bose D, Dhawan H, Kandpal V, Vijay P, Gopinath M (2018) Sustainable power generation from sewage and energy recovery from wastewater with variable resistance using microbial fuel cell. Enzym Microb Technol 118:92–101. https://doi.org/10.1016/j.enzmictec.2018.07.007
38. Shaheen Aziz AP (2015) Utilization of sewage sludge for production of electricity using mediated salt bridge based dual chamber microbial fuel cell. J Bioprocess Biotech 5:1000251. https://doi.org/10.4172/2155-9821.1000251
39. Bose D, Dhawan H, Kandpal V, Vijay P, Gopinath M (2018) Bioelectricity generation from sewage and wastewater treatment using two-chambered microbial fuel cell. Int J Energy Res 42:4335–4344. https://doi.org/10.1002/er.4172
40. Mehariya S, Patel AK, Obulisamy PK, Punniyakotti E, Wong JWC (2018) Co-digestion of food waste and sewage sludge for methane production: current status and perspective. Bioresour Technol 265:519–531. https://doi.org/10.1016/j.biortech.2018.04.030
41. Pandey P, Shinde VN, Deopurkar RL, Kale SP, Patil SA, Pant D (2016) Recent advances in the use of different substrates in microbial fuel cells toward wastewater treatment and simultaneous energy recovery. Appl Energy 168:706–723. https://doi.org/10.1016/j.apenergy.2016.01.056
42. Gude VG (2016) Wastewater treatment in microbial fuel cells e an overview. J Clean Prod 122:287–307. https://doi.org/10.1016/j.jclepro.2016.02.022
43. Cecconet D, Bolognesi S, Molognoni D, Callegari A, Capodaglio AG (2018) Influence of reactor's hydrodynamics on the performance of microbial fuel cells. J Water Process Eng 26:281–288. https://doi.org/10.1016/j.jwpe.2018.10.019
44. Cecconet D, Molognoni D, Callegari A, Capodaglio AG (2018) Agro-food industry wastewater treatment with microbial fuel cells: energetic recovery issues. Int J Hydrog Energy 43:500–511. https://doi.org/10.1016/j.ijhydene.2017.07.231
45. Molognoni D, Chiarolla S, Cecconet D, Callegari A, Capodaglio AG (2018) Industrial wastewater treatment with a bioelectrochemical process: assessment of depuration efficiency and energy production. Water Sci Technol 77:134–144. https://doi.org/10.2166/wst.2017.532
46. Lefebvre O, Uzabiaga A, Chang IS, Kim BH, Ng HY (2011) Microbial fuel cells for energy self-sufficient domestic wastewater treatment-a review and discussion from energetic consideration. Appl Microbiol Biotechnol 89:259–270. https://doi.org/10.1007/s00253-010-2881-z
47. Zhang F, He Z (2013) A cooperative microbial fuel cell system for waste treatment and energy recovery. Environ Technol 34:1905–1913. https://doi.org/10.1080/09593330.2013.770540
48. Abourached C, English MJ, Liu H (2016) Wastewater treatment by microbial fuel cell (MFC) prior irrigation water reuse. J Clean Prod 137:144–149. https://doi.org/10.1016/j.jclepro.2016.07.048
49. Agudelo-escobar LM, Cabrera SE, Rossa CA (2022) A bioelectrochemical system for waste degradation and energy recovery from industrial coffee wastewater. Front Chem Eng 4:814987. https://doi.org/10.3389/fceng.2022.814987
50. Janicek A, Fan Y, Liu H (2014) Design of microbial fuel cells for practical application: a review and analysis of scale-up studies. Biofuels 5:79–92. https://doi.org/10.4155/bfs.13.69
51. Feng Y, He W, Liu J, Wang X, Qu Y, Ren N (2014) Bioresource technology a horizontal plug flow and stackable pilot microbial fuel cell for municipal wastewater treatment. Bioresour Technol 156:132–138. https://doi.org/10.1016/j.biortech.2013.12.104
52. Wu S, Li H, Zhou X, Liang P, Zhang X, Jiang Y, Huang X (2016) A novel pilot-scale stacked microbial fuel cell for ef fi cient electricity generation and wastewater treatment. Water Res 98:396–403. https://doi.org/10.1016/j.watres.2016.04.043
53. Dong Y, Qu Y, He W, Du Y, Liu J, Han X, Feng Y (2015) A 90-liter stackable baffled microbial fuel cell for brewery wastewater treatment based on energy self-sufficient mode. Bioresour Technol 195:66–72. https://doi.org/10.1016/j.biortech.2015.06.026
54. Hiegemann H, Herzer D, Nettmann E, Lübken M, Schulte P, Schmelz KG, Gredigk-Hoffmann S, Wichern M (2016) An integrated 45 L pilot microbial fuel cell system at a full-scale wastewater treatment plant. Bioresour Technol 218:115–122. https://doi.org/10.1016/j.biortech.2016.06.052

55. Ge Z, Wu L, Zhang F, He Z (2015) Energy extraction from a large-scale microbial fuel cell system treating municipal wastewater. J Power Sources 297:260–264. https://doi.org/10.1016/j.jpowsour.2015.07.105
56. Sevda S, Sreekishnan TR, Pous N, Puig S, Pant D (2018) Bioelectroremediation of perchlorate and nitrate contaminated water: a review. Bioresour Technol 255:331–339. https://doi.org/10.1016/j.biortech.2018.02.005
57. Cecconet D, Zou S, Capodaglio AG, He Z (2018) Evaluation of energy consumption of treating nitrate-contaminated groundwater by bioelectrochemical systems. Sci Total Environ 636:881–890. https://doi.org/10.1016/j.scitotenv.2018.04.336
58. Cecconet D, Sabba F, Devecseri M, Callegari A, Capodaglio AG (2020) In situ groundwater remediation with bioelectrochemical systems: a critical review and future perspectives. Environ Int 137:105550. https://doi.org/10.1016/j.envint.2020.105550
59. Cecconet D, Callegari A, Capodaglio AG (2018) Bioelectrochemical systems for removal of selected metals and perchlorate from groundwater: a review. Energies 11:2643. https://doi.org/10.3390/en11102643
60. Deepika J, Meignanalakshmi S, Thilagaraj RW (2015) The optimization of parameters for increased electricity production by a microbial fuel cell using rumen fluid. Int J Green Energy 12:333–338. https://doi.org/10.1080/15435075.2013.847838
61. Cid CA, Stinchcombe A, Ieropoulos I, Hoffmann MR (2018) Urine microbial fuel cells in a semi-controlled environment for onsite urine pre-treatment and electricity production. J Power Sources 400:441–448. https://doi.org/10.1016/j.jpowsour.2018.08.051
62. Ieropoulos I, Greenman J, Melhuish C (2012) Urine utilisation by microbial fuel cells; energy fuel for the future. Phys Chem Chem Phys 14:94–98. https://doi.org/10.1039/c1cp23213d
63. Barbosa SG, Peixoto L, Ter Heijne A, Kuntke P, Alves MM, Pereira MA (2017) Investigating bacterial community changes and organic substrate degradation in microbial fuel cells operating on real human urine. Environ Sci Water Res Technol 3:897–904. https://doi.org/10.1039/c7ew00087a
64. Freguia S, Logrieco ME, Monetti J, Ledezma P, Virdis B, Tsujimura S (2019) Self-powered bioelectrochemical nutrient recovery for fertilizer generation from human urine. Sustainability 11:5490. https://doi.org/10.3390/su11195490
65. Lu S, Li H, Tan G, Wen F, Flynn MT, Zhu X (2019) Resource recovery microbial fuel cells for urine-containing wastewater treatment without external energy consumption. Chem Eng J 373:1072–1080. https://doi.org/10.1016/j.cej.2019.05.130
66. Obata O, Salar-Garcia MJ, Greenman J, Kurt H, Chandran K, Ieropoulos I (2020) Development of efficient electroactive biofilm in urine-fed microbial fuel cell cascades for bioelectricity generation. J Environ Manag 258:109992. https://doi.org/10.1016/j.jenvman.2019.109992

Principle of Bioelectrochemical Systems

Abdul Sattar Jatoi, Rohail Khan, Nabisab Mujawar Mubarak, Arshad Iqbal, and Abdul Karim Shah

Abstract BES represents a contemporary system that combines microbiological and electrochemical approaches to transform chemical energy from organic substrates into electrical power through microbial generation. Through their three main categories—MFCs and MECs, along with their related derivatives—these systems offer sustainable alternatives for energy development, wastewater processing, environmental surveillance, and competent source recovery. The electron transfer abilities of electrode-active microorganisms form the basis for BES operations, enabling either electricity generation or hydrogen production. The various performance aspects of BES are influenced by microbial species, substrate access, electrode components, systemic design, and environmental conditions, including pH and temperature. BES applications span four major areas: renewable energy generation, environmental cleanup services, water desalination, and biotechnological detection systems. The advancement of electrode materials, synthetic biological procedures, and system integration has enhanced BES capability, overcoming material cost inefficiencies and limitations in power generation. This abstract outlines the core elements and working principles as well as their extensive application in developing sustainable power systems and environmental technologies.

Keywords Principle · Bioelectrochemical system · Microbial fuel cell

A. S. Jatoi (✉) · R. Khan · A. Iqbal · A. K. Shah
Department of Chemical Engineering, Dawood University of Engineering and Technology, Karachi, Pakistan
e-mail: abdul.sattar@duet.edu.pk

N. M. Mubarak
Department of Chemical Engineering, College of Engineering, Imam Mohammad Ibn Saud Islamic University (IMSIU), Riyadh, Saudi Arabia

Department of Biosciences, Saveetha Institute of Medical and Technical Sciences, Chennai, India

A. Sattar Jatoi, N. M. Mubarak (eds.), *Bio-Electrochemical Conversion of Waste to Fuel*, Green Energy and Technology,
https://doi.org/10.1007/978-981-95-4310-6_3

1 Introduction

Bioelectrochemical systems (BES) constitute an active research area that combines electrochemical operations with biological techniques to generate energy, facilitate waste removal, and promote environmental improvements [1]. The conversion of electricity or chemical energy through these systems occurs because specific microorganisms can transfer electrons between electrodes [2, 3]. The development of BES technologies began with microbial fuel cell research in the early stages and later progressed to microbial electrolysis cells (MECs) and bioelectrochemical sensors. BES technology serves essential purposes by meeting global energy needs while aiding environmental cleanup and developing sustainable technologies. A basic BES contains four main components: electrodes, microbial cells, electrolytes, and an external circuit [4]. The anode and cathode components connect electron transfer between microorganisms and the system. However, their performance depends heavily on selecting the electrode material between carbon-based and metal oxide electrodes. Under various microbial species and environmental conditions, microbial cells either conduct direct electron transfer (DET) or mediated electron transfer (MET) to transfer electrons to the electrode during their metabolic processes [5]. The electrolyte solution in BES provides essential ionic movement, while the external circuit enables electronic transportation to produce electricity and power water splitting during MEC operations [6]. The primary functionality of BES depends on how electrons are transported through the design and the microbial species present within it. DET employs membrane-bound redox proteins for electronic energy transfer between microbes and electrodes, whereas MET requires mediators or cofactors to facilitate such transfers [7]. Knowledge of these mechanisms is key to maximizing BES performance capabilities. BES systems are organized into groups that serve different functional purposes [8]. The most commonly scrutinized BES technology is the microbial fuel cell, which allows microbes to break down organic matter while generating electricity, thus enabling waste management and power extraction from organic waste materials. Renewable energy storage receives potential from microbial electrolysis cells (MECs) because they utilize microbial metabolism to convert water into hydrogen gas. Bioelectrochemical sensors are utilized in environmental monitoring systems that employ microbial actions to detect specific contaminants and pollutants [9]. Bio-batteries and microbial desalination cells (MDCs) represent new technologies combining BES operations to store energy and desalinate water. The electrical functioning of BES depends heavily on the metabolism of microorganisms, particularly essential species such as *Geobacter sulfurreducens* and *Shewanella oneidensis*. The metabolic process in BES includes anaerobic respiratory mechanisms through which microbes convert organic compounds into energy using electrode-based electron acceptors. Functions of BES depend directly on the availability of carbon sources combined with electron donors and electron acceptors because these nutrients influence the activities of the microbial community. An optimal system performance occurs through mixed cultures because separate species support one another when they work together in the community [10]. The performance of BES depends on three primary factors: electrode material selection, system design structure, and microbial growth conditions. Evaluating electrode materials is critical since efficiency and stability directly impact [11]. Electrochemical systems utilize carbon-based materials, including graphite and carbon nanotubes, which offer large surface areas and high conductivity. Still, metal-based electrodes, such as platinum or gold, exhibit better performance levels. However, they possess higher costs [12]. The performance of such systems directly depends on the microbial growth conditions, including pH, temperature levels, and available nutrients, which control microbial metabolic operations. The cells' connection to the electrodes depends on how well microorganisms interact with electrodes and how electron mediators influence electron transfer [13]. The overall operational efficiency of BES greatly benefits from system design improvements, which optimize the movement of electrons in conjunction with substrate items and microorganisms within the system. BES demonstrates great potential as a power generation technology for bioelectricity production. Microbial fuel cells generate energy through oxidation reactions of microorganism substrates at the anode, allowing electrons to travel externally through an electric circuit to the cathode. Since BES produces less energy than other standard power sources it offers itself as a suitable solution for standalone applications and small-scale systems whose power needs can be met through organic waste conversion. Various factors, including electrode materials, microbial species, substrate concentrations, and system design components, influence the overall energy conversion efficiency in BES. Developments in microorganisms for electrodes represent a critical path to improving the performance and efficiency of BES systems. BES stands out from standard renewable energy sources by utilizing organic waste as a resource for both power generation and waste management. The various applications of BES technologies continue to grow in number. During wastewater treatment, MFCs and MECs combine the pollution treatment of organic pollutants with power and hydrogen production capabilities. BES delivers special value to waste-to-energy implementations due to its ability to operate in two capacities. The process of environmental remediation benefits from BES technology, which allows microbes in BES systems to minimize toxic compounds that physically contaminate soil and heavy metals. Bioelectrochemical sensors enable real-time monitoring of environmental conditions, as these systems detect contaminants such as heavy metals, organic pollutants, and pathogens. BES is a promising tool for remote power generation, as it enables the supply of electricity to sensors, small electronic devices, and communication systems in off-grid areas. The development of BES technologies encounters various obstacles, as they produce energy inefficiently and are expensive to build, while also being prone to microbial changes. Power density represents the major obstacle for BES systems because their electrical output falls below conventional power generation levels [14]. Furthermore, the cost of materials, especially high-performance electrodes and redox mediators, poses a barrier to large-scale deployment. System performance suffers from biofouling as well as microbial activity losses which present stability problems [15].

Productive progress requires improved research in material science together with advancements in microbial engineering and systematic optimization. Investigations in BES technologies have produced two significant advancements involving the manufacture of carbon nanotubes alongside conductive polymers and the employment of synthetic biology to boost microbial functionality [15]. The new technological developments seek to enhance electrical output power and lower system expenses while improving the stability of microbial networks. BES possesses promising potential through future applications which combine BES with solar and wind energy units to enhance net energy efficiency [16]. The ongoing discoveries about microbial metabolism and electron transfers and system designs promote BES potential for sustainable energy solutions and environmental management.

2 Types of Bioelectrochemical Systems

Bioelectrochemical systems (BES) operate in different types, using different mechanisms to achieve specific applications. The five prominent bioelectrochemical systems (BES) consist of microbial fuel cells (MFCs), microbial electrolysis cells (MECs), bioelectrochemical sensors, bio-batteries, and microbial desalination cells (MDCs), which serve distinct functions for energy output, waste management, and environmental screening [16].

The research focuses on microbial fuel cells (MFCs), the leading type of bioelectrochemical system (BES), because they utilize microbial processes to convert organic substances into electricity [17]. The anaerobic environment provides microorganisms with organic matter, which they decompose while capturing electrons that flow toward the anode [18]. The anode receives an electron flow through the external circuit, which combines with cathode protons and oxygen atoms to produce water. Scientists view MFCs as a renewable energy system because the technology converts the chemical energy of organic waste into electricity, which supports the operation of devices and sensors [19]. Microbial metabolism is fundamental to MFC efficiency because specific microorganisms possess the ability to transfer electrons directly to the electrode through direct electron transfer (DET) or mediate electron transfer via mediators. MFC technology demonstrates flexibility across wastewater treatment systems, bioremediation processes, and remote power generation, as it simultaneously cleans wastewater using the electrical energy output [20].

Microbial electrolysis cells (MECs) serve a different purpose than microbial fuel cells (MFCs), as they produce clean hydrogen gas fuels rather than generating electricity. MECs operate through electrolysis due to microbial anode metabolism, which produces electrons and protons [21]. An external voltage allows protons and electrons to reach the cathode, where they combine to form hydrogen gas [21]. The application of MECs demonstrates exceptional effectiveness in producing hydrogen from renewable energy by converting organic waste or wastewater into hydrogen.

MECs serve wastewater management by simultaneously purifying water and generating renewable hydrogen energy [22].

The Bioelectrochemical Sensor is a specific type of BES for environmental substance detection and monitoring purposes. Through changes in metabolic activity, microorganisms reveal environmental changes that these sensors determine by measuring potential and current modifications [23]. Bioelectrochemical sensors are environmental detectors for heavy metals, pesticides, and organic compounds in water or soil. Utilizing specific microorganisms in these sensors establishes a cost-effective, real-time environmental surveillance system due to their sensitivity to environmental contaminants [24]. These sensors operate successfully for environmental diagnostics and pollution detection in both industrial and field applications, providing swift feedback regarding contaminant presence [25].

The energy storage function demands Bio-batteries, which represent a type of BES. Bio-batteries operate similarly to their chemical counterparts, storing energy for future use. However, they convert stored energy using biological materials, consisting of enzymes or living microorganisms, instead of conventional metal electrodes [26]. The technology serves small energy storage needs, including powering portable devices and supporting medical system functions. Bio-batteries offer greater environmental benefits than chemical batteries, as they utilize renewable biological ingredients and degrade through biodegradation during their usable lifespan [27].

Microbial Desalination Cells (MDCs) represent an advanced Bioelectrochemical System (BES) designed to address global freshwater challenges. Utilizing microorganisms in MDC offers a sustainable technology for water desalination through efficient salt removal methods. The microbial metabolic activity at the anode of an MDC generates protons and electrons while the ion transport process decreases the water salinity [28]. The desalination of saline and brackish water sources using MDCs becomes feasible when linked with MFCs or MECs to generate power throughout the desalination process [29]. A unified solution serves as a promising answer to simultaneously address both the water shortage problem and power generation requirements [30].

Several hybrid systems combining BES components demonstrate improved performance alongside a greater number of applications than single systems [31]. Scientists integrate microbial fuel cells and microbial electrolysis cells into a single hybrid system, enabling them to achieve optimal outcomes in energy extraction and hydrogen generation [32]. Active research is currently underway to integrate BES with solar or wind power technologies, aiming to design efficient, multipurpose systems that meet both energy and environmental needs [33].

The choice of a BES system depends on the application's specific needs, as different types of systems are best suited for particular deployments [34]. Microbial fuel cells' low-power operating range suits them for remote wastewater treatment systems, while microbial electrochemical cells produce hydrogen for specific applications [35]. Bioelectrochemical sensors effectively serve environmental surveillance needs, and bio-batteries offer a sustainable power storage method. MDCs deliver possible water scarcity solutions when applied to regions with high salinity content. BES research and development indicate that these systems will provide new solutions across energy production, environmental cleanup, and resource management applications [36].

3 Principles of Electron Transfer in Bioelectrochemical Systems (BES)

Electron transfer is a fundamental operation that makes bioelectrochemical systems (BES) function effectively [37]. Microorganisms in BES systems enable the conversion of chemical energy into electrical energy by transmitting electrons between the cells and electrodes [38]. BES implements either Direct Electron Transfer (DET) or Mediated Electron Transfer (MET) mechanisms to facilitate electron transfer within the system, thereby affecting system performance and efficiency. Knowledge of these processes remains essential for enhancing BES setups in power generation, waste treatment, and environmental applications (Fig. 1).

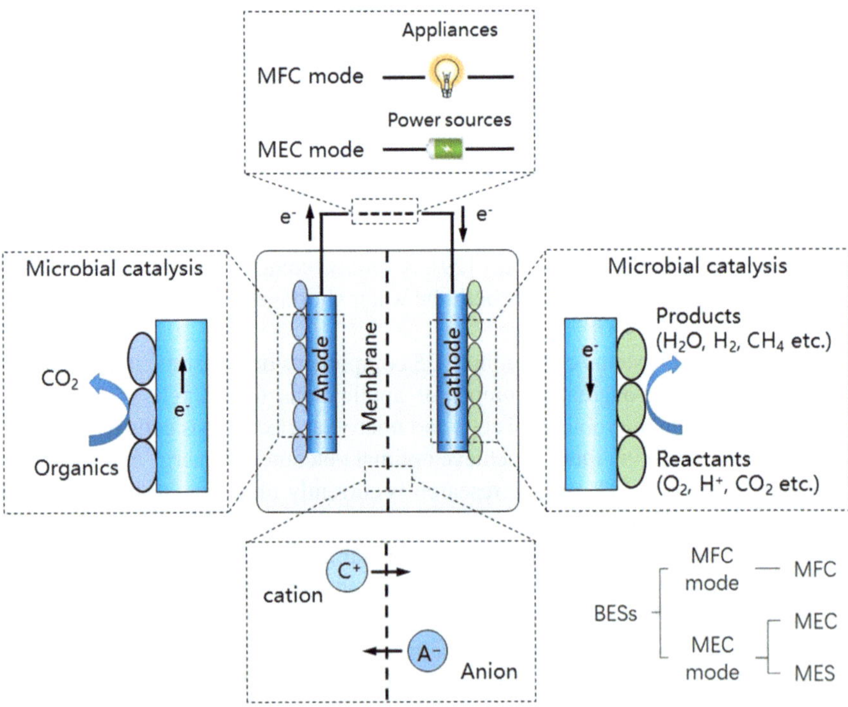

Fig. 1 An overview of the principles of BESs [39]

3.1 Direct Electron Transfer (DET)

The electrode receives electrons from microorganisms through direct electron transfer, eliminating the need for mediator compounds. External bacteria exhibit this phenomenon through specialized membrane proteins that facilitate contact between their redox centers and electrode surfaces. Microbes transfer electron flow from their cellular structures to the anode through proteins found in the outer membrane, specifically cytochromes [40]. The metabolic pathways inside cells transmit electrons to the electrode through electron carrier proteins that produce electricity in MFCs. The execution of DET by microorganisms is influenced by three main elements: the type of microorganism species, the presence of conductive pili or nanowires, and the selection of electrode material [41]. Due to their electron-conducting proteins and pili, *Geobacter sulfurreducens* and *Shewanella oneidensis* are recognized for their effective anode electricity generation in high-performance microbial fuel cells (MFCs) [42].

Among the benefits of DET resides its ability to achieve high power density through direct electron transfer, which prevents losses that mediators could trigger. Not all microorganisms can utilize DET, as only those species that have developed specific electron transfer mechanisms can effectively complete the process. Research devoted to BES continues through investigations seeking better microbial strains and improved electrode materials to enhance DET performance [43].

3.2 Mediated Electron Transfer (MET)

A significant difference exists between DET since mediated electron transfer relies on soluble redox-active molecules and electron mediators, which help carry electrons between microorganisms and electrodes [44]. The electron transfer process utilizes two mediators: flavins, which occur naturally, and methyl viologen or neutral red, which serve as externally added compounds for cell-electrode electron exchange. The microbial cells supply electrons to mediators, which subsequently transport these electrons to electrodes, generating electricity or powering microbial electrolysis cells (MECs) and other types of bioelectrochemical systems (BES) [45].

The introduction of mediators in BES enables the operation of multiple microorganisms that lack direct electron transfer abilities through MET since the mediators facilitate microbial electron transfer to electrodes [46]. Implementing mediators enables the exploitation of diverse microbial species for BES systems, enhancing functionality in different operating environments. Using MET provides an alternative to DET when DET proves inefficient or makes implementation impractical, thereby establishing an alternative mechanism for conducting electrons. Implementing mediators presents several obstacles, as they may impact mediator stability and toxicity properties, and optimal mediator concentration levels require adjustment for maximum efficiency [47].

MECs require mediators to transfer electrons during hydrogen production by facilitating the movement of electrons from the anode to the cathode. Through this process, mediators help the cathode reduce protons effectively, generating hydrogen gas. The selection and management of mediators determine optimal performance because high or inappropriate concentrations of mediators negatively affect microbial metabolism and produce unwanted side effects [48].

4 The Role of Electrode Materials in Electron Transfer

BES electron transfer efficiency directly relates to the material selection of electrodes. Electrode materials form the connection point between external circuitry and microbial cells, while their surface characteristics strongly influence the electron transfer rate. The best materials for BES applications exhibit high electrical conductivity, a large surface area, and suitable chemical reactivity [49].

Carbon-based materials, including graphite, carbon cloth, and carbon nanotubes, are frequently used because they provide high conductivity with large surface area values and biocompatible qualities [50]. These inexpensive materials possess high modification potential and availability, which enables their use in various types of BES designs. Electrode material research has yielded conductive polymers, metal oxide coatings, and composite materials, facilitating enhanced microbial attachment and surface contact to improve electron transfer performance [51].

The layout of electrodes in BES focuses on expanding the contact surface with microorganisms to achieve higher electron conversion ratios throughout each unit of microbial cell mass. Researchers develop improved system electronic transfer techniques by modifying electrodes with conductive nanoparticles or enzymes to increase performance [52].

5 Electrochemical Pathways in BES

The electrochemical pathways controlling electron transfer processes in bioelectrochemical systems depend on microbial metabolisms, the types of substrates used, and the electrode's electronegativity. Bacterial cells extract electrons from organic molecules such as sugars and fatty acids, and the electrodes receive these electrons during the process. Bacteria oxidize organic matter inside MFCs, releasing electrons and protons in the process. Water forms when electrons from the anode use the external circuit to reach the cathode while protons from the electrolyte combine with oxygen. A redox reaction is the basic electrochemical process because anode oxidation co-occurs with cathode reduction [53].

The electron transfer process works efficiently because it considers microbial metabolite redox potential alongside electrode potential and microbial cell membrane electrochemical gradients [54]. Beatification efficiency depends on the predefined BES elements, electrode materials, surface dimensions, and physical structure. The power generation in MFCs and the hydrogen production in MECs can be maximized by properly optimizing critical factors [55].

6 Factors Influencing Electron Transfer Efficiency

The rate of electron transfer within BES depends on the selection of bacterial organisms, as well as substrate accessibility, alongside solution pH drift, environmental temperature variations, and the concentrations of toxic compounds and inhibitory substances [56]. The effectiveness of electron transfer in BES depends heavily on the choice of microbial species, as some microorganisms conduct transfer more rapidly due to their specific metabolic pathways, which contain redox proteins [57]. Electron transfer efficiency in BES depends directly on the availability of electron donor substances, which directly correlates with electron transmission to the electrode.

Environmental factors, including pH and temperature, influence electron transfer efficiency and microbial metabolism. Microbes cannot grow properly within extreme pH conditions, while the electrode surface reactions might become impaired, resulting in decreased efficiency. Temperature changes affect the metabolic functions of microorganisms, causing BES performance to become unstable [58].

The operation of a bioelectrochemical system depends fundamentally on the principles of electron transfer that govern its functionality. The performance of BES varies among microbial species and system design because these systems use direct or mediated electron transfer methods. Effective management of BES technologies depends on enhancing these mechanisms by selecting the correct electrodes and controlling environmental conditions to achieve higher efficiency levels for various applications, including renewable energy and environmental and wastewater management systems. Continued research will lead to the development of improved microbial strains, electrode materials, and system configurations that enhance electronic transfer efficiency in BES systems, thereby establishing them as a sustainable technology for energy solutions.

7 Principles of Electron Transfer in Bioelectrochemical Systems (BES)

The functional success of bioelectrochemical systems (BES) relies on electron transfer as their fundamental operational principle. The conversion of chemical energy into electrical energy (or vice versa) through electron transfer between cells and electrodes is a primary operation of microorganisms within a BES. Direct Electron Transfer (DET) and Mediated Electron Transfer (MET) are primary mechanisms for electron transfer in BES systems, thus affecting the system's operational efficiency. The basis for maximizing BES performance in energy production, alongside wastewater management and environmental applications, depends on understanding these two mechanisms.

7.1 Direct Electron Transfer (DET)

Microorganisms forego using any mediators to transfer electrons directly to electrodes. Bacteria with specialized redox proteins on their outer membrane exhibit this phenomenon, as these proteins directly contact and interact with the electrode surface. The electrochemical process enables outer membrane proteins, particularly cytochromes, to transfer electrons between microbial cells and anode surfaces [59]. The proteins function as electron carriers to transport electrons produced by metabolic pathways between the cell interior and the electrode, generating an electric current in MFCs. The execution of DET by microorganisms depends on three essential factors: the variety of microorganisms, conductive pili or nanowires, and the choice of electrode [60]. Geobacter sulfurreducens and Shewanella oneidensis demonstrate exceptional electron donor capabilities for electrodes because these species possess conductive proteins and pili, enabling high-performance microbial fuel cell (MFC) operation [61].

The power density capabilities of direct electron transfer rise since this method helps reduce losses that would otherwise develop through mediators. The ability to carry out DET exists only for microorganisms that have developed specialized electron transfer mechanisms, as this process is not effective for all microbial species [62]. The improvement of electrodes in conjunction with microorganisms for maximizing DET efficiency remains an active research focus within BES.

7.2 Mediated Electron Transfer (MET)

The main distinction between DET and mediated electron transfer lies in the application of soluble electron mediators and redox-active molecules, which enable communication between microorganisms and electrodes. Two electron transfer mediators exist—naturally produced flavins and externally supplied compounds, including methyl viologen and neutral red, which act as electron shuttles between bacterial cells and electrodes [63]. Microbiological cells transmit electrons to the electrode surface through mediators that accept electrons from the cells, resulting in electricity generation or electrochemical reactions in microbial electrochemical cells (MECs) and other biocatalytic energy systems (BES) types.

MET enables broader participation of microorganisms in BES technology because it allows non-direct electron transfer-capable bacteria to utilize the system.

BES performance optimization benefits from using mediators because they enable the utilization of multiple microbial species across various environments [64]. The MET technology provides an alternative electronic pathway for systems where DET is ineffective or creates obstacles that challenge electron flow. Using mediators in power generation systems introduces new operational challenges because mediator breakdown occurs while exposure to toxic elements is present and optimizing proper mediator dosing is necessary.

During hydrogen production in MEC technology, mediators must perform the essential task of transmitting electrons from the anode to the cathode. The efficient reduction of cathode protons occurs through mediation processes, resulting in the production of hydrogen gas. The optimal performance of microbial systems depends heavily on these two factors, as the amount and selection of mediators determine microbial metabolic efficiency and unwanted side reactions.

7.3 The Role of Electrode Materials in Electron Transfer

Electron transfer efficiency in BES depends substantially on which electrode material gets selected. The electron transfer rate between microbial cells and the external circuit depends heavily on the properties of electrode materials, which serve as the connecting point between the two entities. Applications of BES perform optimally when using materials that possess high electrical conductivity, a large surface area, and suitable chemical reactivity.

Graphite, carbon cloth, and carbon nanotubes are the top choices for BES technologies due to their high conductivity, large surface area characteristics, and good biocompatibility. The materials demonstrate cost efficiency and easy modifiability, accommodating different BES project types. Advanced discoveries in electrode science now use combined materials, including conductive polymers, metal oxide coatings, and composite structures, which enhance electron transfer by improving microbial attachment and surface conductivity.

The contact area between microbial cells and electrodes in BES devices reaches maximum size, leading to higher electron transfer across each unit of microbial biomass. Research teams investigate various electrode improvement methods, including coating electrodes with conductive nanoparticles and enzymes, to enhance electron transfer processes and improve system efficiency.

7.4 Electrochemical Pathways in BES

The flow of electrochemical pathways that regulate electron transfer is influenced by three key factors: microbial metabolism, substrate type, and electrode potential. The microbial cells derive electrons from organic carbon substances, including sugars and fatty acids, which they supply to electrodes functioning as electron acceptors. The anode environment in MFCs is inhabited by bacteria that oxidize organic matter, thereby producing electrons and protons. The electrons route from the anode through an external wiring structure to the cathode for water production, following the combination of protons and oxygen in the electrolyte solution. A redox reaction is the electrochemical mechanism separating the oxidation process into the anode region and the reduction reaction into the cathode region.

The transfer of electrons in this process depends on three main elements: the redox potential of microbial metabolites, the electrode potential, and the electrochemical gradients of the microbial cell membrane. An efficient design of the BES electrodes, incorporating material composition, surface dimensions, and layout, can improve the pace and effectiveness of the redox transformation procedures. Enhancing main factors and subsidiary conditions allows MFCs and MECs to achieve their highest possible power level and hydrogen generation capacity.

8 Factors Influencing Electron Transfer Efficiency

Various factors, including microbial species composition, substrate availability, pH values, temperatures, and exposure to inhibitors or toxic materials, influence the operation of electron transfer in BES. The essential aspect of electron transfer efficiency in BES depends directly on the selected microbial species, as particular microorganisms exhibit faster electron transfer speeds through their natural metabolic pathways and redox proteins. The availability of substances that donate electrons affects electron transfer directly through their relationship with the number of electrons transferred to the electrode.

Environmental conditions, including pH value and temperature measurements, influence microbial electron transfer efficiency. The effect of extreme pH conditions reduces microbial population growth, creating difficulties during redox reactions, which in turn decrease overall system performance. Temperature variations control the speed of microorganism metabolism, generating instability within bioelectrochemical systems.

Bioelectrochemical systems operate through fundamental principles of electron transfer to deliver their operational results. The microbial system design and the selected species determine which mechanism between direct and mediated electron transfer will function more effectively. Developing efficient BES applications depends on enhancing these mechanisms, utilizing suitable electrode materials, and implementing effective environmental management practices. BES shows promise for sustainable energy and environmental solutions because ongoing research will enhance microbial strains, electrode materials, and system configurations (Table 1).

Table 1 Effect of different factor

Factor	Description	Effect on electron transfer efficiency	Example
Microbial species	The type of microorganism determines its ability to transfer electrons	High-efficiency species increase electron transfer rates	*Geobacter sulfurreducens*, *Shewanella oneidensis*
Substrate type and concentration	The nature and amount of organic matter available as an electron donor	Optimal substrates enhance microbial metabolism and electron release	Acetate, glucose, wastewater organics
Electrode material	Surface area, conductivity, and biocompatibility of electrodes	Better materials improve microbial adhesion and electron flow	Carbon cloth, carbon nanotubes, graphene coatings
pH	It affects microbial activity and redox reactions at electrode surfaces	Extreme pH values reduce microbial performance	Optimal range: Typically pH 6.5–7.5
Temperature	Influences the metabolic rate of microorganisms	Higher metabolic rates at optimal temperatures increase electron transfer	Optimal range: 25 °C–40 °C
Presence of mediators	Use of electron shuttles (natural or synthetic)	Mediators can enhance electron transfer when the direct electron transfer (DET) is weak	Flavins (natural), neutral red (synthetic)
Electrode configuration	Arrangement and distance between electrodes	Shorter distances reduce internal resistance, improving efficiency	Flat plate, 3D structured electrodes
Biofilm formation	Thickness and conductivity of microbial biofilms on electrodes	Well-developed biofilms improve direct electron transfer	Thin, dense biofilms preferred
Oxygen concentration (at anode)	Oxygen competes as a terminal electron acceptor	High oxygen levels decrease electron flow to the anode	Anaerobic or low-oxygen conditions are preferred
Ionic strength of electrolyte	Conductivity of the electrolyte solution	Higher ionic strength reduces internal resistance	Adding salts like NaCl to the electrolyte

8.1 Energy Harvesting and Conversion in Bioelectrochemical Systems (BES)

The energy harvesting and conversion operations in bioelectrochemical systems (BES) are incredibly appealing because this technology transforms chemical energy into electrical power with the help of biological processes. BES technology excels because it obtains energy from renewable organic substances, including waste products, resolving dual energy and ecological issues. BES energy harvesting functions through the microorganism substrate degradation process, which enables electron transfer through an electrode system that generates electrical power. This segment analyzes the execution methods, convertibility structures, and utility of energy recovery in BES, as well as the technical improvements and operational hurdles that guide current research.

8.2 Microbial Fuel Cells (MFCs) as Energy Harvesting Devices

The Microbial Fuel Cell functions as the central harvesting component of a BES and stands as the most widely researched electrical energy generator that utilizes organic chemical energy. Microorganisms, such as bacteria, conduct the oxidation of organic compounds (glucose, acetate) inside MFCs within anaerobic anode chambers, releasing electrons and protons. The transferred electrons use an external circuit to reach the cathode, where they unite with oxygen and protons to produce water. The movement of electrons along the external circuit produces electric power that can operate technical equipment, automatic sensors, and expansive energy systems under particular conditions.

MFCs are so powerful in converting organic waste and wastewater into usable electricity. At the same time, they clean up polluted water surfaces, making them highly viable for powering environmental cleanup operations. The quantity of energy that MFCs generate depends on the microorganisms and substrates utilized in combination with their system design and electrode selection. The power output of MFC devices depends on specific microorganism species, such as *Geobacter sulfurreducens* and *Shewanella oneidensis*because these organisms effectively transfer electrons to electrodes.

8.3 MFCs Operate Based on Two Essential Functional Parameters: Power Density and Efficiency

The practical application of MFCs in energy collection depends heavily on power density, which refers to the electrical power output per unit of microbial biomass and substrate consumption rate. The power-generating abilities of MFCs remain lower than those generated from conventional fuel sources and solar energy systems. MFCs produce continuous electrical power from organic waste without requiring external energy, making them ideal for isolated locations and small-scale energy needs.

Various system parameters undergo intensive research to enhance the operational efficiency of MFCs. Researchers focus on implementing novel electrode materials that provide enhanced surface areas, optimal conductivity, and improved biocompatibility. Due to their superior surface area and high conductive properties, carbon-based materials, including graphite, carbon cloth, and carbon nanotubes, are widely used in electrochemistry. The effects between microorganisms and electrodes perform better when researchers implement conductive polymers and nano-structured electrodes. Fueling the electricity production ability of microbes can be enhanced by selecting or genetically altering microbial strains, as this practice offers improved electron transfer efficiency in electricity generation.

8.4 Microbial Electrolysis Cells (MECs) and Energy Conversion

Energy conversion using MECs stands out as an advanced technology that combines energy harvesting capabilities with a special focus on producing valuable fuel through hydrogen generation. Microbial electrolysis cells employ an electrochemical process similar to MFCs, but they utilize an external voltage to reduce H^+ ions at the cathode for hydrogen gas production. The microorganisms inside MECs break down organic materials in the anodic chamber, creating electrons and protons. Energy first passes electrons from the protons through the electrolyte fluid, reaching the cathode terminus, where external voltage applications force the combination of protons and electrons into hydrogen gas (H_2).

REA scores highly as a technology platform for generating hydrogen from renewable sources because it works as an environmentally friendly alternative fuel. Hydrogen gas produced by MECs enables storage as an energy carrier, serving as a solution that supports clean power generation and energy storage. Hydrogen production in MECs achieves higher efficiency through the optimisation of microorganism type, applied voltage, and electrode materials, as well as operational factors such as pH and temperature. The optimization of operational factors will enable MECs to produce higher hydrogen yields at lower energy requirements, thus providing improved competitiveness compared to other hydrogen technology variants, such as water electrolysis.

8.5 Bioelectrochemical Systems for Waste-to-Energy Applications

BES's primary benefit in energy harvesting is its ability to convert waste into energy. Bioelectrochemical Systems, including MFCs, MECs, and others, function as devices for extracting energy from organic waste materials such as food waste, agricultural residues, and wastewater, as well as different biomass types. Bioelectrochemical systems offer an effective waste-to-energy solution for rural areas, wastewater treatment facilities, and remote locations, as they efficiently manage waste while generating energy through a single system.

Wastewater treatment facilities benefit from BES technology because it utilizes organic pollutants to generate electricity during treatment operations, thereby creating renewable energy and reducing environmental impacts associated with wastewater disposal. The MFC or MEC relies on wastewater organic matter acting as an electron source, which a microbe population decomposes to generate either electricity or hydrogen fuel. Industry wastewater treatment plants can implement this process to obtain dual benefits from waste elimination and sustainable energy generation.

8.6 Bio-Batteries and Long-Term Energy Storage

Bio-batteries serve as an innovative technology inside BES to store energy. Bio-batteries utilize the exact battery mechanisms as conventional batteries but employ biological materials, including enzymes and microorganisms, to generate electrical energy through chemical reactions. Bio-batteries' sustainability advantage arises from using biologically renewable materials instead of toxic battery constituents in conventional electronic cells. Portable electronics, medical devices, and other power-related devices are ideal applications for bio-battery technology. The development of bio-batteries continues in the initial stages. Still, enzyme engineering and improvements in microbial fuel cells promise to transform these batteries into viable choices compared to conventional power sources.

9 Applications of Bioelectrochemical Systems

Bioelectrochemical systems (BES) applications span diverse energy management sectors and environmental applications for generating sustainable processes using microorganisms. Energy production through microbial fuel cells (MFCs) converts organic waste into electricity, making them suitable for wastewater treatment, remote power generation, and environmental monitoring tasks. Cells with Microbial Electrolysis function through BES technology to produce hydrogen gas as an energy carrier by utilizing microbial energy from waste sources. Bioelectrochemical sensors powered by BES act as environmental detection devices that evaluate toxic substances, including heavy metals and organic contaminants, in water and soil-based materials. BES serves as a basis for microbial desalination cells (MDCs) to deliver an environmentally friendly solution for water desalination through microbial activity targeting water scarcity problems. Research groups are developing bio-batteries from biological materials to generate electricity for medical uses and portable electronics. BES combines environmental sustainability with energy generation through its innovative approach to creating clean power systems, waste management, and environmental monitoring capabilities (Fig. 2).

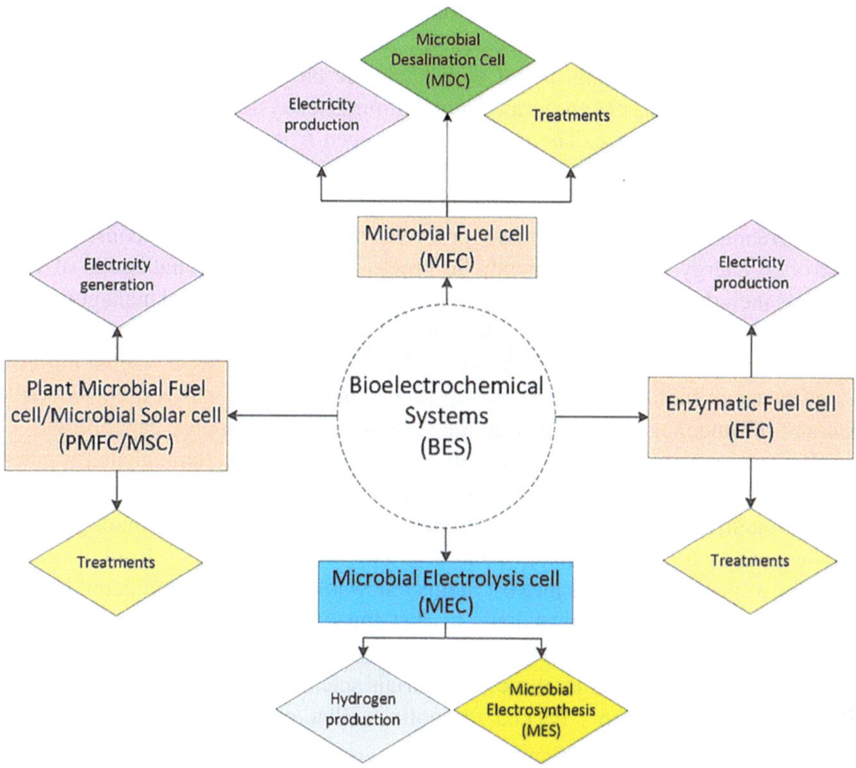

Fig. 2 Application of bioelectrochemical systems [65]

10 Challenges and Limitations of Bioelectrochemical Systems (BES)

Several significant challenges and operational limitations prevent widespread commercial use of bioelectrochemical systems (BES) despite their ability to produce sustainable energy, treat water sources, and recover essential materials. The power density of microbial fuel cells (MFCs) represents the main challenge because their output electricity typically fails to support power-intensive applications. The combination of substandard electron transfer operations, poor microbial performance, and resistance factors across the system prevents efficient power delivery. The expense of state-of-the-art electrodes and their limited durability are material-related constraints for widespread adoption because such advanced materials typically require platinum or nanostructured carbon composites. The performance of BES depends substantially on selecting the proper microorganisms, as electrochemically active microorganisms need to form stable biofilms. Still, they have limited diversity and poor stability when the conditions change. The upsizing of laboratory reactors presents two core concerns: efficiency decreases due to high internal resistance, substrate distribution inconsistencies, and elevated maintenance work.

Uncontrolled biofouling, along with electrode clogging, represents a primary obstacle, as it restricts electron transfer and diminishes system efficiency while eliciting increased resistance. The transport of oxygen from the anode to the cathode compartment lowers both Coulombic efficiency and microbial efficiency, particularly when proper compartmentalization through membranes is absent. Real-world waste streams cause substrate variety problems because they create inconsistency in microbial energy output and metabolic processes. The operational needs of BES systems include accurate control of pH, temperature regulation, and management of electrolyte conductivity; however, they face challenges in maintaining these conditions when operating extremely large or decentralized systems. Establishing efficient bacterial biofilms through startup procedures typically takes a considerable amount of time, negatively affecting the system's prompt response capabilities.

BES deployment faces restrictions due to regulatory and technological obstacles, which demand infrastructure configuration changes when integrated with traditional systems such as wastewater treatment infrastructure. BES technologies face challenges in gaining widespread awareness and investment support from the public domain, as people show less interest in this renewable technology compared to wind and solar options, which slows down market entry. The successful implementation of bioelectrochemical systems requires collaboration between microbiologists and engineers who work together through materials science innovations to develop economically viable systems backed by policies that promote partnership growth between governments and private entities.

11 Conclusion

Bioelectrochemical systems (BES) offer a groundbreaking approach to addressing global energy challenges while enhancing environmental remediation and resource recovery through microorganism-based electrochemical processes. BES-based technology features MFCs and MECs, among other cell designs, which utilize biological processes to generate electricity, produce fuel, and simultaneously treat wastewater and contaminated water. Assessing transfer, along with microbial metabolic processes and electrode interface construction fundamentals, determines the performance and scalability capabilities of BES technology. The combination of progress made in material development and microbial engineering, along with reactor optimization, has not yet resolved the fundamental limitations of BES systems, including power-generating capacity limitations, cost efficiency issues, and design complexity. Researchers continue their work to enhance BES performance, leading to lower prices and broader industrial opportunities. The increasing support from interdisciplinary researchers and public funding, together with policy backing, positions BES as a powerful solution for clean energy generation, environmental sustainability, and the development of a circular bioeconomy in the future.

References

1. Hassan RY, Febbraio F, Andreescu S (2021) Microbial electrochemical systems: principles, construction and biosensing applications. Sensors 21:1279
2. Shanbhag MM, Manasa G, Mascarenhas RJ, Mondal K, Shetti NP (2023) Fundamentals of bio-electrochemical sensing. Chem Eng J Adv 16:100516
3. Jatoi AS, Ahmed J, Bhutto AA, Shaikh MS (2024) Sustainable bio-energy generation via the conversion of spent wash using dual chamber microbial fuel cell. Waste Disp Sustain Energy 6:219–231
4. Jatoi AS, Hashmi Z, Anjum A, Bhatti ZA, Siyal SH, Mazari S et al (2023) Overview of bio-electrochemical approaches for sulfur reduction: current and future perspectives. Biomass Convers Biorefinery 13:12333–12348
5. Jatoi AS, Mallah NB, Ahmed J, Shah SA, Sabzoi N, Ibrahim SM (2025) Enhancing the performance of microbial fuel cell via use of new type of electrode for the bioenergy generation along with the treatment of distillery effluent. Biomass Convers Biorefinery 15:3151–3164
6. Jatoi AS, Akhter F, Mazari SA, Sabzoi N, Aziz S, Soomro SA et al (2021) Advanced microbial fuel cell for waste water treatment—a review. Environ Sci Pollut Res 28:5005–5019
7. Mubarak NM, Sattar A, Mazari SA, Nizamuddin S (2023) Advanced nanomaterials and nanocomposites for bioelectrochemical systems. Elsevier, Amsterdam
8. Jatoi AS, Hashmi Z, Mazari SA, Mubarak NM, Karri RR, Ramesh S et al (2022) A comprehensive review of microbial desalination cells for present and future challenges. Desalination 535:115808
9. Gautam R, Nayak JK, Ress NV, Steinberger-Wilckens R, Ghosh UK (2023) Bio-hydrogen production through microbial electrolysis cell: structural components and influencing factors. Chem Eng J 455:140535
10. Savla N, Pandit S, Verma JP, Awasthi AK, Sana SS, Prasad R (2021) Techno-economical evaluation and life cycle assessment of microbial electrochemical systems: a review. Curr Res Green Sustain Chem 4:100111
11. Chen S, Zhang T, Tappertzhofen S, Yang Y, Valov I (2023) Electrochemical-memristor-based artificial neurons and synapses—fundamentals, applications, and challenges. Adv Mater 35:2301924
12. Mahmoud RH, Gomaa OM, Hassan RY (2022) Bio-electrochemical frameworks governing microbial fuel cell performance: technical bottlenecks and proposed solutions. RSC Adv 12:5749–5764
13. Gupta S, Patro A, Mittal Y, Dwivedi S, Saket P, Panja R et al (2023) The race between classical microbial fuel cells, sediment-microbial fuel cells, plant-microbial fuel cells, and constructed wetlands-microbial fuel cells: applications and technology readiness level. Sci Total Environ 879:162757
14. Bazina N, Ahmed TG, Almdaaf M, Jibia S, Sarker M (2023) Power generation from wastewater using microbial fuel cells: a review. J Biotechnol 374:17–30
15. Roy H, Rahman TU, Tasnim N, Arju J, Rafid MM, Islam MR et al (2023) Microbial fuel cell construction features and application for sustainable wastewater treatment. Membranes 13:490
16. Gul H, Raza W, Lee J, Azam M, Ashraf M, Kim K-H (2021) Progress in microbial fuel cell technology for wastewater treatment and energy harvesting. Chemosphere 281:130828
17. Arun J, SundarRajan P, Pavithra KG, Priyadharsini P, Shyam S, Goutham R et al (2024) New insights into microbial electrolysis cells (MEC) and microbial fuel cells (MFC) for simultaneous wastewater treatment and green fuel (hydrogen) generation. Fuel 355:129530
18. Shrivastava A, Sharma RK (2022) Lignocellulosic biomass based microbial fuel cells: performance and applications. J Clean Prod 361:132269
19. Wang J, Ren K, Zhu Y, Huang J, Liu S (2022) A review of recent advances in microbial fuel cells: preparation, operation, and application. Biotech 11:44

20. Meylani V, Surahman E, Fudholi A, Almalki WH, Ilyas N, Sayyed R (2023) Biodiversity in microbial fuel cells: review of a promising technology for wastewater treatment. J Environ Chem Eng 11:109503
21. Nawaz A, Ul Haq I, Qaisar K, Gunes B, Raja SI, Mohyuddin K et al (2022) Microbial fuel cells: insight into simultaneous wastewater treatment and bioelectricity generation. Process Saf Environ Prot 161:357–373
22. Garbini GL, Barra Caracciolo A, Grenni P (2023) Electroactive bacteria in natural ecosystems and their applications in microbial fuel cells for bioremediation: a review. Microorganisms 11:1255
23. Khandaker S, Das S, Hossain MT, Islam A, Miah MR, Awual MR (2021) Sustainable approach for wastewater treatment using microbial fuel cells and green energy generation–a comprehensive review. J Mol Liq 344:117795
24. Ma J, Zhang J, Zhang Y, Guo Q, Hu T, Xiao H et al (2023) Progress on anodic modification materials and future development directions in microbial fuel cells. J Power Sources 556:232486
25. Kurniawan TA, Othman MHD, Liang X, Ayub M, Goh HH, Kusworo TD et al (2022) Microbial fuel cells (MFC): a potential game-changer in renewable energy development. Sustainability 14:16847
26. Sonawane JM, Mahadevan R, Pandey A, Greener J (2022) Recent progress in microbial fuel cells using substrates from diverse sources. Heliyon 8:e12353
27. Dwivedi KA, Huang S-J, Wang C-T, Kumar S (2022) Fundamental understanding of microbial fuel cell technology: recent development and challenges. Chemosphere 288:132446
28. Dwivedi KA, Huang S-J, Wang C-T (2022) Integration of various technology-based approaches for enhancing the performance of microbial fuel cell technology: a review. Chemosphere 287:132248
29. Rojas-Flores S, De La Cruz-Noriega M, Nazario-Naveda R, Benites SM, Delfín-Narciso D, Rojas-Villacorta W et al (2022) Bioelectricity through microbial fuel cells using avocado waste. Energy Rep 8:376–382
30. Huang Y, Zhao Y, Tang C, Yadav AK, Abbassi R, Kang P et al (2023) A glance of coupled water and wastewater treatment systems based on microbial fuel cells. Sci Total Environ 892:164599
31. Atawi IE, Al-Shetwi AQ, Magableh AM, Albalawi OH (2022) Recent advances in hybrid energy storage system integrated renewable power generation: configuration, control, applications, and future directions. Batteries 9:29
32. Choudhury S (2022) Review of energy storage system technologies integration to microgrid: types, control strategies, issues, and future prospects. J Energy Storage 48:103966
33. Bamisile O, Cai D, Adun H, Dagbasi M, Ukwuoma CC, Huang Q et al (2024) Towards renewables development: review of optimization techniques for energy storage and hybrid renewable energy systems. Heliyon 10:e37482
34. Xu Y, Leonforte F, Del Pero C (2024) District multi-energy systems: a comprehensive review of configurations, technologies, and performances. Build Environ 253:111318
35. Agajie TF, Ali A, Fopah-Lele A, Amoussou I, Khan B, Velasco CLR et al (2023) A comprehensive review on techno-economic analysis and optimal sizing of hybrid renewable energy sources with energy storage systems. Energies 16:642
36. Panbechi B, Hajinezhad A, Moosavian SF, Fattahi R (2025) Enhancing the enviro-economic viability of biogas-solar hybrid systems: strategies for sustainable energy development in Iran. Energ Strat Rev 58:101671
37. Mohanakrishna G, Al-Raoush RI, Abu-Reesh IM (2021) Integrating electrochemical and bioelectrochemical systems for energetically sustainable treatment of produced water. Fuel 285:119104
38. Wang S, Adekunle A, Raghavan V (2022) Exploring the integration of bioelectrochemical systems and hydroponics: possibilities, challenges, and innovations. J Clean Prod 366:132855

39. Li Z, Fu Q, Su H, Yang W, Chen H, Zhang B et al (2022) Model development of bioelectrochemical systems: a critical review from the perspective of physiochemical principles and mathematical methods. Water Res 226:119311
40. Zhao P, Zhang H, Sun X, Hao S, Dong S (2022) A hybrid bioelectrochemical device based on glucose/O2 enzymatic biofuel cell for energy conversion and storage. Electrochim Acta 420:140440
41. Debnath K, Dutta S (2023) Bio-electrochemical system analysis and improvement: a technical review. Clean Circ Bioecon 6:100052
42. Włodarczyk B, Włodarczyk PP (2025) Enhancing microbial fuel cell operation with molasses decoction from yeast production. Desalin Water Treat 322:101192
43. Mochammad W, Fujimura S, Shichida M, Nagamine M, Kiyoshi K, Inoue K (2025) Tofu whey treatment and electricity generation by stacked microbial fuel cells. J Environ Chem Eng 13:116353
44. Li S, Duan L, Gao Q, Zhang H (2025) Reduction of methane emission from microbial fuel cells during sulfamethoxazole wastewater treatment. Chin Chem Lett 36:110997
45. Liang L, Lin Y, Ren R, Zhang Q, Qu J, Liu Y et al (2025) Recent advances in antibiotic removal using independent and coupled microbial fuel cells (MFCs): mechanisms and cells (MFCs): mechanisms and practical applications. Process Eng 71:107226
46. Ojha R, Pradhan D (2025) The potential of microbial fuel cell for converting waste to energy: an overview. Sustain Chem Environ 9:100196
47. Baskar S, Jayaprabakar J, Rajan JA, Kumar NRU et al (2025) Possible bio cathode materials for usage in microbial fuel cells towards energy generation and wastewater treatment to sustain environment: a review. Results Eng 25:104161
48. Ye Y, Yan X, Jiang Y, Wang S, Liu D, Ren Y et al (2025) Optimized feeding schemes of heterotrophic anodic denitrification coupled with cathodic phosphate recovery from wastewater using a microbial fuel cell. Sci Total Environ 981:179590
49. Güneş S, Ayol A, Tezer O (2025) Biodegradation and power generation from olive oil mill wastewater using soil-based microbial fuel cells. Int J Hydrog Energy 116:410–419
50. Li C, Zheng M, Yang X, Li C, Lu J, Song Z et al (2025) Electrostatic self-assembly of copper oxide nano catchers boosting the efficiency of oxygen reduction in microbial fuel cell. J Power Sources 643:237059
51. Özyörü Zİ, Uğur Nigiz F (2025) Electricity production from dairy wastewater using phosphotungstic acid-poly(vinylidene fluoride) membrane supported microbial fuel cell. Int J Hydrog Energy 142:842–851
52. Douma MNK, Ondel O, Tsafack P, Mieyeville F, Kengnou NA (2025) Microbial fuel cell: investigation of the electrical power production of cow dung and human faeces using 3D-printed reactors. Bioresour Technol Rep 29:102036
53. Sreelakshmi C, Kiruthika S, Jeyalakshmi R (2025) Optimization modeling and economics assessment on simultaneous struvite and bioelectricity production from waste nutrient solution in the microbial fuel cell. Biochem Eng J 218:109705
54. Wang C-T, Das B, Saladaga IA (2025) Microalgae biocathode coupled polyvinylalcohol proton exchange membrane for performance of recirculation honeycomb microbial fuel cells. Bioresour Technol Rep 29:102037
55. Ndive JN, Eze SO, Nnabuife SG, Kuang B, Rana ZA (2025) Dual-chamber microbial fuel cell for azo-dye degradation and electricity generation in textile wastewater treatment. Waste Manage Bull 3:100195
56. Mkilima T, Zharkenov Y, Abduova A, Kudaibergenov N, Fazylov K, Toleubayeva S et al (2025) Bioelectrochemical degradation of pollutants in wastewater using a dual-chamber microbial fuel cell with graphene-modified electrodes and electroactive bacteria. Case Stud Chem Environ Eng 11:101184
57. Tapan NA, Günay ME, Gürbüz T (2024) Routes to optimum conditions of plant based microbial fuel cells by reinforcement learning. Int J Hydrog Energy 142:813–824

58. Singh S, Kirrolia AS, Pannu D, Bishnoi NR (2025) Exploring the synergy between microbial fuel cells and constructed wetlands in dairy waste water treatment: a comprehensive review. J Water Process Eng 71:107211
59. Xu J, Zhang X, Jin D, Wu P (2025) Anammox-coupled microbial fuel cell (Anammox-MFC) for wastewater treatment: a critical review. J Water Process Eng 71:107285
60. Li B, Wang X, Wu L (2025) Yolk-shell structured ultra-small Fe3C nanoparticles encapsulated in S, N-codoped porous carbon as effective catalysts for the oxygen reduction reaction in microbial fuel cells. J Power Sources 644:237058
61. Fattah IMR, Alom J, Zaman JU, Ban S, Veza I, Kalam MA et al (2025) Hydrogel-derived materials for microbial fuel cell. J Power Sources 625:235688
62. Dehghanian M, Mahdavi MA, Gheshlaghi R, Darband GB, Absalan Y (2025) Synthesis of TiC@C–TiO2 as an emerging efficient nanostructure for anode modification in high-power microbial fuel cells. Int J Hydrog Energy 116:591–600
63. Liu X, Zhou E, Fan Y, Wang F, Xu D (2025) Riboflavin-mediated extracellular electron transfer enhances microbiologically influenced corrosion of 316L stainless steel by enterococcus faecalis. Bioelectrochemistry 165:108982
64. Zhu H, Ma H, Zhao Z, Xu L, Li M, Liu W et al (2025) Electron transfer tuning for persulfate activation via the radical and non-radical pathways with biochar mediator. J Hazard Mater 486:136825
65. Bajracharya S, Sharma M, Mohanakrishna G, Dominguez Benneton X, Strik DPBTB, Sarma PM et al (2016) An overview on emerging bioelectrochemical systems (BESs): technology for sustainable electricity, waste remediation, resource recovery, chemical production and beyond. Renew Energy 98:153–170

Design and Operation of Bioelectrochemical System

Swapna Gade, Ratibha Desai, and Yuvraj Patil

Abstract The MFC design supports bioelectricity generation in part because of the electrochemical gradient that exists between the anode and cathode. Mechanism for producing electrons and protons by microbial oxidation of organic compounds. It is carried out in simple, soft environmental conditions and operates using waste organic material as a substrate. Bioelectrochemical systems are more laborious since the BES process uses both biological and electrical means. Various models were employed during the optimization of the reactor design. A study on various model design reactors reveals that the components across different reactors are nearly identical. This chapter aims to supply ideas for the plan and assignments of BS to study various components that affect the design process.

Keyword Reactor process optimization · BES

1 Introduction

Microbial fuel cells are devices that convert biochemical energy into electricity. Its basic behavior is based on the redox reaction of the substrate produced by bacteria. Two types of bioelectrical systems (BESs) that harness energy from renewable organic materials are microbial fuel cells (MFCs) and microbial electrolysis cells (MECs). These bioelectrochemical systems (BESs), which utilize microorganisms

S. Gade (✉)
Chemical Engineering and Process Development Division, National Chemical Laboratory, Pune, India

R. Desai
Marathwada Mitra Mandals Institute of Technology, Pune, India

Y. Patil
School of Health Sciences and Technology, Dr. Vishwanath Karad MIT World Peace University, Pune, India

A. Sattar Jatoi, N. M. Mubarak (eds.), *Bio-Electrochemical Conversion of Waste to Fuel*, Green Energy and Technology,
https://doi.org/10.1007/978-981-95-4310-6_4

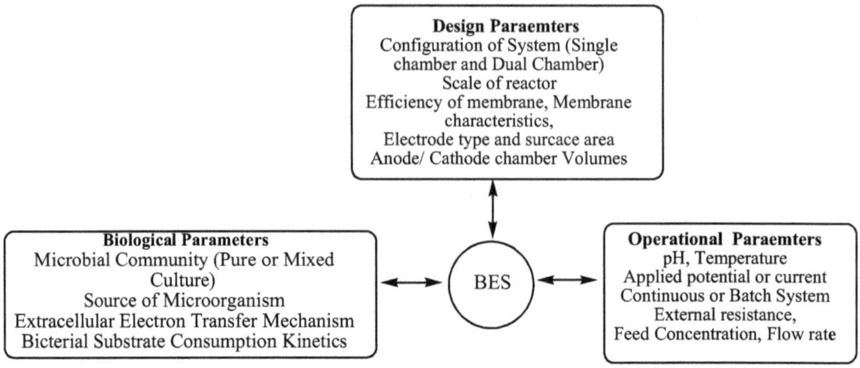

Fig. 1 Parameters that influence the BES performance

to facilitate oxidation or reduction processes by moving electrons to and from an electrode, have garnered considerable attention lately as a potentially revolutionary technological advancement. BES, like any other electrochemical cell (e.g., battery), consists of an anode, cathode, and a separating membrane. Microorganisms catalyze at least one or both of the electrodes in a BES reaction. BES is a versatile technology that combines electrochemistry and living biological systems to generate electricity (microbial fuel cell, MFC), produce hydrogen (microbial electrolysis cells, MEC), synthesize value-added chemicals (microbial electro-synthesis, MES), desalinate (microbial desalination cell, MDC), and remove contaminants (microbial remediation cell, MRC) [1, 2]. Among other biological and design variables, factors affecting the cell potential (Ecell) of a bioelectrochemical system (BES) include the type of bacterial population, the anodes and cathodes utilized, and the substrate. To lower the electron acceptor at the cathode [3, 4], the BES's total cell potential (Ecell) decides whether electricity can be generated when positive (E0cell>) and whether additional external power is required. These two situations are illustrated in the BES scheme in Fig. 1. Figure 1b illustrates a more general scheme for the other BES, specifically for a typical microbial fuel cell (one with a positive overall cell potential).

Bioelectrochemical systems are more challenging as the BES process uses both microorganisms and biological and electrical means. Various models were employed during the optimization of the reactor design. A study on various reactor model designs revealed that the components across different reactors are nearly identical. This chapter aims to provide ideas for planning and assigning BS and to examine various components that affect the design process.

2 Components of BES

In the BES, microorganisms are attached to the surface of the positive electrode, where they are catalyzed by the oxidation of organic compounds and form living biofilms. Therefore, interaction between the anode and microorganisms is necessary for the operation of a BES, such as a microbial fuel cell (MFC). Primarily, the electrode plays a crucial role in the MFC, where it is involved in an electrochemical reaction and undergoes complete reduction. A wide range of materials demonstrated significant efficiency in biological and wastewater treatment.

2.1 Anode Materials

Living biofilms identify the anode of a microbial fuel cell (MFC). Metabolic electrons in biofilms are transferred to the anode, which affects the material at the interface and alters the electron transmittance (Fig. 2). The anode material selected is crucial for enhancing the biofilm's development and electron transfer efficiency [5]. Moreover, anode creation and synthesis should be flexible, utilizing inexpensive and readily available materials.

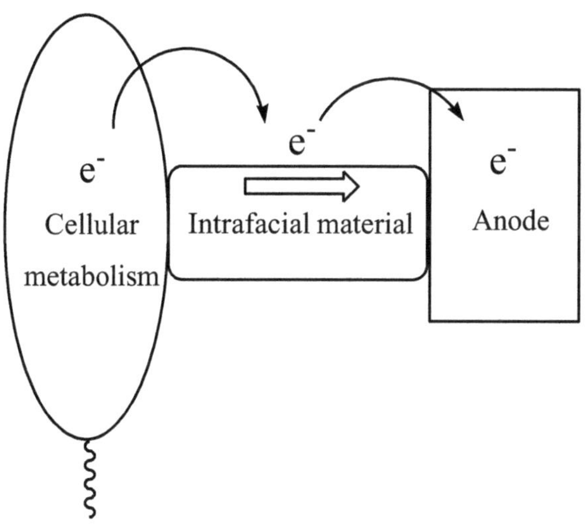

Fig. 2 Electrons generated metabolically flow to the anode

2.1.1 Nanoporous Carbon-Based Electrodes

As anode materials in microbial fuel cells (MFCs), nanoporous carbon-based electrodes have attracted considerable interest owing to their excellent conductivity, outstanding corrosion resistance, and better biocompatibility [6].

2.1.2 Carbon Nanotubes

Microbial fuel cells (MFCs) can greatly benefit from the high conductivity and mechanical stability of carbon nanotubes (CNTs) as electrode materials. Furthermore, CNTs have outstanding biocompatibility with bacteria, which promotes the growth of biofilms on their surface. External supply from the bacterial film of the electronic C-type (OM C-Cyts) to the solid electron of the electron (EET), comprising an electrode [7], may be strengthened by the CNT network, either in practice or in theory. The CNT facilitated the EET process by establishing an electronic channel between the CN and OM C-Cyts, as determined by molecular mechanics modeling. The presence of the reduction function group for the United Nations can play a decisive role in establishing this electronic channel. CNT geometry is critical to the performance of bioanodes [8]. This study has concluded that longer and more freely packaged CNTs are highly suitable for effective EET processes. It also enables microbial electrodes to interact less than the amount of amorphous carbon. The simple addition of CNT powder has enhanced mass transfer, reduced anode resistance, and promoted mass movement within the Biofilm10, all of which have improved the performance of Geobacter sulfur. The generated electrodes provided numerous surfaces for bacterial adhesion, enabling the CN functional group to facilitate electron transfer (EET). CNT hydrogels are considered one of the most important factors in the practical application of MFCs, as they exhibit excellent stability in long-term measurements. Microgroscale MFCs are highly attractive for use in devices with ultra-low power, so there have been several attempts to produce them. MWCNT reduced the contact resistance of the modified electrode and provided a large specific surface for bacterial immobilization. Using the MWCNT anode and air cathode, the modified MFC-microsize (75 μL) exhibits significantly improved performance compared to previous work, achieving a current density of 880 mA cm^{-2}. However, the design of the air cathode was left to distribute oxygen by the anode to boost the efficiency of the bianode. Furthermore, bacteria impeded the high surface area MWCNT offers. Therefore, the long-term operation of the device requires optimization of electrode manufacturing. The porous bioanode of the 3D (3D) CNT has shown to be a promising candidate for the development of electrode, and it is a promising strategy to increase the productivity of Bioanod [9] by comparing a competent weekend compared to the 212 G. For example, the CNT doped by nitrogen showed better performance than body and carbon tissue. The doping of nitrogen increased the active area in bionodes and slightly reduced the internal resistance of the electrodes. In general, the bacterial membrane is considered to be negatively charged in a neutral solution [6]. From the membranes that are affected by this sound, the bacteria prefer to attach to a hard surface that is positively charged. This element inspired researchers to create high-end bioanodes on a positive surface. For example, positive mesoporous polysulfone support single-walled bio-ode CNT (SWCNT) has developed a dense biofilm to attract Shewanella cells [10]. Negative increased current generation and sewage treatment results from the development of thick biofilm. Depending on the load between the anode and bacterial cells, the initial treatment of acid as a mountain can lead to a much more positive surface, thereby improving the labor force.

2.1.3 Graphene

Among two-dimensional (2D) nanomaterials, graphene stands out for its exceptional electrical conductivity, high mechanical stability, and superior biocompatibility [11]. Moreover, compared to other carbon nanostructures, graphene has a greater specific surface area. These advantageous properties have led to graphene's widespread use in a range of energy devices, including microbial fuel cells (MFCs), where it serves as a practical electrode material. Graphene-based MFCs typically demonstrate superior performance compared to those utilizing activated carbon and carbon nanotubes (CNTs) [12]. The initial investigation into graphene-based bioanodes was conducted several years ago [13], utilizing an MFC seeded with *E. coli*. The graphene functioned as the bioanode and was attached to a stainless steel mesh. When compared to the control, which was the stainless steel mesh anode, the modified anode showed a noticeably higher current generation. This enhanced performance can be attributed primarily to the anode's high surface area, which facilitated substantial biofilm development. A vacuum-stripped graphene scaffold bioanode generated a current 78 times greater than that of the control carbon cloth in a microbial fuel cell (MFC) utilizing Pseudomonas aeruginosa as the biocatalyst [14]. *P. aeruginosa* is known to release exogenous mediators, such as phenazine, during its metabolic processes [15]. Graphene can promote phenazine production, thereby accelerating mediated electron transfer (EET) in *P. aeruginosa* [14]. Moreover, the vacuum-stripped graphene electrode displayed an enhanced charge transfer rate and a large surface area. Similarly, graph-coated carbon fabrics enhanced the performance of *P. aeruginosa* in MFCS, as the graph-stimulating effect on phenazine generation [14, 16] represents graphs on stainless steel networks as a reasonably inexpensive bioanode source for MFCs. Particularly concerning electrical conductivity and high specific surface areas, the advantage of combining graphics sponge and stainless steel increases MFC power, allowing for the maximum power density of 1.32 m^2 to be utilized. Using an ice cream template approach, the bioanode derived from the 3D graph sponge produced a maximum power density of 427 W m^1, higher than that of MFC23 based on carbon Fell [17]. This remarkable current generation suggests that 3D graphs are promising bioanode materials and can enhance the utility of microbial fuel cells (MFCs). The addition of polypyrrole (PPy) nanotubes reduced the load transmission resistance of the electrodes, resulting in a significant EET process [17] (EET) of electron transfer (EET) after analysis using electrochemical impedance spectroscopy (ICE). Furthermore, compared to control carbon paper anodes that showed low biofilm development, scan electron microscope (SEM) images (SEM) showed that the biofilm largely covers the entire surface of the PPY nanorola node. This finding means that modified bioanodes improve bacterial adhesion by providing larger active surfaces and functional groups. In PPY hydrogel-inoculated microbial fuel cells (MFCs), bioanodes made with carbon nanotubes (CNTs) showed increased power generation.

2.1.4 Conductive Polymers

Conductive polymers are particularly useful as electrode materials in microbial fuel cells (MFCs) due to their excellent electrical conductivity, biological compatibility, high chemical stability, and ease of manufacturing and processing. Among these, polypyrrole (PPy) has been the most extensively researched conductive polymer to date. PPy stays stable in water and conducts electricity well. Its synthesis is straightforward, and its surface characteristics, including porosity, can be readily modified for diverse applications. The reactive self-degraded template method [10] was used to construct the PPy nanotube membrane bioanode, resulting in a six-fold increase in current production in a Shewanella MFC compared to the control. Analysis using electrochemical impedance spectroscopy (EIS) revealed that PPy nanotubes enhanced the electron transfer process by significantly reducing the electrode's charge transfer resistance. As opposed to the restricted biofilm formation seen on the control carbon paper anode, the PPy-nanotube anode's entire surface was effectively covered by biofilms, according to scanning electron microscopy (SEM) images. According to this finding, the modified bioanode's increased active surface area and functional groups promote better bacterial adhesion. In MFCs included in mixed bacterial cultures, bioanodes made with carbon nanotubes (CNTs) and PPy-Hydrogel showed higher power. This was attributed to improved electron transfer and decreased internal resistance [18]. The porous and hydrophilic properties of the modified bioanode facilitated the efficient metabolic processes of the associated microorganisms by promoting the development of a dense biofilm and allowing unhindered mass transport. It is known that anthraquinone-2,6-sulfonic acid (AQDS) is a highly successful mediator in the extracellular electron transfer (EET) process [19]. In Shewanella decolorationis S12 microbial fuel cells (MFCs), PPy modified with AQDS functioned as an efficient bioanode, generating a current density that was thirteen times greater than that of the control bioanode. This improvement was ascribed to AQDS's facilitation of electron transfer and PPy's enhancement of conductivity and biofilm formation, which successfully addressed the limitations of mass transfer and charge transfer.

2.1.5 Metal Nanoparticles

Carbon-based materials are frequently utilized as bioanode materials in microbial fuel cells (MFCs), as was previously mentioned. Compared to most metals, these materials have two or three orders of magnitude lower electrical conductivity, indicating a significant limitation [20]. A viable tactic for improving MFC performance may be the use of metal-based bioanodes. Many different types of metal nanoparticles have been frequently used as bioanode materials in MFC applications, including non-poster metals such as nickel (Ni), copper (Cu), and titan (Ti), and precious metals such as gold (Au) and palladium (Pd). On bare Au surfaces, electrogenic bacteria have difficulty forming a thick biofilm [21]. However, it has been demonstrated that applying Au coatings to carbon-based electrode surfaces can significantly enhance the activity of electrogenic bacteria, such as Shewanella oneidensis MR-1. The sputtered electrode exhibited enhanced electrochemical performance relative to the control carbon paper, a phenomenon likely linked to the improved performance of the microbial fuel cell (MFC). Developed a bioanode by applying gold nanoparticles onto carbon paper through a layer-by-layer assembly technique [22], resulting in a significant enhancement of MFC performance compared to the unmodified carbon paper. The presence of the gold layer on the modified electrode contributed to a larger active surface area and increased conductivity, facilitating efficient electron transfer processes. Various microorganisms can synthesize metal nanoparticles when exposed to metal ions [23]. For instance, when exposed to palladium ions, the sulfate-reducing bacterium Desulfovibrio desulfuricans produced membrane-bound palladium nano-assemblies, which enhanced electron transfer between the cells and the electrode due to their superior electrical conductivity. An efficient method for improving electron transfer processes in MFCs has been the integration of nanomaterials into bacterial systems. For example, the increased electrical conductivity of Shewanella resulted from the addition of nickel and iron oxide/sulfide nanoparticles, which significantly accelerated the rate of electron transfer [24]. While it is commonly accepted that copper and silver are not biocompatible with bacteria [25], recent findings suggest that this antibacterial property may not apply to electrogenic bacteria [20]. Both copper and silver have been employed in the fabrication of bioanodes for MFCs, with the bioanodes modified with these metals achieving maximum current densities of 1.5 mA cm^2 and 1.1 mA cm^2, respectively [20]. These current densities are comparable to that of a benchmark graphite bioanode, which produces 1 mA cm^2. The similar performance of these costly materials compared to carbon-based bioanodes raises an important question regarding the rationale behind using expensive materials when their performance does not surpass that of more economical carbon alternatives.

Titanium (Ti) presents numerous benefits compared to traditional carbon and metal electrodes, including exceptional biocompatibility, resistance to corrosion, and favorable dimensional stability [26]. Despite these advantages, the application of Ti as bioanodes in microbial fuel cells (MFCs) remains limited, likely due to inadequate electrical connectivity between bacterial cells and the titanium surface. The technique to improve the use of Ti as a bioanode involves heating the metal to create an oxide layer on its surface [27]. This oxide layer enhances electrocatalytic performance by facilitating efficient electron transfer between the bacterial cells and the cathode. Compared to routine carbon-based materials, stainless steel is the preferred material for bioanode applications in microbial fuel cells (MFCs). Mechanical strength, good conductivity, and stability under the most extreme experimental conditions characterize stainless steel. For long-term use, stainless steel is the preferred electrode due to its high corrosion resistance and affordability. Intensive testing by Pocaznoi et al. [28] shows that stainless steel acts as an efficient bioanode for MFCs compared to carbon-based electrodes. The relatively small stainless steel surface area compared to traditional carbon-containing materials, such as carbon fabrics, is one of its main drawbacks. MFC used bioanodes made of stainless steel foam to tackle this issue. In this case, the maximum power density was generated at 80 A/m^2. The foam structure provided a three-dimensional, porous structure that facilitated effective biofilm formation and mass transport. Many pretreatment procedures were presented to improve the effectiveness of bioanodes made of stainless steel. A heat treatment is an effective method for improving surface properties of stainless steel. The stainless steel was preheated to 600 °C for 5 min to form an iron oxide layer on its surface, significantly improving the electrochemical interaction between the biofilm and the bioanode. Iron oxide layers were developed to protect the electrodes from rust. The heat-treated stainless steel electrodes demonstrated the potential of MFCs as inexpensive and scalable bioanodes, generating a high power density of 1.5 mW cm^2. Similarly, the iron oxide layer after flame oxidation on the surface of stainless steel was verified by X-ray base spectroscopy [29]. Compared to flame oxidation electrodes with controls (carbon structure and untreated stainless steel), the former exhibited generation of electricity generation. Analysis of the bacterial community revealed that Geobacter occurs more frequently at flame oxidation electrodes than during the control. A higher generation of MFCs may be a result of the stimulating growth of Geobacter on modified electrodes. This study showed that the bioanode material determines the bacterial population of electrode-bound biofilms.

2.1.6 Chemical Cathodes

Carbon-based materials are frequently used as MFC cathodes due to their wide surface area, high stability, low cost, and superior biocompatibility. Installing MFC, which is not effective or efficient, requires substantial enhancements. Since the high cost of the Pt catalyst precludes its actual application, it is necessary to explore inexpensive materials for cost-effective electrodes. Producing a low cathode method that attaches metal oxide nanoparticles to a carbon support is a promising approach. When Pt cathode [30] was compared to the cathodes Mn_2O_3 and Fe_2O_3, it demonstrated superior ORR results created a single chamber MFC [31] ORR cathode for iron-aminopyrine (Fe-AAPyr). Activated carbon and Pt cathodes performed worse than the Fe-AAPyr cathode. Following extended use of the MFC, a biofilm formation was found on the Fe-AAPyr cathode. The formation of biofilms typically impairs cathode performance by obstructing catalytic active sites. Interestingly, although the biofilm developed due to extended contact with wastewater, the Fe-AAPyr cathode maintained its performance without any degradation. However, biofilm development on the Pt cathode demonstrated a notable reduction in ORR activity, which may have been brought on by catalyst inactivation. Demonstration of the Fe-AAPyr shows that it is extremely stable for long-term MFC operation without compromising its ORR activity in harsh wastewater conditions.

2.1.7 Biocathodes

The ORR capacity of certain aerobic bacteria can occasionally surpass that of traditional abiotic cathode catalysts [32]. BioCathode is an example of a microbial cathode that can perform reduction reactions, such as the oxygen reduction reaction (ORR) typically observed in microbial fuel cell (MFC) cathodes. Biological materials are superior to non-biological catalysts because they are more cost-effective, longer-lasting, and can withstand sulfide poisoning. Furthermore, Biocathode can stick to oxygen more than Pt. Only when dissolved oxygen levels are typically depleted to 0.12 mg/L, aerobic bacteria are utilized. Like bioanodes, carbon-based materials such as carbon cloth, carbon paper, and graphite films are often used as biocathode materials. It has also been shown that stainless steel is a suitable material for biological material structures. An example of a microbial cathode that can perform reduction reactions, such as the oxygen reduction reaction (ORR), at an MFC cathode is biological substances. The benefits of biocathodes over non-biological catalysts include affordability, sustainability, and the ability to withstand sulfide poisoning. Furthermore, Biocathode has a stronger affinity for oxygen than Pt. Aerobic bacteria typically have the largest oxygen compound capacity for dissolved oxygen concentrations of 0.12 mg/L. Carbon-based materials, such as graphite felt, carbon paper, and carbon fabrics, are often used as biocathode materials, just like bioanodes. Moreover, stainless steel has demonstrated its potential as an effective material for the development of biological substances. Pure cultures and microbial consortiums are utilized in mixed cultures to produce biological casthodes [33]. By employing graphite felt, carbon paper, and stainless steel masking boxes, we systematically investigate the influence of cathode materials on the formation of microbial fuel cell (MFC) biocathodes. The cathode was inoculated with anaerobic mud. The biocathode made from graphite outperformed those constructed from carbon paper and stainless steel networks in terms of ORR activity. The efficiency of the biocathode can be significantly enhanced by increasing the specific surfaces available for microbial attachment. To boost the MFC performance by enhancing the ORR activity of the biocathode and providing a larger area for biofilm development, graphite granules combined with activated carbon grinding and activated carbon powder were introduced into the cathode chamber. The internal resistance of the MFC system was markedly decreased due to the enhanced biological cathode.

2.2 *Membranes*

A membrane in MFC often separates the anode and cathode. The membrane can stop substrate crossover and oxygen diffusion between the MFC's chambers. Membrane use, however, significantly raises the cost of MFC design and prevents commercial applications. Additionally, it raises internal resistance, which can occasionally permit oxygen to spread and impair MFC [34] performance. There are two kinds of movies at MFC. Specifically, the negative ionic membrane and the cation membrane.

2.2.1 Cation Exchange Membranes

MFCs typically utilize cation exchange membranes (CEMs), such as Ultrx CMI 7000, Hyflon, Nafion, and Zirfons, as membrane separators due to their high proton conductivity. The most widely used membrane in MFCs is Nafion, which features a negatively charged, hydrophilic sulphonate group that makes it highly proton-conductive. Thin Nafion membranes have less ohmic resistance [35], which has shown to be more effective than MFC thickness. However, the high oxygen permeability of the thin membrane and substrate crossing reduces the Coulomb efficiency of the MFC system. To address this issue, the membrane thickness must be optimized to enhance the performance of MFCs. A significant issue related to the Nafion-based MFC [36] is the reduction in pH levels within the anode compartment. Protons produced at the anode by the oxidation of the substrate may result in a pH drop, which is essential for Nafion membranes as we know that low pH [36] inhibits bacterial respiration at the anode, which impairs MFC performance. A difference in charge between the anode and cathode is also produced by the Nafion membrane's capacity to move cations (such as Na) alongside protons.

2.2.2 Anion Exchange Membranes

Typically, the anion exchange membrane (AEM) generates a higher current than the MFC's cathode exchange membrane (CEM). From the cathode OH ions that are transported via AEM consume the protons produced by the anode, significantly preventing the pH drop at the positive pole. To this end, AEM lowers the cathode's transport and resistance, as well as the cathode's resistance caused by the transported cation.

3 Operational Parameters

MFC performance improvement involves both biological and engineering factors, such as the impact of critical parameters like pH, temperature, supply speed, shift voltage, and organic load.

3.1 *Effect of pH*

In the microbial fuel cell (MFC), a proton is produced by the anode reaction. It moves to the cathode electrode, where the formation of oxygen or water occurs with the help of the recovery process. When an MFC is operated for an extended period, protons from the microbial oxidation of organic molecules accumulate at the anode, resulting in acidification. This is made worse by the slow and imperfect diffusion and migration of protons through the membrane. The cathode side experiences alkalinity from this proton and continues to be consumed in the oxygen recovery reaction, which is insufficient in the positive oxidation process. The mechanism produces a pH concentration gradient in the membrane, imposing electrochemical and thermodynamic limitations on MFC performance. The Nernst equation indicates that the likelihood of an oxygen reduction reaction increases with a decrease in pH value, and therefore suggests that an increase in pH value in the cathode compartment can significantly hinder generation. Therefore, the decrease in the work pH will enhance the oxygen restoration process and ultimately improve the current output power of microbial fuel cells (MFCs).Furthermore, bacteria typically require a pH around neutral to grow optimally, and they adapt their activity to variations in both internal and external pH levels. The pH can affect several major physiological parameters, including bacterial growth conditions, such as ion concentration, membrane potential, and proton concentration, as well as biofilms [37–40]. Therefore, pH is a key element that can alter substrate metabolic activity, and the process of electron and proton formation occurs within the anodic pH microenvironment. The performance and stability of biofilm can affect the capacity of the acidified anode to reduce bacterial activity [38, 39, 41].

3.2 *Effect of Temperature*

Microbial fuel elements are highly susceptible to temperature fluctuations because they can impact mass transfer and system parameters, such as solution conductivity, activation energy, and mass transfer coefficients. In addition, optimal temperatures of 55, 69, 68, and 69–85 °C affect the properties and distribution of microbial communities, as well as thermodynamic properties such as free energy and electrode potential. There is a lack of comprehensive information on how temperature affects microbial fuel cells, despite the significant growth in research in this area in recent years. MFC performance in terms of CPC removal and electrical production is greatly influenced by temperature, which can also have a high output and removal of CPC [42–55]. The reduction in ohmic resistance caused by improved microbial metabolism and membrane permeability, as well as increased conductivity of liquid solutions, can lead to higher power density at temperatures of 67, 72, 74, 79, and 80 °C. The growth of biofilms at the anode, which affects anodic biocatalytic activity, can be used to investigate this microbial activity. Research on how temperature affects biofilm formation reveals that the starting temperature of an MFC significantly influences both the time required for startup and the initial biofilm formation. Higher temperatures quickly produce a stable biofilm and MFC functioning, which in turn results in improved performance [48, 52, 56, 57]. Biofilms have been shown to exhibit the largest biological electrical catalyst activity between 30 °C and 45 °C [48, 52, 56]. From this, the work of MFC in this temperature range is more profitable for obtaining higher charges, specifically at 55, 69, 75, and 91. Lower or higher operating temperatures lead to reduced biofilms due to biofilm separation and irreversible denaturation processes, which can cause inactivation of bacterial metabolic activity and MFC performance [58, 59]. The initial temperature required for biofilm growth affects the number of initial biofilm growth steps and the distribution of the biofilm matrix, as various electric bacteria have different temperature ranges and leak at different temperatures. Bacteria were able to transform their metabolism to function at various temperatures after the biofilm developed, ranging from 71.82 to 84. This clearly shows how important the initial temperature of biofilm formation is; according to this, MFC can function at various temperatures without a significant reduction in productivity.

3.3 *Organic Loading Rate (OLR)*

The speed and load speed (SLR) of the organic load (OLR) are the parameters that influence the precipitation's characteristics in the early stages. These parameters determine the mass of microorganisms that can be used, as well as the ability of the unit volume reactor and the deformation of the organic substrate, respectively. The power density and Coulomb efficiency of microbial fuel cells (MFCs) are affected by the substrate conversion rate, where substrate loading rate (SLR) and organic loading rate (OLR) are essential factors in determining MFC performance. To achieve the best results in terms of removing organic matter and producing electricity, MFCs need to operate at an ideal SLR or OLR. The performance of MFC was examined to investigate the effects of SLR and OLR. The result shows an increase in electrical production and decomposition of the substrate. As a result, more organic substances turn into energy. In addition, as the OLR increases, the internal resistance of the cells decreases, which is associated with the increase in ionic force of the increased concentration of the concentration of increased concentrations from the increase in the concentration of volatile fatty acids? [43, 60] Competition between bacteria and other microorganisms that produce electricity when high OLRs, which show high conditions related to the surface of the positive electrode surface. This competition may result in a greater removal of organic matter unrelated to current generation, which would decrease Coulombic efficiency while increasing substrate degradation. Even if low OLR and high internal resistance are present, the coulomb efficiency is improved because fewer substrates are used to produce methane [43, 60].

3.4 Understanding the Feed Rate

To understand the operation of the microbial fuel cell (MFC) in continuous mode, it involves fluid mechanics that affect the performance of the MFC and the mechanics of the biological microbial community. When cleaning wastewater, research parameters such as flow rate, hydraulic retention time (HRT), and shift voltage are required before effectively implementing MFC. The increase in flow rate has been associated with an increase in output power; however, excessively high flow rates can reduce the density of power. This indicates that optimal performance is achieved when there is sufficient time to establish its own niche within a microbial cluster, which can enhance the hydrolysis of nutrients and the substrate. In addition, it was noted that the higher the flow rate, the more negative effects on the elimination of chemical needs for oxygen (CPC) and cooling efficiency. As the flow rate increases, the time required to convert microorganisms reduces the intensity of dynamic biological fluids. It is a very crucial factor that affects microbial adhesion and development of biofilm. Principles can contribute to the formation of thick and dense bio-feet when the threshold and motion speed of stretching strength increase, which can improve electronic transmission and MFC output. The formation of these reliable biofilms may be associated with the increase in baptism by the reaction to the increase in biomass resulting from strong separation.

4 Conclusion

This book chapter presents the design and operation of bioelectrochemical systems (BES), with a focus on their application in power-to-gas technology. This chapter covers all aspects of the value chain, from reactor design and operation to process modeling and electric grid integration. To provide a comprehensive understanding, five BES reactors were operated over an extended period and monitored from both electrical and chemical perspectives. Key operational conditions, such as pH, temperature, organic loading rate, and feed rate, are also discussed in detail. These factors play a crucial role in optimizing the performance and efficiency of Building Energy Systems (BES) systems.

References

1. Pant D, Singh A, Van Bogaert G, Irving Olsen S, Singh Nigam P, Diels L, Vanbroekhoven K (2012) Bioelectrochemical systems (BES) for sustainable energy production and product recovery from organic wastes and industrial wastewaters. RSC Adv 2(4):1248–1263
2. Harnisch F, Schröder U (2010) From MFC to MXC: chemical and biological cathodes and their potential for microbial bioelectrochemical systems. Chem Soc Rev 39(11):4433–4448
3. Rabaey K, Angenent L, Schroder U, Keller J (2009) Bioelectrochemical systems. IWA Publishing, London
4. Rabaey K, Rozendal RA (2010) Microbial electrosynthesis—revisiting the electrical route for microbial production. Nat Rev Microbiol 8(10):706–716
5. Xie X, Criddle C, Cui Y (2015) Design and fabrication of bioelectrodes for microbial bioelectrochemical systems. Energy Environ Sci 8:3418–3441
6. Kalathil S, Pant D (2016) Nanotechnology to rescue bacterial bidirectional extracellular electron transfer in bioelectrochemical systems. RSC Adv 6:30582–30597
7. Liu C, Chen J-J, Huang Y-X, Sun X-F, Sheng G-P, Li D-B, Xiong L, Zhnag Y-Y, Zhao F, Yu H-Q (2014) Experimental and theoretical demonstrations for the mechanism behind enhanced microbial electron transfer by CNT network. Sci Rep 4:3732
8. Erbay C, Yang G, Figueiredo PD, Sadr R, Yu C, Han A (2015) Three-dimensional porous carbon nanotube sponges for high-performance anodes of MFCs. J Power Sources 298:177–183
9. Ci S, Wen Z, Chen J, He Z (2012) Decorating anode with bamboo-like nitrogen-doped carbon nanotubes for MFCs. Electrochem Commun 14:71–74
10. Kalathil S, Nguyen VH, Shim J-J, Khan MM, Lee J, Cho MH (2013) Enhanced performance of a MFC using CNT/MnO2 nanocomposite as a bioanode material. J Nanosci Nanotechnol 13:7712–7716
11. Filip J, Tkac J (2014) Is graphene worth using in biofuel cells? Electrochim Acta 136:340–354
12. Yuan H, He Z (2015) Graphene-modified electrodes for enhancing the performance of MFCs. Nanoscale 7:7022–7029
13. Zhang Y, Mo G, Li X, Zhang W, Zhang J, Ye J, Huang X, Yu C (2011) A graphene modified anode to improve the performance of MFCs. J Power Sources 196:5402–5540
14. He Z, Liu J, Qiao Y, Li CM, Yang Tan TT (2012) Architecture engineering of hierarchically porous chitosan/vacuum-stripped graphene scaffold as bioanode for high performance MFC. Nano Lett 12:4738–4741
15. Rabaey K, Boon N, Hofte M, Verstraete W (2005) Microbial phenazine production enhances electron transfer in biofuel cells. Environ Sci Technol 39:3401–3408
16. Xie X, Yu G, Liu N, Bao Z, Criddle CS, Cui Y (2012) Graphene-sponges as high-performance low-cost anodes for MFCs. Energy Environ Sci 5:6862–6866
17. Ren H, Tian H, Gardner CL, Ren T-L, Chae J (2016) A miniaturized MFC with three dimensional graphene macroporous scaffold anode demonstrating a record power density of over 10,000 W m^{-3}. Nanoscale 8:3539–3547
18. Tang X, Li H, Du Z, Wang W, Ng HY (2015) Conductive polypyrrole hydrogels and carbon nanotubes composite as an anode for MFCs. RSC Adv 5:50968–50974
19. Adachi M, Shimomura T, Komatsu M, Yakuwaa H, Miyaa A (2008) A novel mediator-polymer-modified anode for MFCs. Chem Commun 17:2055–2057
20. Baudler A, Schmidt I, Langner M, Greiner A, Schröder U (2015) Does it have to be carbon? Metal anodes in MFCs and related bioelectrochemical systems. Energy Environ Sci 8:2048–2055
21. Crittenden SR, Sund CJ, Sumner JJ (2006) Mediating electron transfer from bacteria. Langmuir 22(23):9473–9476
22. Guo W, Pi Y, Song H, Tang W, Sun J (2012) Layer-by-layer assembled gold nanoparticles modified anode and its application in MFCs. Colloids Surf A Physicochem Eng Asp 415:105–111
23. Kalathil S, Lee J, Cho MH (2011) Electrochemically active biofilm-mediated synthesis of silver nanoparticles in water. Green Chem 13:1482–1485

24. Jiang X, Hu J, Lieber AM, Jackan CS, Biffinger JC, Fitzgerald LA, Ringeisen BR, Lieber CM (2014) Nanoparticle facilitated extracellular electron transfer in MFCs. Nano Lett 14:6737–6742
25. Grass G, Rensing C, Solioz M (2011) Metallic copper as an antimicrobial surface. Appl Environ Microbiol 77:1541–1547
26. Sun M, Zhang F, Tong Z-H, Sheng G-P, Chen Y-Z, Zhao Y, Chen Y-P, Zhou S-Y, Liu G, Tian Y-C, Yu H-Q (2010) A gold-sputtered carbon paper as an anode for improved electricity generation from an MFC inoculated with Shewanellaoneidensis MR-1. Biosens Bioelectron 26:338–343
27. Zhou X, Chen X, Li H, Xiong J, Li X, Li W (2016) Surface oxygen-rich titanium as anode for high performance MFC. Electrochim Acta 209:582–590
28. Pocaznoi D, Calmet A, Etcheverry L, Erable B, Bergel A (2012) Stainless steel is a promising electrode material for anodes of MFCs. Energy Environ Sci 5:9645–9652
29. Yamashita T, Ishida M, Asakawa S, Kanamori H, Sasaki H, Ogino A, Katayose Y, Hatta T, Yokoyama H (2016) Enhanced electrical power generation using flame-oxidized stainless-steel anode in MFCs and the anodic community structure. Biotechnol Biofuels 9:62
30. Martin E, Tartakovsky B, Savadogo O (2011) Cathode materials evaluation in MFCs: A comparison of carbon, Mn2O3, Fe2O3 and platinum materials. Electrochim Acta 58:58–66
31. Santoro C, Serov A, Narvaez Villarrubia CW, Stariha S, Babanova S, Artyushkova K, Schuler AJ, Atanassov P (2015) High catalytic activity and pollutants resistivity using Fe-AAPyr cathode catalyst for MFC application. Sci Rep 5:16596
32. Kim BH, Lim SS, Wan Daud WR, Gadd GM, Chang IS (2015) The biocathode of microbial electrochemical systems and microbially-influenced corrosion. Bioresour Technol 190:395–401
33. Butti SK, Velvizhi G, Sulonen MLK, Haavisto JM, Koroglu EO, Cetinkaya AY, Singh S, Arya D, Modestra JA, Krishna KV, Verma A, Ozkaya B, Lakaniemi A-M, Puhakka JA, Mohan SV (2016) Microbial electrochemical technologies with the perspective of harnessing bioenergy: maneuvering towards upscaling. Renew Sustain Energy Rev 53:462–476
34. Leong JX, Wan Daud WR, Ghasemi M, Liew KB, Ismail M (2013) Ion exchange membranes as separators in MFCs for bioenergy conversion: a comprehensive review. Renew Sust Energ Rev 28:575–587
35. Jung DH, Lee CH, Kim CS, Shin DR (1998) Performance of a direct methanol polymer electrolyte fuel cell. J Power Sources 71:169–173
36. Kim JR, Cheng S, Oh S-E, Logan BE (2007) Power generation using different cation, anion, and ultrafiltration membranes in MFCs. Environ Sci Technol 41:1004–1009
37. Zhang L, Li C, Ding L, Xu K, Ren H (2011) Influences of initial pH on performance and anodic microbes of fed-batch microbial fuel cells. J Chem Technol Biotechnol 86:1226–1232
38. Yuan Y, Zhao B, Zhou S, Zhong S, Zhuang L (2011) Electrocatalytic activity of anodic biofilm responses to pH changes in microbial fuel cells. Bioresour Technol 102:6887–6891
39. Patil SA, Harnisch F, Koch C, Hübschmann T, Fetzer I, CarmonaMartínez AA, Müller S, Schröder U (2011) Electroactive mixed culture derived biofilms in microbial bioelectrochemical systems: the role of pH on biofilm formation, performance and composition. Bioresour Technol 102:6887–6891
40. Jung S, Mench MM, Regan JM (2011) Impedance characteristics and polarization behavior of a microbial fuel cell in response to short-term changes in medium pH. Environ Sci Technol 45:9069–9074
41. Erable B, Etcheverry L, Bergel A (2009) Increased power from a two-chamber microbial fuel cell with a low-pH air-cathode compartment. Electrochem Commun 11:619–622
42. Jadhav GS, Ghangrekar MM (2009) Performance of microbial fuel cell subjected to variation in pH, temperature, external load and substrate concentration. Bioresour Technol 100:717–723
43. Martin E, Savadogo O, Guiot SR, Tartakovsky B (2010) The influence of operational conditions on the performance of a microbial fuel cell seeded with mesophilic anaerobic sludge. Biochem Eng J 51:132–139

44. Wang X, Feng YJ, Qu YP, Li DM, Li H, Ren NQ (2008) Effect of temperature on performance of microbial fuel cell using beer wastewater. Environ Sci 29:3128–3132
45. Hong SW, Chang IS, Choi YS, Chung TH (2009) Experimental evaluation of influential factors for electricity harvesting from sediment using microbial fuel cell. Bioresour Technol 100:3029–3035
46. Ahn Y, Logan BE (2010) Effectiveness of domestic wastewater treatment using microbial fuel cells at ambient and mesophilic temperatures. Bioresour Technol 101:469–475
47. Guerrero AL, Scott K, Head IM, Mateo F, Ginesta A, Godinez C (2010) Effect of temperature on the performance of microbial fuel cells. Fuel 89:3985–3994
48. Patil SA, Harnisch F, Kapadnis B, Schröder U (2010) Electroactive mixed culture biofilms in microbial bioelectrochemical systems: the role of temperature for biofilm formation and performance. Biosens Bioelectron 26:803–808
49. Liu Y, Climent V, Berná A, Feliu JM (2011) Effect of temperature on the catalytic ability of electrochemically active biofilm as anode catalyst in microbial fuel cells. Electroanalysis 23:387–394
50. Behera M, Murthy SSR, Ghangrekar MM (2011) Effect of operating temperature on performance of microbial fuel cell. Water Sci Technol 64:917–922
51. Wang XL, Wu C, Zhang JQ, Chi QL, Tian SS (2011) Acclimation stage on the performance of microbial fuel cells subjected to variation in COD, temperature, and electron acceptor. Adv Mater Res 183-185:2346–2350
52. Michie IS, Kim JR, Dinsdale RM, Guwy AJ, Premier GC (2011) Operational temperature regulates anodic biofilm growth and the development of electrogenic activity. Appl Microbiol Biotechnol 92:419–430
53. Tang Y, He Y, Yu P, Sun H, Fu J (2012) Effect of temperature on electricity generation of single-chamber microbial fuel cells with proton exchange membrane. Adv Mater Res 393-395:1169–1172
54. Liu ZH, Li XM, Zheng FL, Yang Q (2012) Operating characteristics of microbial fuel cell using sludge, China. Environ Sci 32:268–273
55. Campo AG, Lobato J, Canizares P, Rodrigo MA, Morales FJF (2012) Short-term effects of temperature and COD in a microbial fuel cell. Appl Energy. https://doi.org/10.1016/j.apenergy.2012.02.064
56. Michie IS, Kim JR, Dinsdale RM, Guwy AJ, Premier GC (2011) The influence of psychrophilic and mesophilic start-up temperature on microbial fuel cell system performance. Environ Sci Technol 4:1011–1019
57. Cheng S, Xing D, Logan BE (2011) Electricity generation of single-chamber microbialfuel cells atlow temperatures. Biosens Bioelectron 26:1913–1917
58. Liu L, Tsyganova O, Lee DJ, Suc A, Chang JS, Wang A, Ren N (2012) Anodic biofilm in single-chamber microbial fuel cells cultivated under different temperatures. Int J Hydrog Energy. https://doi.org/10.1016/j.ijhydene.2012.03.084
59. Guerrero AL, Scott K, Head IM, Mateo F, Ginestá A, Fernández FJH, Godínez C (2010) Low temperature performance of microbial fuel cells. Chem Eng Trans 21:463–468
60. Velvizhi G, Mohan SV (2012) Electrogenic activity and electron losses under increasing organic load of recalcitrant pharmaceutical wastewater. Int J Hydrogen Energy 37:5969–5978

Conversion Processes and Pathways in Bio-Electrochemical Systems

Abdul Sattar Jatoi and Nabisab Mujawar Mubarak

Abstract Bio-electrochemical systems (BES) create a sustainable frontier through innovative pathways that transform biological substances and chemicals into electricity, combined with biofuels and value-creating products. The system operates within the intersection of microbiology and electrochemistry, enabling electroactive microorganisms to conduct reduction–oxidation reactions on electrodes under specific conditions. The main conversion processes, mechanistic pathways, and integration patterns operating within BES systems are analyzed by discussing substrate processing, electron transfer processes, and coupling between biochemical and electrochemical systems. This section explains the difference between anodic oxidation and cathodic reduction reactions. Also, it describes direct and mediated electron transfer pathways, as well as metabolic and enzymatic networks that drive the conversion processes. The article evaluates BES types, including microbial fuel cells (MFCs), microbial electrolysis cells (MECs), and microbial electrosynthesis systems (MES), based on their efficiency levels, system designs, and application boundaries. The techniques for technological optimization are discussed through analysis of performance metrics, including Coulombic efficiency, power density, and substrate conversion rates. The overview concludes with a discussion of existing challenges related to electrode design, microbial community management, and system scalability, as well as suggestions for future research directions. The synthesis provides a comprehensive foundation for advancing BES development as a robust technology for energy recovery, complementing environmental stewardship.

A. S. Jatoi (✉)
Department of Chemical Engineering, Dawood University of Engineering and Technology, Karachi, Pakistan
e-mail: abdul.sattar@duet.edu.pk

N. M. Mubarak
Department of Chemical Engineering, College of Engineering, Imam Mohammad Ibn Saud Islamic University (IMSIU), Riyadh, Saudi Arabia

Department of Biosciences, Saveetha School of Engineering, Saveetha Institute of Medical and Technical Sciences, Chennai, India

A. Sattar Jatoi, N. M. Mubarak (eds.), *Bio-Electrochemical Conversion of Waste to Fuel*, Green Energy and Technology,
https://doi.org/10.1007/978-981-95-4310-6_5

Keywords Electrochemical · Microbial fuel · Biochemical · Energy

1 Introduction

The rising demand for sustainable waste management and efficient energy solutions drives research into biological-electrochemical integration technologies since their practical applications increase interest. Modern technologies have established bio-electrochemical systems (BES) as a ground-breaking solution to bridge experimental research in microbial metabolism with electrochemical processes [1]. BES utilizes electroactive, electrogenic bacteria to power electrical processes, where these bacteria exchange electrons between themselves and electrodes, transforming chemical energy from organic substrates into electricity and essential biochemical products. Due to their biological and electrochemical nature, BES enable several conversion processes, including power generation, hydrogen production, and the recovery of nutrients through biofuel and chemical production [2].

Bio-electrochemical systems are a diverse technology group that merges MFCs with MECs and MDCs, including MES. Various configurations exist to perform specific conversion operations that match the characteristics of the substrate with those of the target product under particular environmental conditions [3]. MFCs function to transform organic waste into electricity, but MECs need an external voltage to convert wastewater into hydrogen gas. Microbial electric synthesis requires carbon dioxide and microbial catalysts at the cathode to produce marketable organic compounds such as acetate, ethanol, or methane. Every system within this family relies on the fundamental bio-electrochemical interface, allowing microbial metabolic activity to intersect with electrochemical redox reactions [4].

Electron generation and proton production remain at the center of BES operations because targeted substrates are transformed into electrons while generating protons during conversion processes involving metabolic pathways [5]. The transformation occurs through microbial populations that utilize enzymes to break down various substances, including carbohydrates, proteins, fats, and lignocellulosic biomass, to produce basic intermediates. The intermediates produce reducing equivalents, such as NADH and $FADH_2$, through catabolic processes before they move to external circuits via electron transfer mechanisms [6]. Electrons move through the anode and cathode sections before being used to decrease terminal electron acceptors such as oxygen, carbon dioxide, and nitrate, according to system configurations. The BES's total performance depends on three fundamental parameters: metabolic efficiency, electron transfer capacity, and electrochemical stability [7].

BES stands apart from ordinary biological or electrochemical methods through its ability to enable electrons to move in both directions between microbial cells and electrodes. Bacterial electrons are transferred to MFC and MEC anodes, leading to the oxidation of organic compounds and the donation of electrons. During MES operation, microbes function as electrotrophs, receiving electrons at the cathode and reducing carbon dioxide into new organic molecules [8]. Quantifying electron transport mechanisms that function through cytochromes embedded in bacterial membranes, conductive pili, and soluble redox shuttles helps system designers maximize the operational capabilities of their systems [9].

The conversion efficiency in BES depends on four primary groups of factors: substrate nature and concentration, microbial community structure, electrode physical and chemical properties, and operating parameters, including pH, temperature, and electrolyte conductivity [10]. Different operating variables determine both the speed of substrate breakdown and the degree of electricity recovery, as well as the yield of products. BES evaluation utilizes quantitative metrics to assess system performance across various configurations, including Coulombic efficiency, energy recovery, and substrate conversion rate measurements [11].

BES systems play a crucial role in extracting valuable resources from waste materials. Various types of wastewater, including municipal waste, agricultural discharge, industrial byproducts, and food manufacturing wastewater, contain high amounts of biodegradable organic matter and nutrients [12]. The integrated BES system treats waste streams while generating electricity, producing hydrogen gas or methane, or creating nutrient-rich bio-fertilizers. MFC applications in wastewater treatment result in a reduction of BOD in the influent water while producing electrical power. Through its operation, MECs produce hydrogen gas that serves as a clean energy vector [13]. BES systems achieve higher sustainability and economic viability when their multiple functions are combined to support environmental projects and energy outputs.

The wide-scale implementation of BES technology remains limited by numerous operational and technical barriers [14]. The application of BES technologies faces multiple barriers due to several obstacles, including low power density production, limited operational stability, high electrode resistance, slow microbial electrode colonization rates, and difficulties in scaling up [15]. Improving BES operations requires a thorough understanding of the biochemical and electrochemical conversion pathways that occur throughout BES operations. Research teams studying systems biology, metagenomics, synthetic biology, and materials science now build improved and enhanced BES capabilities. Genetic modification of electroactive bacteria improves their electron transfer speed and the precision with which they select compounds [16]. BES performance increases when nanostructured electrodes blend with redox-active materials because they enhance conductivity and improve microbial attachment [17].

The future development of BES plays a crucial role in integrating with emerging technologies from various scientific fields. Anaerobic digestion technology, integrated with photobioreactors or membrane filtration systems alongside BES, allows for system advantages that individual systems cannot achieve effectively [18]. Implementing machine learning systems and artificial intelligence through real-time monitoring enables improved operations for BES under varying environmental conditions. The realization of BES's maximum potential depends on a complete understanding of the conversion mechanisms and pathways within these systems [19].

This chapter details the complex procedures that enable the BES function. Our study thoroughly explains basic bio-electrochemical conversions that combine BES system types with electrochemical operation fundamentals. We examine the biochemical processes involved in substrate breakdown and microbial metabolic pathways that lead to product generation. The research examines direct and mediated electron transfer methods concerning microbial functions during electrode reactions. The study examines the reactions occurring at the anode and cathode in various BES configurations, resulting in energy transformation and chemical conversions [20]. This chapter continues with sections on integration, performance measurement methods, and specific case studies that demonstrate BES applications in energy generation, environmental remediation, and resource conversion projects. The final section of our study examines the existing research boundaries and new development requirements for this emerging scientific discipline.

The detailed examination of this book provides a fundamental understanding of bio-electrochemical system conversion methods and pathways while demonstrating their capacity to address global sustainability challenges.

2 Types of Bio-Electrochemical Systems and Their Conversion Capabilities

Microorganisms working with electrodes form a diverse technology group called bio-electrochemical systems (BES), which convert chemical energy into usable electrical energy and marketable products [21]. The microbial fuel cell (MFC) is the leading bio-electrochemical system (BES) technology, which enables electroactive microorganisms to generate electricity from the oxidation of organic substrates at the anode before the electrons are transferred through an external circuit to the cathode for current production [22]. Organic-rich wastewater treatment through MFC technology becomes advantageous because it combines energy generation and wastewater cleanup operations. The performance evaluation of these systems depends on measuring their power density and Coulombic efficiency rating. The lower voltages produced by MFC systems limit their usage to small-scale power generation and their incorporation into larger power units [23].

The purpose of microbial electrolysis cells (MECs) revolves around the manufacture of hydrogen, methane, and alternative fuel products. An external voltage of less than 1 V drives hydrogen evolution at the cathode of MECs, similar to how these systems operate like MFCs. The technology demonstrates high efficiency as a biohydrogen production method for various substrates, including wastewater, fermentation outputs, and agricultural waste materials [24]. The hydrogen product from MECs has storage potential as a clean fuel, making MECs suitable for developing a hydrogen economy. Methanogenic MECs can generate biomethane through the reduction of carbon dioxide and proton combination, facilitated by the controlled operation of microbial consortia, thereby enabling carbon sequestration and energy recovery protocols [25]. The microbial electrosynthesis system functions as a BES that operates through electron donation from the cathode, reversing the MFC electron processes. The microbial electrosynthesis system enables electroactive microbes, including *Sporomusa ovata* and *Clostridium ljungdahlii*, to produce valorized compounds such as acetate, ethanol, or butyrate by utilizing electrons supplied from the electrode [26]. These systems demonstrate high potential value for both carbon recycling and bio-based chemical production since they enable the conversion of waste CO_2 into marketable compounds. Maintaining high yield and selectivity from MES technologies requires accurate settings of both culture conditions and electrode potentials [27].

Enzymatic biofuel cells (EBFCs) constitute an advanced BES technology that performs electrode reactions by employing isolated redox enzymes instead of complete microbial cells. EBFCs possess operational reliability and specialized reaction control capabilities, allowing their implementation in biosensor systems, wearable devices, and implantable medical instruments. Enzyme deactivation alongside cofactor dependency remains the main hurdle slowing down their large-scale development for long-term operations.

The photosynthetic BES (Photo-BES) utilizes phototrophic organisms, such as cyanobacteria, algae, or photosynthetic bacteria, that use solar power for electron transport. The technology demonstrates the potential to produce electricity directly from sunrays, accompanied by a CO_2 reaction, while generating oxygen during the overall process that functions similarly to natural photosynthesis. Photo-BES primarily exists in experimental stages, although its future potential suggests sustainable carbon capture and solar power harvesting capabilities.

The different types of BES structures vary according to their complexity levels, microbial requirements, preferred substrates, electricity production, hydrogen or methane output, and the production of organic acids. Scientists today develop combined BES platforms that integrate multiple system types, such as MEC–MFC or solar units, to enhance operational flexibility and overall system performance. Two critical factors determine which BES configuration to use: the target application, which can be either wastewater treatment or decentralized energy generation, and whether it involves carbon recycling or biomanufacturing. The development of BES research will yield innovations that enhance the functionality, economic viability, and operational efficiency of these technological systems.

Type of BES	Electron flow direction	Main output/ product	Substrate type	Microorganisms used	External energy required	Applications
Microbial fuel cell (MFC)	Anode → cathode	Electricity	Organic waste, wastewater, and acetate	*Geobacter* spp., *Shewanella* spp.	No	Wastewater treatment, remote sensors, and small-scale power
Microbial electrolysis cell (MEC)	Anode → cathode (with voltage)	Hydrogen, methane	Organic matter, fermentation effluent	Mixed anaerobic consortia, methanogens	Yes (~0.2–1.0 V)	Biohydrogen production, energy recovery from waste
Microbial Electrosynthesis system (MES)	Cathode → microbe	Acetate, ethanol, butyrate	CO_2, syngas	*Sporomusa ovata*, *clostridium* spp.	Yes	CO_2 capture and conversion, biomanufacturing
Enzymatic biofuel cell (EBFC)	Enzyme → electrode	Electricity (in small devices)	Glucose, lactate, ethanol	Isolated enzymes (e.g., glucose oxidase)	No	Biomedical devices, biosensors, wearable electronics
Photosynthetic BES (photo-BES)	Light-driven electron transfer	Electricity, O_2, biomass	CO_2, water	Algae, cyanobacteria, and purple bacteria	Light (solar)	Solar-powered bioproduction, carbon sequestration

3 Substrate Utilization and Bioconversion Pathways

BES utilize substrate metabolism pathways as their metabolic core structure to produce electrons and protons, which generate electricity and facilitate chemical synthesis [28]. The microbial communities inside BES can transform basic compounds, such as acetate and glucose, as well as complex waste materials, including lignocellulosic biomass, food leftovers, and wastewater organic matter [29]. The microbial degradation process involves macromolecule hydrolysis, the fermentation of volatile fatty acids and alcohols, and the production of acetate, accompanied by the generation of hydrogen. This leads to electrogenic bacteria, such as Geobacter or Shewanella, generating electrons for anode transfer [30]. Substrate complexity determines whether multiple-stage degradation and conversion operations must be carried out in cooperation by syntrophic microbial consortia to sustain continuous electron flux [31]. The performance of electron transfer depends on the metabolic pathways, which include the tricarboxylic acid (TCA) cycle and anaerobic respiration that use the electrode as an electron acceptor. The performance of bio-electrochemical systems depends on the characteristics of the substances they convert, including their ability to biodegrade, their concentrations, the C/N ratio, and their toxic potential. Multiple factors related to thermodynamics and speed limit the total effectiveness of biochemical conversion systems [32]. Selecting a substrate knowingly plays a crucial role in determining microbial processes, electron yield, and product output, which can be electricity in microbial fuel cells, hydrogen in microbial electrolysis cells, or chemicals in microbial electrosynthesis systems. This means that substrate decisions must be optimized for BES systems to function effectively.

4 Electron Transfer Pathways and Mechanisms

Bio-electrochemical systems (BES) require fundamental electron transfer pathways and mechanisms to maintain their operations through which microorganisms move electrons resulting from substrate oxidation either to electrode surfaces or from electrodes to microbial cells. The two established transfer mechanisms include direct electron transfer (DET) and mediated electron transfer (MET) [33]. The bacterium Geobacter and Shewanella utilize outer-membrane cytochromes, conductive pili (nanowires), and tightly bound redox proteins to directly transfer electrons to electrodes in DET systems with high efficiency [34]. Bacterial cells move electrons between their components and electrodes by utilizing solvent-exchangeable redox mediators, including naturally produced compounds (flavins and phenazines) or external additives (neutral red and methylene blue). For extended BES function, DET provides better stability since it avoids the complications associated with soluble mediators; however, MET enhances slow electron conduction between the electrode and the microorganism [35]. These electrical pathways exhibit variable

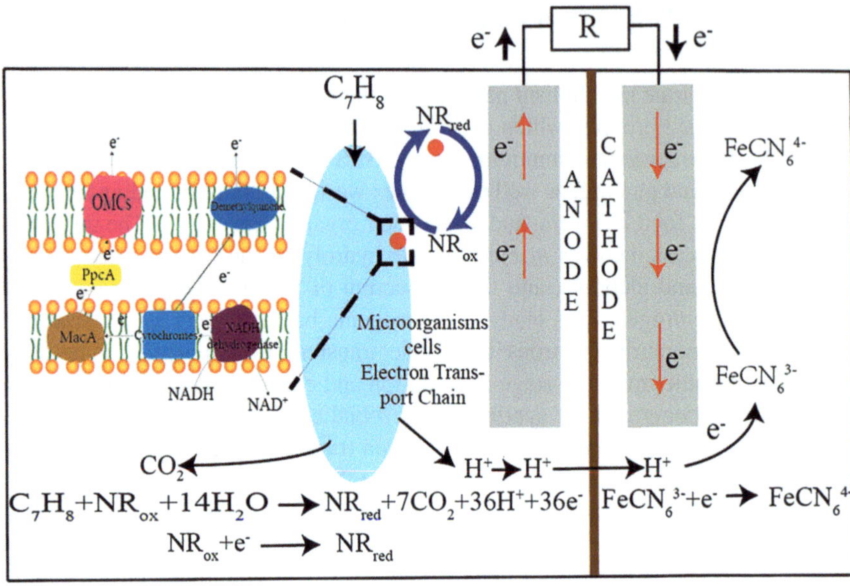

Fig. 1 Electron transfer pathways and mechanisms in bio-electrochemical systems [38]

performance due to their dependence on microbial species, biofilm structure, electrode material properties, surface composition, and environmental factors such as pH and ionic strength [36]. Modern studies have identified multi-heme cytochromes that work in conjunction with electron-hopping complexes to connect microbes within biofilm matrices, thereby forming intricate electron transfer systems [37]. Improving Coulombic efficiency, power output, and product yields in BES applications depends on optimizing electron transfer processes through microbial selection, genetic engineering, and electrode surface modifications (Fig. 1).

5 Cathodic and Anodic Reactions

Electron flow and resulting electrode products are major operational processes of bio-electrochemical systems (BES). Microorganisms break down organic substances, such as acetate and glucose, or wastewater pollution during anodic processes by establishing electron and proton separations [39]. The anode accepts electrons through DET or MET pathways as microbes release electrons, but protons build up into a solution gradient [40]. The primary anodic reactions often involve the oxidation of organic matter to produce carbon dioxide, protons, and electrons, such as in the oxidation of acetate:

$$CH_3COO^- + 2H_2O \rightarrow 2CO_2 + 7H^+ + 8e^-.CH_3COO^-$$
$$+2H_2O \rightarrow 2CO_2 + 7H^+ + 8e^-.$$

Based on system design, the cathode part receives electrons through an external circuit, which gets transferred to different terminal electron acceptor materials, including oxygen, carbon dioxide, or nitrate [41]. Microbial fuel cells (MFCs) normally perform the reduction of oxygen to water as their cathodic reaction:

$$O_2 + 4H^+ + 4e^- \rightarrow 2H_2O.O2 + 4H^+ + 4e^- \rightarrow 2H2O.$$

In Microbial Electrolysis Cells (MECs), where an external voltage is applied, the cathodic reaction can involve hydrogen evolution from protons:

$$2H^+ + 2e^- \rightarrow H_2.2H^+ + 2e^- \rightarrow H2.$$

The reducible carbon dioxide products in microbial electrosynthesis systems (MES) consist of acetate, methane, and various organic acids. The selected electron acceptor, microbial communities, and applied potential determine the nature of the cathodic reaction. MES electrode efficiency depends on electrode materials, biofilm development, and changes in pH and ionic strength because optimizations try to reduce internal resistance while improving the electrical connection between electrodes. BES technologies enable energy development and achieve nutrient recovery while conducting chemical synthesis by integrating opposing anodic and cathodic reactions [42].

6 Integration of Biochemical and Electrochemical Pathways

The basic operation of bio-electrochemical systems (BES) relies on combining biochemical reactions with electrochemical processes to convert organic substances into valuable energy and compounds [43]. The BES function is due to microbial communities that connect electrode electrochemistry with the biological breakdown of organic substances and the transfer of electrons. Microbial cells conduct electron transfers at the bio-electrochemical interface by donating electrons to the anode when breaking down substrates or accepting electrons from the cathode for electrosynthesis and hydrogen production reactions [44]. The biochemical process of microbial decomposition involves hydrolysis, fermentation, and acetogenesis, enabling microbes to produce acetate and hydrogen intermediates before converting these compounds into electrons for transport to the anode. At the same time, the electrochemical process allows electrons to travel along the external circuit, accelerating both energy production and reduction reactions occurring at the cathode, including oxygen reduction and carbon dioxide reduction. The successful operation of BES requires perfect integration between biochemical compounds and electrochemical processes [45]. Microbial fuel cells (MFCs) require ideal electron conduction between microbial cellular processes and the anode, as well as a rapid oxygen reduction reaction at the cathode. The external voltage in microbial electrolysis cells drives the cathode proton reduction process to produce hydrogen gas by

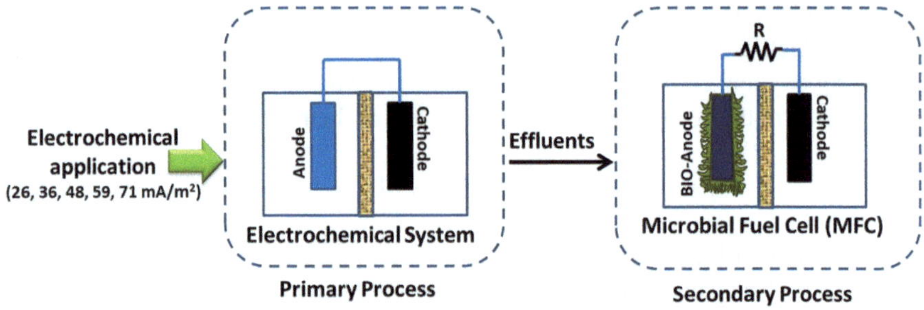

Fig. 2 Integration of biochemical and electrochemical pathways [46]

enhancing process coupling. The operational success of integrated pathways in BES technology depends on four main categories: electrode material selection, microbial populations held within the system, electron transport options, and operational system conditions (Fig. 2). Research on BES integration focuses on determining the optimal biochemical-to-electrochemical process ratios, as this maximizes efficiency and stability and enhances product yield [47].

7 Conversion Efficiency and Performance Metrics

The performance assessment of bio-electrochemical systems (BES) requires a precise evaluation of conversion efficiency and performance metrics for their deployment in energy generation, waste management, and chemical production systems [48]. The ratio between useful product output and initial input is the basis for calculating conversion efficiency in research processes. MFCs achieve operational efficiency by converting chemical energy from organic substrates into electrical output [49–51]. The widely employed performance indicator known as Coulombic efficiency calculates the electrical output percentage compared to all theoretical electrochemical electron releases from substrate breakdown. Two essential system performance indicators exist: power density, which measures the output power per unit of system size or surface area, and current density, which determines the quantity of electron flow per unit of electrode surface area [52]. MECs utilize hydrogen yield or hydrogen production rate metrics for performance assessment because they quantify the hydrogen production from each consumed unit of organic substrate. The product yield combined with faradaic efficiency is crucial in evaluating performance outcomes in microbial electrosynthesis systems (MES) [53]. The performance analysis of BES significantly depends on internal resistance, as high levels of internal resistance limit electron flow, resulting in reduced overall efficiency. The energy recovery potential of systems and their voltage output, specifically in microbial fuel cells (MFCs), serves as a crucial metric determining the practicality of BES for green energy applications [54]. Implementing BES technologies depends on optimizing efficiency metrics, which requires strategic design decisions, the selection of electrode materials, microbial strain engineering, and optimized operational methods.

8 Challenges and Future Perspectives

Several obstacles limit the commercialization and widespread adoption of bio-electrochemical systems (BES) despite their success in development. BES achieves insufficient power generation and poor conversion efficiency in microbial fuel cells (MFCs), making them unfit for substantial commercial energy production. This inefficiency primarily develops due to high internal resistance, suboptimal electrode materials, and sluggish electron transfer kinetics between microbial cells and electrodes. The optimization of BES performance faces complications due to the combination of multiple organic compounds in waste streams, which necessitate specialized microbial alliances and intricate biochemical processes. Independent electrode biofilms lead to performance decreases over time, requiring regular operation and maintenance activities. Raw microbial capabilities can be enhanced through genetic engineering combined with strain optimization to improve their ability to transfer electron signals more efficiently. BES faces significant challenges in scalability because laboratory systems currently prove difficult to scale up for industrial use due to barriers related to price, operational performance, and system interconnectivity. The future development of BES technology relies on electrode materials that incorporate nanomaterials, bioelectrode engineering, and synthetic biology approaches, which enhance electron transfer efficiency and system performance. The long-term advancement of sustainable waste-to-energy technology requires the integration of BES with renewable energy systems, which develop hybrid capabilities to process complex waste materials, such as food waste, wastewater, and lignocellulosic biomass. The next phase of the study will focus on enhancing system performance through advanced modeling techniques and automated control mechanisms, as well as on developing new performance assessments tailored to various application needs, such as energy recovery and chemical manufacturing. Advanced BES systems will succeed in the future by surpassing their existing technical barriers to produce maximum sustainable power generation alongside resource extraction and environmental improvement.

9 Conclusion

Bio-electrochemical systems (BES) demonstrate effectiveness as a multi-purpose technology due to their ability to generate energy from waste materials through microbial-electrochemical interactions. Biochemical and electrochemical networking efficiently transforms organic materials into useful electricity generation outputs, hydrogen production, or chemical manufacturing products. Larger-scale deployment of BES meets resistance because researchers must overcome four primary hurdles: low power generation, high internal electrical resistance, complicated substrate requirements, and electrode fouling issues. The future of better efficiency systems depends heavily on active development in microbial strain optimization, electrode material technology, and system development. BES technology will progress through combined systems that integrate renewable power generation with enhanced biocatalysts and innovative engineering approaches, thereby deploying sustainable environmental and clean energy solutions. Ongoing research and innovation will enable BES to develop a clean energy framework, a waste management solution, and a circular economy model for the future, thereby strengthening a sustainable global infrastructure.

References

1. Sathya PM, Mohan H, Park J-H, Seralathan K-K, Cho M, Oh B-T (2024) Bio-electrochemical degradation of carbamazepine (CBZ): a comprehensive study on effectiveness, degradation pathway, and toxicological assessment. J Environ Manag 360:121161
2. Kadam R, Jo S, Cha J, Yang H, Park J, Jun HB (2024) Influence of increasing anode surface area on nitrite-absent ammonium oxidation in a continuous single-chamber bio-electrochemical system. Chemosphere 353:141579
3. Zhou W, Peng S, Yuan J, Gao Y (2025) Application of bio-electrochemical systems for phosphorus resource recovery: Progress and prospects. J Environ Manag 375:124220
4. Tang J, Li Z, Xiao X, Liu B, Huang W, Xie Q et al (2025) Recent advancements in antibiotics removal by bio-electrochemical systems (BESs): from mechanisms to application of emerging combined systems. Water Res 268:122683
5. Li C, Yun M, Zhu X, Ma J, Zhang Z, Kong W et al (2025) High-value conversion of bio-dried products derived from municipal organic wastes: activated carbon preparation and electrochemical assessment. Fuel 387:134323
6. Singh NK, Kumar M, Sharath Kumar R, Renju, Khan VA (2025) Microalgal bio-electrochemical system and pretreatment technologies for biohydrogen production. In: Jaiswal KK, Singh B, Jaiswal AK, Gagnon DY (eds) Microalgal biofuels. Woodhead Publishing, Sawston, pp 203–232
7. Wang B, Liu Y, Zhang X, Wen G, Chen X, Wen C et al (2024) Enhanced ammonia nitration by bio-electrochemical systems with constructed wetlands. Bioresour Technol 410:131296
8. Kanellos G, Monokrousou E, Tremouli A, Lyberatos G (2024) Treatment of two-phase olive mill waste and degradation of phenolic compounds with simultaneous bio-electrochemical conversion of CO2 to CH4 using a microbial electrolysis cell. Sustain Chem Pharm 39:101589
9. Renju, Singh R (2024) (Bio)electrochemical system: a systematic approach from agricultural waste to sewage wastewater treatment with nutrients and hydrogen recovery. J Clean Prod 457:142387
10. Ding L, Zamalloa C, Lin H, Hu B (2025) Bio-electrochemically assisted sulfide, phosphorus, and nitrogen remediation in continuous anaerobic digestion of dairy manure with improved biogas production. Chemosphere 376:144288
11. Rahman S, Al Shuaili T, Al-Mamun A, Khudaish E, Sana A, Baawain MS et al (2025) Cost effective reduced graphene oxide/polyaniline composite coated SSM cathode for bio-electrochemical desalination: advancing desalination via cathodic improvement. Desalination 597:118399

12. Gupta R, Sethi S, Bharshankh A, Sahu R, Biswas R (2023) A perspective on bio-electrochemical system (BES) as a tool for boosting the performance of the nonperforming anaerobic units. Cleaner Circ Bioecon 6:100066
13. Yadav S, Ahmad A, Gulati C, Ghangrekar MM, Dubey BK (2024) Zn-Al@LDH infused hydrochar as cathode catalyst for upgrading tetracycline degradation and hospital wastewater treatment: a synergy of Fenton-like and bio-electrochemical systems. J Environ Chem Eng 12:113874
14. Cao Z, Yan L, Duan X, Li Z, Wang X (2025) Adsorption and nitrogen removal of NH4+-N based on Mn (II)/α-MnO2 cycle in bio-electrochemical system. Bioresour Technol 431:132628
15. Addagada L, Goel M, Shahid MK, Prabhu SV, Chand S, Sahoo NK et al (2023) Tricks and tracks in resource recovery from wastewater using bio-electrochemical systems (BES): a systematic review on recent advancements and future directions. J Water Process Eng 56:104580
16. Wang S, Hadji-Thomas A, Adekunle A, Raghavan V (2024) The exploitation of bio-electrochemical system and microplastics removal: possibilities and perspectives. Sci Total Environ 930:172737
17. Zhu Q, Li X, Nie Z, Wang Y, Dang T, Papadakis VG et al (2023) In-situ microbial protein production by using nitrogen extracted from multifunctional bio-electrochemical system. J Environ Manag 347:119050
18. Sun J, Pan F, Zhu H, Wu Q, Pan C, Lu F (2024) Enhancing low-temperature anaerobic digestion of low-strength organic wastewater through bio-electrochemical technology. Int J Hydrog Energy 58:1062–1074
19. Zhang Y, Su J, Li T, Yin L, Xian H, Wang K (2025) Advances in electrooxidation pathways for the production of carboxylic acids from bio-alcohols. Chem Eng J 508:160894
20. Kadam R, Jo S, Khanthong K, Lee J, Jang H, Park J (2023) Potential of nitrite-absent anaerobic ammonium oxidation by mixed culture ANAMMOX granules in a single chamber bio-electrochemical system. Chemosphere 345:140494
21. Ramírez Á, Muñoz-Morales M, López-Fernández E, Fernández-Morales FJ, Llanos J (2024) Advancing circular economy: critical insights into waste biomass derived carbon electrodes for (bio)electrochemical water treatment. Curr Opin Electrochem 46:101492
22. Zhang B, Shi S, Tang R, Qiao C, Yang M, You Z et al (2023) Recent advances in enrichment, isolation, and bio-electrochemical activity evaluation of exoelectrogenic microorganisms. Biotechnol Adv 66:108175
23. Godbole V, Singh S, Ramamurthy PC, Khan NA, Bisht M, Pal MK et al (2023) Electroactive microbe communication: a crucial aspect for energy generation in bio-electrochemical systems. J Environ Chem Eng 11:110646
24. Siwal SS, Zhang Q, Saini AK, Gupta VK, Roberts D, Saini V et al (2021) Recent advances in bio-electrochemical system analysis in biorefineries. J Environ Chem Eng 9:105982
25. Jia J, Sun Y, Huo L, Zhao L, Liu Z, Liu Z et al (2025) Bio-carbon composite for supercapacitor electrodes: harnessing hydrochar frameworks and bio-tar polymerization. Fuel Process Technol 269:108178
26. Boto ST, Bardl B, Harnisch F, Rosenbaum MA (2023) Microbial electrosynthesis with clostridium ljungdahlii benefits from hydrogen electron mediation and permits a greater variety of products. Green Chem 25:4375–4386
27. Bañeras L, Cabeza Á, Perona-Vico E, Lopez-Abelarias M, Puig S, De Wever H (2024) Microbial models for biocathodic electrochemical CO2 transformation: a comprehensive review on pure cultures. Bioresour Technol Rep 25:101766
28. Suboktagin S, Ullah MW, Sethupathy S, Keerio HA, Alabbosh KF, Khan KA et al (2025) Microbial cell factories for bioconversion of lignin to vanillin — challenges and opportunities: a review. Int J Biol Macromol 309:142805
29. Güneş S, Ayol A, Tezer O (2025) Biodegradation and power generation from olive oil mill wastewater using soil-based microbial fuel cells. Int J Hydrog Energy 116:410–419

30. Rathinavel N, Samuel JOE, Veleeswaran A, Nallathambi S, Ponnuchamy K, Muthusamy G et al (2025) Exploring the key factors enhancing the microbial fuel cell performance. Process Saf Environ Prot 193:385–402
31. Han J, Hamza F, Guo J, Sayed M, Pyo S-H, Xu Y (2025) Advanced technological approaches and market status analysis of xylose bioconversion and utilization: Xylooligosacharides and xylonic acid as emerging products. Biotechnol Adv 79:108509
32. Radeef AY, Najim AA (2024) Microbial fuel cell: the renewable and sustainable magical system for wastewater treatment and bioenergy recovery. Energy 360:100001
33. Kumar A, Pandit S, Sharma K, Agrawal S, Kuhad RC, Mathuriya AS et al (2024) Microbial degradation of cellulose extracted from wheat bran for bioelectricity production using microbial fuel cell. Process Saf Environ Prot 190:574–585
34. Gasparrini W, Lee SH, Woolston BM (2025) Pathways to sustainability: a quantitative comparison of aerobic and anaerobic C1 bioconversion routes. Curr Opin Biotechnol 93:103310
35. Bazina N, Ahmed TG, Almdaaf M, Jibia S, Sarker M (2023) Power generation from wastewater using microbial fuel cells: a review. J Biotechnol 374:17–30
36. Pandya RS, Kaur T, Bhattacharya R, Bose D, Saraf D (2024) Harnessing microorganisms for bioenergy with microbial fuel cells: powering the future. Water-Energy Nexus 7:1–12
37. Surendra KC, Angelidaki I, Khanal SK (2022) Bioconversion of waste-to-resources (BWR-2021): valorization of industrial and agro-wastes to fuel, feed, fertilizer, and biobased products. Bioresour Technol 347:126739
38. Chen H, Yu Y, Yu Y, Ye J, Zhang S, Chen J (2021) Exogenous electron transfer mediator enhancing gaseous toluene degradation in a microbial fuel cell: performance and electron transfer mechanism. Chemosphere 282:131028
39. He Z, Huang Y, Manohar AK, Mansfeld F (2008) Effect of electrolyte pH on the rate of the anodic and cathodic reactions in an air-cathode microbial fuel cell. Bioelectrochemistry 74:78–82
40. Ye Y, Yan X, Jiang Y, Wang S, Liu D, Ren Y et al (2025) Optimized feeding schemes of heterotrophic anodic denitrification coupled with cathodic phosphate recovery from wastewater using a microbial fuel cell. Sci Total Environ 981:179590
41. Ghasemi S, Gheshlaghi R, Mahdavi MA, Abazarian E (2024) Evaluation of low-cost carbon/metal electrodes as cathodes and anodes in sediment microbial fuel cells. Fuel 373:132349
42. Guo C, Zhang X, Zhao W, Han Y, Pei L, Hou Y et al (2023) Study on the performance of anodic desulfurization and cathodic denitrification coupled electricity generation in microbial fuel cell. J Water Process Eng 54:103932
43. Dehghanian M, Mahdavi MA, Gheshlaghi R, Darband GB, Absalan Y (2025) Synthesis of TiC@C–TiO2 as an emerging efficient nanostructure for anode modification in high-power microbial fuel cells. Int J Hydrog Energy 116:591–600
44. Christwardana M, Prayoga BA, Kuntolaksono S, Joelianingsih J, Maulana AY (2025) Enhanced electron transfer in microbial fuel cells using calcined eggshell anode with optimized performance. Bioresour Technol Rep 29:102088
45. Xie L, Tanaka F, Yagi T, Hashimoto H, Ikeru K, Igarashi T et al (2024) Multi-anode enhanced the bioelectricity generation in air-cathode microbial fuel cells towards energy self-sustaining wastewater treatment. Environ Res 243:117744
46. Mohanakrishna G, Al-Raoush RI, Abu-Reesh IM (2021) Integrating electrochemical and bioelectrochemical systems for energetically sustainable treatment of produced water. Fuel 285:119104
47. Opoku PA, Shaowei X, Jingyu H (2025) Bioenergy generation, organic matter, and simultaneous nitrogen and phosphorus removal in comparative pyrite and graphite-based constructed wetland-microbial fuel cells at different anode scales. J Water Process Eng 69:106882
48. Jatoi AS, Akhter F, Mazari SA, Sabzoi N, Aziz S, Soomro SA et al (2021) Advanced microbial fuel cell for waste water treatment—a review. Environ Sci Pollut Res 28:5005–5019

49. Jatoi AS, Hashmi Z, Anjum A, Bhatti ZA, Siyal SH, Mazari S et al (2021) Overview of bio-electrochemical approaches for sulfur reduction: current and future perspectives. Biomass Convers Biorefinery 13:1–16
50. Jatoi AS, Hashmi Z, Mazari SA, Mubarak NM, Karri RR, Ramesh S et al (2022) A comprehensive review of microbial desalination cells for present and future challenges. Desalination 535:115808
51. Jatoi AS, Mallah NB, Ahmed J, Shah SA, Sabzoi N, Ibrahim SM (2025) Enhancing the performance of microbial fuel cell via use of new type of electrode for the bioenergy generation along with the treatment of distillery effluent. Biomass Convers Biorefinery 15:3151–3164
52. Naderi A, Kakavandi B, Giannakis S, Angelidaki I, Rezaei Kalantary R (2023) Putting the electro-bugs to work: a systematic review of 22 years of advances in bio-electrochemical systems and the parameters governing their performance. Environ Res 229:115843
53. Kumar G, Saratale RG, Kadier A, Sivagurunathan P, Zhen G, Kim S-H et al (2017) A review on bio-electrochemical systems (BESs) for the syngas and value added biochemicals production. Chemosphere 177:84–92
54. Srikanth S, Kumar M, Puri SK (2018) Bio-electrochemical system (BES) as an innovative approach for sustainable waste management in petroleum industry. Bioresour Technol 265:506–518

Microbial Communities and Their Role in Bioelectrochemical Systems

Rahul Palani, Sai Nandhini Ravi, and Karthik Pothiyappan

Abstract Microbial communities play a vital role in bioelectrochemical systems (BES), enabling sustainable energy generation, wastewater treatment, and bioremediation. These systems utilise the metabolic processes of electroactive microorganisms, facilitating electron transfer between electrodes and substrates. The diversity, composition, and interactions of microbial consortia have a significant impact on the efficiency and performance of biogas energy systems (BES). Key groups, such as Geobacter, Shewanella, and Pseudomonas, are recognised for their ability to conduct extracellular electron transfer (EET), which contributes to power generation and pollutant degradation. The microbial community structures in BES are shaped by environmental factors, electrode materials, and operational parameters, which in turn affect biofilm formation and electron transfer pathways. Understanding these complex microbial interactions is essential for optimising BES performance and stability. Advances in omics technologies, including metagenomics, transcriptomics, and proteomics, have provided more profound insights into microbial dynamics and functional roles in BES. Future research aims to engineer microbial communities through synthetic biology and adaptive evolution to elevate electron transfer efficiency and system robustness. This review examines the current understanding of microbial communities in Bioelectrochemical systems, their functional mechanisms, and strategies to enhance their application in energy and environmental sustainability. A comprehensive understanding of these microbial interactions will drive the development of more efficient and scalable bioreactor systems (BES) technologies.

R. Palani · S. N. Ravi (✉)
Department of Biotechnology, Vel Tech Rangarajan Dr. Sagunthala R&D Institute of Science and Technology, Chennai, Tamil Nadu, India

K. Pothiyappan
Faculty of Engineering, Department of Food Technology, Karpagam Academy of Higher Education, Coimbatore, Tamil Nadu, India

A. Sattar Jatoi, N. M. Mubarak (eds.), *Bio-Electrochemical Conversion of Waste to Fuel*, Green Energy and Technology,
https://doi.org/10.1007/978-981-95-4310-6_6

Keywords Microbial communities · Bioelectrochemical systems · Extracellular electron transfer · Biofilm formation

1 Introduction

Fossil fuels result from the compression of dead plant and animal matter, which layers of rock and mud have compressed. Upon longer exposure to pressure, this plant and animal matter gets converted into valuable resources such as coal, natural gas, and petroleum oil. These fossil fuels have been utilised for years to generate electricity and power modern transportation technologies, including cars and buses [1]. Although fossil fuels are efficient, they have an irreversible impact on the environment, increasing toxic greenhouse gases that are a primary cause of global warming and ozone layer depletion. With the growing demand for electricity generation, greenhouse gas emissions have also increased exponentially. Recently, the focus has been on sustainable energy technologies to replace electricity production from conventional resources such as fossil fuels [2]. As a result, technologies utilising renewable energy sources, such as solar energy, ocean, wind, biomass-derived energies, and Energies from hydrogen, have been preferred to decrease the damage caused by greenhouse gases exponentially and contribute to addressing the overall energy demand crisis [3–5]. Due to their ability to produce electricity without causing any harm. Damage to the environment: Renewable energy (RE) technologies are gaining popularity due to increasing awareness of clean energy among the public [6]. Even though renewable energy (RE) technologies have drastically reduced the reliance on fossil fuels for energy production, challenges persist in stabilising and implementing these RE technologies. Examples of the crisis include counterproductive energy storage in solar and wind energy systems, low conversion rates, and resource-intensive processes in biofuel energy systems. New methods and approaches are being researched to produce electricity efficiently and increase environmental sustainability, thereby addressing the growing need for power demand.

A bioelectrochemical system (BES) operates as a tool that effectively converts chemical energy into various useful resources, including electricity, hydrogen, and value-added chemicals. Over the past few years, the utilisation of BES has gained interest due to the possible applications and uses it brings to the table. Several studies have demonstrated the potential of BES to utilise organic matter to generate clean and sustainable electricity. Generally, BES combines Biological processes, enzymes, and various microorganisms to produce electricity and other valuable byproducts [7]. Due to their vast applications, the potential of microorganisms in BES to utilise wastewater as a source has been studied. It appears that BES can treat waste wastewater and produce electricity. At the same time, these types of BES systems also aid in wastewater treatment. Bioelectrochemical systems are broadly classified into Microbial fuel cells (MFCs), microbial electrolysis cells (MECs), microbial desalination cells (MDCs), microbial reverse-electrodialysis cells (MRCs), microbial solar cells (MSCs), enzymatic biofuel cells (EBCs), and microbial electrosynthesis cells. All these systems utilise various principles to achieve their potential uses [8]. One of the most commonly used BES types is the microbial fuel cell (MFC). Microbial fuel cells hold promise compared to other methods available for wastewater treatment and energy recovery from organic matter. The primary objective of MFC is to recover electricity while also providing wastewater treatment, which is achieved by utilising electrogenic bacteria as biocatalysts for oxidising organic matter in the anodic chambers and an external circuit to harness the electricity generated during this process.

2 Fundamentals of Bioelectrochemical Systems and Microbial Interaction

Bioelectrochemical systems (BESs) are systems that induce electrode reactions using microorganisms as biocatalysts. Prominent examples of BESs include MFCs and MECs, which often merge electrochemical and microbial processes to yield bioelectricity, hydrogen, and other value-added chemicals [9]. A typical BES comprises two electrodes, such as a cathode and an anode, connected by an external conductive wire to complete the electric circuit. MFCs are primarily used for producing electricity; meanwhile, MECs are employed to drive chemical reactions for generating hydrogen or other valuable chemicals using electricity. For electricity and valuable chemical production, wastewater can be utilised as a bacterial source and energy for MECs and MFCs. Throughout the process, organic matter from the wastewater is eliminated and utilised as a source of energy for the microorganisms, thereby treating the wastewater and producing electricity [10].

BES utilises distinctive microorganisms that are crucial in transporting electrons to an external solid electron acceptor, such as the electrodes from the electron transport chain inside the cell. This exchange reaction with extracellular solids is known as "extracellular electron transfer" (EET). Typically, in MFCs, in the absence of oxygen, the electrically active bacteria in the anode compartment oxidise organic matter in the wastewater, while the anaerobically present bacteria release protons, electrons, and CO_2 as they break down organic matter. The electrons produced by microorganisms are delivered to the anode through various possible pathways and reach the cathode through an external circuit. A cation exchange membrane is placed between the cathode and anode sections to allow the protons to travel between the two electrodes of the MFCs. In the cathode region, reactions occur between protons and electrons and oxygen, typically from air, resulting in the formation of water and thus completing the circuit. In MECs, there is a slight difference compared to MFCs, where during the reaction process in the cationic chamber, oxygen is absent. Due to the absence of oxygen, a reaction occurs between protons ($H+$) and electrons, resulting in the production of hydrogen as a byproduct of the reaction [3]. Bioelectrochemical system technology has been modified and given various names based on the exclusive function each variant satisfies. Some of the variants of BES include the microbial desalination cell (MDC), microbial remediation cell (MRC), microbial chemical cell (MCC), microbial solar cell (MSC), and microbial electrosynthesis (MES), among others. The principles and details of each are explained in the cited literature [4]. Some of the main fundamentals of BES include the transfer of electrons to the electrodes from the microorganisms. Bioelectrochemical systems employ various techniques, including the extracellular electron transfer mechanism and outward electron transfer.

2.1 Extracellular Electron Transfer Mechanisms (EET)

EET is a key process associated with BES that combines insoluble acceptors and donors with microorganisms; based on this principle, various applications can be carried out. Upon researching the fundamentals of electron transfer, two unique classifications of transfer methods were identified: direct extracellular electron transfer (DET) and indirect extracellular electron transfer (IET). DET amid microorganisms and electrodes has been suggested to occur in two ways. One is through direct electron transfer in microbes, which occurs due to the direct attachment of outer membrane proteins to the electrode. This enables electrons to be transferred directly from the cell interior to the external electrode surface [5]. The second method of direct electron transfer occurs between microorganisms and electrodes through cell extensions, such as nanowires or pili. These extensions connect the microorganism to the electrode, facilitating the transfer of electrons between them. With nanowires, the range by which the microbes can make a connection to the electrode is extended to nearly tens of microns. So, the cells can make even distant connections possible.

Indirect or mediated electron transfer from microorganisms to the electrode surface occurs through electron-transporting compounds, which facilitate the shuttling of electrons back and forth between microbes and electrodes. These compounds are naturally present or could be microbial secretions such as flavins. These mediators facilitate electron transport from and to the electrode due to their cyclic redox properties, which enables long-range electron transfer that supports sustainable and efficient electricity generation. Additionally, these EET mechanisms enable microorganisms to form multi-layered biofilm structures.

2.1.1 Outward Electron Transfer

Outward electron transfer occurs via the transfer of electrons from intercellular electron transport chains to pericellular electron acceptors on the outside. Through this unique occurrence, electroactive microbes perform respiration [11]. In BESs, one advantage in microbial electron transfer is their ability to transfer electrons to minerals. It is then Used to collect electrons by placing an insoluble solid electrode as a terminal electron acceptor and extracting electrons as current [12].

3 Types of Microorganisms Involved in Bio Electrochemical Systems

Bioelectrochemical systems (BES) have been reported to work with different types of biocatalysts, each providing distinct outputs; the classification of a common group of microbes utilised is described in Fig. 1.

3.1 Exoelectrogenic Bacteria

Microbes capable of exocellular electron transfer are classified as exoelectrogens. They have different names, such as electrochemically active bacteria [13], anode-respiring bacteria [14], and electricigens [15]. Studies conducted on many different pure cultures of bacteria can produce power in anodic. Four out of five classes of the Proteobacteria, Firmicutes, and Acidobacteria phyla have demonstrated the generation of electrical current. Yeast pichia anomala in MFC produces electricity by utilising redox enzymes on its outer membrane [16]—the phototrophic oxygenic cyanobacterium synechocystis sp. PCC 6803 was identified as generating electricity through conductive appendages called nanowires [17]. Electrogens utilise exocellular electron transfer (EET) in microbial fuel cells (MFCs) through three key mechanisms. First, in respiration, bacteria transfer electrons from terminal oxidases to solid metal oxides, such as Fe(III), reducing them to Fe(II). Secondly, direct interspecies electron transfer occurs without the need for intermediates. This was observed in *Pelotomaculum thermopropionicum* and *Methanothermobacter*

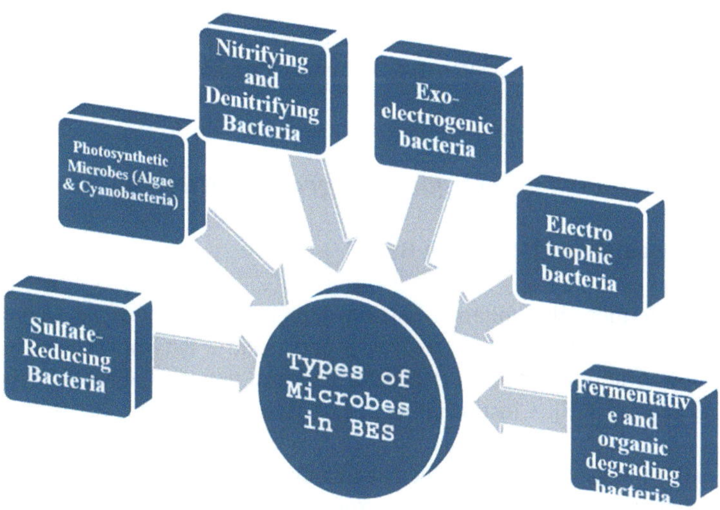

Fig. 1 Types of microbes

thermautotrophicus via conductive appendages. Thirdly, biocathodes with electrogens speed up the oxygen reduction process, enabling bacterial growth by utilising electrons from the anode [18]. These biocathodes also facilitate the reduction of nitrate and hydrogen evolution [19]. The ability of microbes to both donate and accept electrons highlights EET as a natural microbial process. Additional studies on dissimilatory metal-reducing bacteria have identified three electron transfer (EET) mechanisms: direct electron transfer (DET) via cytochromes in the outer membrane, mediator-based shuttling, and electrically conductive nanowires [20]. Initially, nanowires were characterised by their conductance across their diameter. Moreover, recent studies show that *Shewanella oneidensis* MR-1 nanowires exhibit nonlinear electrical transport along their length [21].

3.2 Electrotrophic Microorganisms

Electrotrophic microorganisms utilise cathode-based electrons for metabolism, playing a significant role in microbial electrochemical systems (MES), such as microbial electrosynthesis (MES) and microbial electrolysis cells (MECs). Acidithiobacillus ferrooxidans exemplifies electrotrophy by utilising reversed electron flow mechanisms to reduce NAD^+ while adapting to both soluble Fe^{2+} and solid conductive surfaces [22]. Shewanella oneidensis and Geobacter sulfurreducens maintain electron flow via outer membrane cytochromes (OMCs) and menaquinone (MQ), whereas Sporomusa ovata reduces CO_2 into acetate and butyrate through cytochrome-dependent electron transfer. Gram-positive electrotrophs, such as Clostridium ljungdahlii and Acetobacterium woodii, depend on electron bifurcation for ATP generation and CO_2 fixation, which is distinct from the mechanisms of gram-negative bacteria [23]. These microorganisms drive bioelectrosynthesis and support microbial fuel cells (MFCs) and microbial electrochemical cells (MECs), contributing to sustainable energy and carbon capture [24].

3.3 Fermentative and Organic-Degrading Microbes

3.3.1 Fermentative Microbes in BES

The role of fermentative microbes is to disintegrate complex organic compounds into simpler substrates, such as hydrogen and volatile fatty acids, which in turn serve as electron donors for electroactive bacteria, thereby enhancing bioelectrochemical activity [25]. Clostridium spp. and Enterobacter spp. Produce essential intermediates, such as acetate and butyrate, which support interspecies electron transfer (IET) with electrogens. Electron transfer from fermentative bacteria occurs through indirect pathways, utilising hydrogen or formate as carriers, and this process can also occur via direct pathways through conductive pili or biofilms [26].

3.3.2 Organic-Degrading Microbes in BES

Organic-degrading microbes, including Geobacter, Shewanella, Pseudomonas, and Desulfovibrio, break down pollutants and participate in extracellular electron transfer (EET) [27]. Pseudomonas aeruginosa produces pyocyanin as a redox mediator while sulfate-reducing bacteria (*Desulfovibrio* spp.) and denitrifying bacteria (*Paracoccus* spp.) contribute to organic degradation and bioelectricity generation [28]. The efficiency of organic degradation in BES depends on the microbial consortia, substrate availability, and reactor conditions. BES applications include the degradation of petroleum hydrocarbons, polycyclic aromatic hydrocarbons (PAHs), and industrial dyes.

3.4 Sulfate-Reducing Bacteria (SRB)

Sulfate-reducing bacteria (SRB) play a crucial role in bioelectrochemical systems (BES) by purposing sulfate as a terminal electron acceptor, enabling organic matter degradation and bioelectricity generation. SRB, which are found in anoxic environments, perform dissimilatory sulfate reduction by converting sulfate (SO_4^{2-}) into hydrogen sulfide (H_2S), making them valuable for sulfate removal, hydrogen production and bioremediation of metal-contaminated environments. These SRBs can perform both direct and indirect pathways of electron transfer mechanisms. Direct electron uptake (DEU) occurs via cytochromes in the outer membrane and nanowires, a mechanism similar to that found in Geobacter and Shewanella species. In contrast, indirect electron uptake (IEU) relies on intermediates such as hydrogen or formate, which are electrochemically produced at the cathode. The ability of certain species, such as Desulfovibrio piger and Desulfopila corrodes, to directly accept electrons from solid electrodes makes them promising candidates for biocathodes in bioelectrochemical systems (BES) applications. These properties enables SRB applications in wastewater treatment, where SRB-based biocathodes achieve high sulfate removal efficiency while generating electricity. Additionally, SRB also contributes to microbial electrosynthesis (MES) by utilising cathodic electrons to reduce CO_2 into valuable chemicals, such as acetate and ethanol. [29].

3.5 Nitrifying and Denitrifying Bacteria

3.5.1 Nitrifying and Denitrifying Bacteria in Bioelectrochemical Systems (BES)

Nitrification and denitrification in BES involve distinct bacterial communities that work together to remove nitrogen from the system. Ammonia-oxidising bacteria (AOB), such as Nitrosomonas, convert ammonium (NH_4^+) to nitrite (NO_2^-), followed by nitrite-oxidising bacteria (NOB), such as Nitrobacter, which oxidise NO_2^- to nitrate (NO_3^-) [1]. This aerobic process requires either increased aeration or a separate nitrifying unit in the cathode to facilitate oxidation. Heterotrophic bacteria, such as Paracoccus and Pseudomonas, carry out denitrification by using organic carbon as an electron donor to reduce NO_3^- to N_2. Autotrophic bacteria denitrify by relying on hydrogen, iron, or sulphur compounds for energy, using CO_2 or bicarbonate as a carbon source. In BES, autotrophic bacteria denitrify and facilitate nitrate reduction in cathodic biofilms, even under conditions of reduced carbon availability. Simultaneous nitrification and denitrification (SND) occur within a single BES biofilm due to oxygen gradients. Meanwhile, the Nitrosomonas and Nitrospira present in the outer aerobic layer facilitate the nitrification process in the BES. In contrast, the inner anaerobic layers, enriched with *Paracoccus* and *Pseudomonas*, perform denitrification, making the BES highly efficient for nitrogen removal.

3.6 Photosynthetic Microbes (Algae and Cyanobacteria)

3.6.1 Microalgae in BES

BES equipped with microalgae enhances biomass production, wastewater treatment, and energy generation. Microbial fuel cells with chlorella demonstrated sustainable biomass and energy production [30]. MFCs also improved nitrogen removal efficiency in wastewater treatment [16]. Ceramic MFC catholytes enhanced microalgae lysis under light/dark cycles, increasing electricity generation. Reviews highlight the role of MFCs in wastewater treatment, biomass production, and energy generation, identifying key factors that influence their performance and future research directions [31].

3.6.2 Cyanobacteria in BES

Microalgae-MFCs are a form of biophotovoltaic BES that utilises microalgae as cathodic catalysts to generate bioelectricity with exoelectrogenic microbes. These systems also produce oxygen, biomass, and value-added chemicals. Large-scale oceanic algal clusters outperformed commercially available graphite and silicon electrodes in energy output and storage capacity (Table 1). Algae-based systems sustainably recycle nutrients and convert carbon dioxide from the atmosphere into carbon-based fuels, thereby aiding in the continuous, self-sufficient operation of MFCs [32].

Table 1 Types of microbes involved in BES

S. No.	Type of microorganism	Examples	Type of BES	Role in BES
1.	Exoelectrogenic bacteria (exoelectrogens)	*Geobacter* sp.	Single-chamber MFC	Direct extracellular electron transfer (EET), biofilm formation
		Geobacter sp.	Double-chamber MFC	Direct EET, used for phenolic wastewater treatment
		Shewanella sp.	Single-chamber MFC	DET and IET mechanisms, adaptation to nutrient-limited environmental
		Shewanella sp.	Double-chamber MFC	Used in wastewater treatment, potential in petrochemical applications
2.	Electrotrophic microorganisms	*Methanobacterium* sp.	Microbial electrolysis cell (MEC)	Accepts electrons for CO, reduction to methane.
		Sporomusa sp.	MEC	Converts CO into acetate using electrons
3.	Fermentative and organic-degrading microbes	*Clostridium* sp.	Single-chamber MFC	Degrades complex organic matter to simpler substrates for electrons
		Clostridium sp.	Double-chamber MFC	Used in active sludge systems, it contributes to Electrogenic metabolism
4	Sulfate-reducing bacteria (SRB)	*Desulfovibrio* sp.	Two-chamber MFC	Sulphur metabolism, electron exchange, biofilm-associate D corrosion
	Nitrifying and denitrifying bacteria	*Desulfovibrio* sp.	MEC	Sulfate reduction and hydrogen generation
		Nitrosomonas sp.	Two-chamber MFC	Converts ammonia to nitrite in nitrogen cycle applications
		Paracoccus sp.	Constructed wetland MFC (MFC-CW)	Nitrate reduction and nitrogen removal
5	Photosynthetic microbes (algae and cyanobacteria)	*Synechocystis* sp.	Single-chamber MFC	Uses light energy to generate electrons for bioelectrochemical reactions
		Chlorella sp.	MEC	Enhances electrogenic performance through photosynthesis

4 Mechanisms and Pathways

Within the bioelectrochemical system, electrodes in the anode compartment facilitate microbial growth and the production of biofilms. Biofilm production is a significant phenomenon that occurs as a result of microbial metabolism, where electrons are generated within the system. The produced electrons reach the electrodes in different ways, such as direct electron transfer (DET) through membrane-bound protein structures, such as cytochromes [33]. Another method of electron transfer involves the use of pili or nanowires [34]. Lastly, electron transfer occurs through the involvement of redox mediators secreted by the bacteria or added to the system to facilitate electron transfer [35]. Mechanism and pathways involved in BES was illustrated in Fig. 2.

4.1 Direct Electron Transfer (DET)

With many studies focused on achieving deeper insights into the DET to the anode, little is known about the molecular mechanism involved in direct electron transfer from microbes to the cathode in BESs. G. sulfurreducens is a notable exoelectrogen, verified and validated as an effective cathode catalyst for redox processes [36]. The discovery of G. sulfurreducens's ability to gain electrons directly from the electrode is considered one of the most outstanding achievements in microbial electrosynthesis [37]. Electron transfer through membrane cytochromes, periplasmic and nanowires

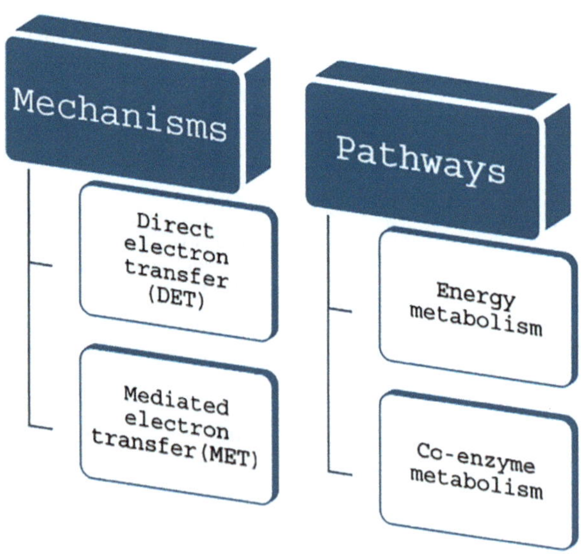

Fig. 2 Mechanisms and pathways involved in BES

represents direct electron transfer, the most apparent mechanism in extracellular electron transfer(EET). The capacity of Living microbes for electron transfer is released during oxidative substrate deterioration, which has been utilised at the anode in various bioelectrochemical systems (BESs). The key mechanisms involved in microbial ET to the electrodes are membrane-spanning cytochromes [38] Or electrically conductive nanowires/pili [39], followed by contribution of exogenous redox or self-excreted mediators [40], and via the reduced primary metabolites oxidation [41], All dependent on the microbial catalysts employed and their efficiency in achieving EET.

In Geobacter sulfurreducens, the DET mechanism is primarily facilitated by conductive pili (nanowires) and cytochromes in the outer membrane (OMCs), which act as electron conduits between the cells and the electrode. A typical approach in natural environments involves the use of c-type cytochromes in microbial electron uptake from an electron donor; for example, the OMC Cyc2 of Acidithiobacillus ferrooxidans has been observed to receive electrons from the iron(II) oxidation reaction [42]. The periplasmic cytochrome cyt579, a chemolithoautotrophic iron oxidiser found in Leptospirillum ferriphilum, has been suggested to be a primary iron oxidoreductase. DET can also occur through interspecies electron transfer, where one microorganism transfers electrons to another via direct physical contact, allowing for cooperative energy conservation and metabolic processes within microbial communities. A distinct indirect mechanism of microbial electron intake from the cathode involves molecular hydrogen—water electrolysis, where the electrochemical reduction of protons at the cathode results in the production of hydrogen. Microbes subsequently oxidise this hydrogen, which is linked to reducing a final electron acceptor such as nitrate, sulfate, or CO_2. A diverse range of microbes, including acetogens and chemolithotrophs, can utilise hydrogen as an electron donor [43]. The ET to microbes has been utilised and analysed mainly for denitrification reactions in wastewater treatment [44].

4.2 Mediated Electron Transfer (MET)

Indirect electron transfer mechanisms that do not require physical contact between the cathode and microbial cells occur via redox mediators that can be secreted by the microbes or exogenously added. Microorganisms can employ several mediators to induce electron uptake from the cathode. Electron shuttles are often utilised to facilitate electron transfer from microbes to cathodes, and some examples include 2,4-anthraquinone disulfonate, neutral red, methyl viologen, and cobalt sepulchrate [45]. Self-secreted mediators, such as flavins and quinones, are commonly analysed in Shewanella species, and these can enhance electron transfer efficiency in BES applications. These mediators function by accepting electrons from the microbial electron transport chain (ETC) and directly transferring them to the electrode, thereby facilitating energy generation and bio-production processes.

4.3 Energy Metabolism

Within the electrochemical system, bacteria can obtain energy through two distinct methods. First is the traditional respiration chain, which is the main energy generation pathway. Insoluble metals operate as electron acceptors for electrochemically active strains and take part in the extracellular electron transport chain in nature. MFC systems were inoculated with Geobacter and Shewanella, and these microbes obtain energy continuously through the degradation of organic acids, utilising them in favour of cell growth and metabolism. The redox balance was sustained to the anode through the release of electrons, and this substitutes insoluble metal as the final electron acceptor [46]. Another way of obtaining energy is electron transfer. As electrons navigate from the cathode within the cell's environment, accompanied by proton-coupled transport, the hypothesis is that the released protons generate a proton motive force (PMF), which drives the synthesis of ATP [47]. In comparison with aerobic respiration, the energy produced by oxidative phosphorylation is not sufficient for cell growth at elevated rates under aerobic conditions. The primary reason is the low ATP supply levels within the cell. Due to the role of electron acceptors as intermediate metabolites, the oxygen supply levels were inadequate [48]. Usually, cell appendage synthesis is an energy-demanding process; a shuttle electron can provide more energy through strongly impactful electron transfer under similar parameters [49]. Sometimes, bacteria can receive sufficient energy to sustain cell growth and metabolism in MEC systems, even though the system does not possess an original electron transfer chain [50].

4.4 Co-Enzyme Metabolism

The intracellular reducing power level is identified by the concentration of co-enzymes such as NADH or NADPH, which are also involved in many oxidation–reduction reactions. The intracellular steady state is disrupted by the redox reactions that inevitably occur alongside electron production and consumption, which interfere with the release and capture of electrons in bioelectrochemical systems; these events can ultimately lead to metabolic flux and energy production imbalance [51]. In electrochemical systems, the anode and cathode undergo oxidation and reduction reactions, respectively. In MECs, the cathode electrode acting as the sole electron donor, Geobacter, can catalyse the fumarate reduction process. This reduction process is catalysed by fumarate dehydrogenase (NADH-dependent). In electrochemically inactive bacteria, past experiments have shown that electrons can be transferred to ferric iron from the cathode via NADH production, utilising ferric citrate as the electron acceptor. E. coli carries out this process, which can be achieved without the use of membrane electron transport carriers [52]. An analysis of a theory revealed that intracellular reducing power (NADH) can be enhanced through biological electrolytic synthesis, and an elevated concentration of NADH could impact intracellular reducing and energy reactions [53].

5 Microbial Fuel Cells: Roles of Microbial Communities

Microbial fuel cells are a new type of bioreactor that could potentially yield tremendous savings in the industry [54]. However, controversial issues like the efficiency and commercialisation of MFC hinder their future applications. The most advantageous point of an MFC is the reduction of energy costs in wastewater treatment. Logan assumed that a population of 100,000 people would generate 16.4×10^6 (L/year) of wastewater, which can run a 2.3 MW power plant producing power from wastewater treatment [55].

A microbial fuel cell typically consists of several components, divided into two chambers: one containing a cathode and the other containing an anode. These chambers are isolated by a proton exchange membrane (PEM). The anodic chamber contains microbes with a suitable substrate, which is then anaerobically broken down, releasing electrons that are transferred to the cathode through an external circuit from the anode. Protons are specifically passed through the exchange membrane. Products created from microbial action in the anodic compartment then travel to the cathode and interact with oxygen to form water [56].

6 Factors Affecting Microbial Performance in Bioelectrochemical Systems

The microorganisms functioning in BES are descendants of indigenous microbes, and they are significantly impacted by external environmental factors, including pH, temperature, oxygen level, moisture content, conductivity, nutrient ions, growth factors, contaminant types, and concentrations [57]. On this note, the system is subjected to bioelectrochemical treatment and various enhancement measures, including adjustments to pH, temperature, or moisture. Magnetic field acclimation of anodic microorganisms and substrate used. Changes in these conditions could significantly affect the efficiency of contaminant degradation, electricity generation and waste removal from wastewater in BES (Fig. 3).

6.1 Effect of pH on Microbial Activity in BES

Microbial actions are reduced when the pH is outside the desirable range, as pH has a significant effect on macromolecules such as proteins and nucleic acids. This change in pH alters biological activity and cell membrane charge, greatly disrupting nutrient absorption. Microorganisms in BES adapt to distinct pH conditions, making it crucial to identify favourable pH values for both electrogenic and electrogenic. The anolyte pH directly controls electron and proton generation, influencing substrate-specific microbial metabolism. High pH (>9) reduces protons, hindering

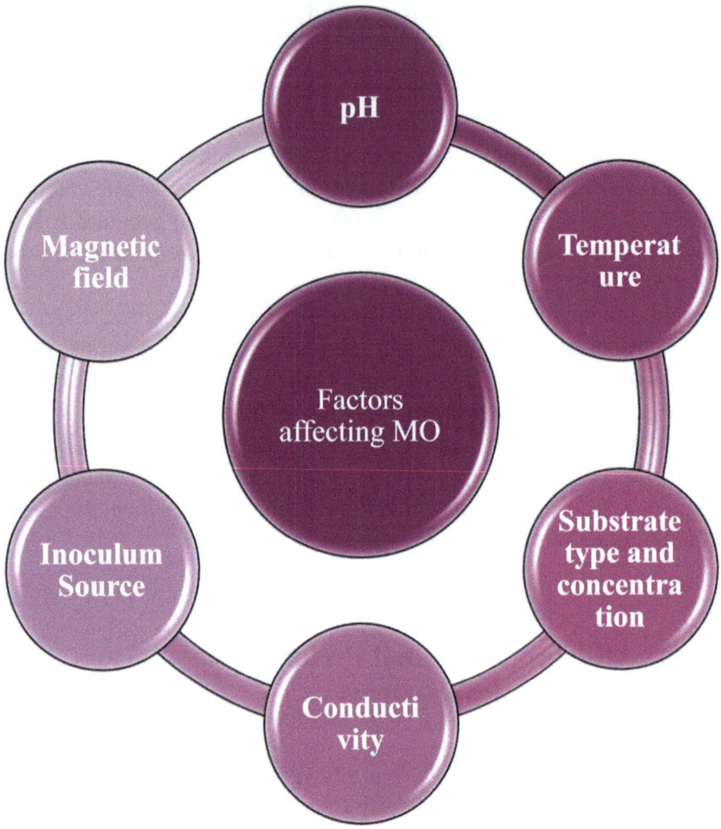

Fig. 3 Factors affecting microbes in BES

electron transfer and favours methanogen growth. Meanwhile, a slightly lower pH (6.2–7.5) inhibits methane yield, but a pH of 5.5 disrupts hydrolysis and carboxylation, thereby affecting the degradation balance. pH also significantly alters the internal resistance of BES by modifying the forms of ions, such as heavy metals. Acidification of the anode enhances heavy metal desorption, whereas cathode alkalization leads to cation precipitation, which reduces conductivity and increases resistance [58]. Furthermore, a higher pH difference between the cathode and anode increases proton flux through the PEM, reducing system resistance.

6.2 *Effect of Temperature on Microbial Performance*

One crucial parameter for microbial growth is temperature, affecting the growth rate, activity, and distribution in BESs. Many microorganisms thrive at temperatures between 35–40 °C, optimising enzyme activity, biofilm formation, substrate disintegration, and electricity production. Studies have shown that raising the temperature enhances BES power output, following the Arrhenius law, which states that it is optimal between 30 °C and 45 °C. However, higher temperature impairs microbial enzyme activity and BES performance. In a study conducted for MFC, the optimal temperature for anaerobic degradation of petroleum hydrocarbons was found to be 40 °C, which doubles the degradation rate and power density compared to 30 °C. However, at 50 °C, the rate dropped to a quarter of its value. In Geobacter, temperature has little to no effect due to its broad adaptation [8]. The starting temperature had a significant impact on BES's long-term performance. Systems initiated at higher temperatures exhibit greater electrical stability, but systems started at lower temperatures may hinder start-up.

6.3 Effect of Conductivity on Microbial Activity

In BES, the anode acts as a constant source of electron acceptors, enhancing the metabolic rate of anaerobic microorganisms that transfer more electrons by using organic contaminants. Yet, BESs often experience hindrance from high internal resistance, particularly in soil- and sediment-based systems, which limits electrical conductivity, suppresses microbial metabolism, and decreases coulombic efficiency. Much-needed metal ions (Na^+, K^+, Cu^{2+}, Zn^{2+}) support microbial growth by maintaining a dynamic balance through physiological and metabolic processes [59]. However, excessive concentration becomes toxic, potentially rendering bacteria nonviable.

6.4 Effect of Substrate Type and Concentration

Experiments have shown that operational microbes in BESs are directly linked to the substrates, with the types and concentrations of the administered substrate significantly affecting the microbial community structure. BESs that adapt to various carbon sources exhibit different responses to changes in substrate. The microbes enriched with a glucose-rich medium performed better than those enriched with acetate or butyrate; meanwhile, acetate-enriched BESs cannot utilise glucose after a sudden substrate shift [60]. In addition, system performance is inversely related to substrate complexity but directly proportional to substrate bioavailability.

7 Challenges and Future Directions

Even though there have been significant advancements in BES performance over the past two decades, several challenges stand in the way of commercialising BES. Those challenges include reduced power output, the high cost of the electrodes used in the systems, the cost of materials and separators, and the difficulties encountered during the scaling-up process. MFCs function as a voltage provider with elevated resistance, but their power output is ultimately lower than that of chemical fuel cells. Current flowing through an external circuit dissipates energy through multiple losses, reducing the actual cell potential below its expected value. These losses are categorised as activation, concentration, ohmic, and coulombic losses [61]. Activation occurs due to the energy required to activate electrode reactions; concentration loss occurs due to the depletion of reactants near the electrodes. Ohmic losses are caused by the resistance of the electrodes, external circuits, electrolytes, and contacts. Coulombic loss occurs due to reactions that do not contribute to the current production and energy losses from the substrate. One of the significant challenges is the scaling up of BES. Most studies are conducted on a lab scale with chamber volumes ranging from millilitres to litres. Researchers documented the maximum power density of 2.15 kW/m^3 from a 0.335 mL chamber and 1.55 kW/m^3 from a 2.5 mL chamber.

Organelle-based biofuel cells offer advantages over organisms, such as bacteria, including easier electron transfer, improved substrate transportation, and greater stability compared to isolated metabolic enzymes. These features make them highly promising despite their low power outputs. Meanwhile, mitochondria-based biofuel cells have enabled self-powered biosensors with ultra-low detection limits, which are helpful for in situ sensing of explosives and toxins. One additional main application is cost-effective drug screening, in which bioelectrocatalysis alters the mechanisms involved in inhibition by harnessing some light. Enhancing organelle-based biofuel cells is crucial for generating micropower, improving biosensor sensitivity, and expanding their use in quantitative analysis.

8 Conclusion

Bioelectrochemical systems (BES) utilise microbial activity to drive electron transfer, providing definite applications in energy generation and environmental sustainability. This review has focused on the fundamentals of BESs, including key microbial contributors, their metabolic pathways, and mechanisms that facilitate electricity production and electron transfer. It also aims to provide an understanding of the factors affecting BES performance, the challenges it faces, and potential research directions. Although BES technology has advanced in many aspects, such as the ability to transfer electrons, the stability of microbes and the system's scalability still require further improvement. Future research should focus heavily on optimising microbial communities, genetic manipulation to enhance microbial electroactivity, and the development of novel materials to improve BES efficiency in electricity generation. Improvements in this field could significantly impact sustainable energy production and waste management, which drives the upcoming applications of bioelectrochemical systems.

References

1. Ortiz-Martínez VM, Salar-García MJ, de los Ríos AP, Hernandez-Fernandez FJ, Egea JA, Lozano LJ (2015) Developments in microbial fuel cell modeling. Chem Eng J 271:50–60
2. Rizzi F, van Eck NJ, Frey M (2014) The production of scientific knowledge on renewable energies: worldwide trends, dynamics can challenges and implications for management. Renew Energy 62:657–671
3. Kadier A, Kalil MS, Abdeshahian P, Chandrasekhar K, Mohamed A, Azman NF et al (2016) Recent advances and emerging challenges in microbial electrolysis cells (MECs) formicrobial production of hydrogen and value-added chemicals. Renew Sust Energ Rev 61:501–525. https://doi.org/10.1016/j.rser.2016.04.017
4. Kumar P, Chandrasekhar K, Kumari A, Sathiyamoorthi E, Kim B (2018) Electro-fermentation in aid of bioenergy and biopolymers. Energies 11(2):343–343. https://doi.org/10.3390/en11020343
5. Harnisch F, Schroder U (2010) From MFC to MXC: chemical and biological cathodes and their potential for microbial bioelectrochemical systems. Chem Soc Rev 39(11):4433–4448. https://doi.org/10.1039/c003068f
6. Qazi A, Hussain F, Abd Rahim N, Hardaker G, Alghazzawi D, Shaban K, Haruna K (2019) Towards sustainable energy: a systematic review of renewable energy sources technologies, and public opinions. IEEE Access 7:63837–63851
7. Potter MC (1915) Electrical effects accompanying the decomposition of organic compounds. II. Ionisa Ghent tion of the gases produced during fermentation. Proc R Soc Lond A 91:465–480
8. Logan BE, Hamelers B, Rozendal R, Schröder U, Keller J, Freguia S et al (2006) Microbial fuel cells: methodology and technology. Environ Sci Technol 40(17):5181–5192
9. Logan BE, Rossi R, Ragab A, Saikaly PE (2019) Electroactive microorganisms in bioelectrochemical systems. Nat Rev Microbiol 17(5):307–319. https://doi.org/10.1038/s41579-019-0173-x
10. Nevin KP, Woodard TL, Franks AE, Summers ZM, Lovley DR (2010) Microbial electrosynthesis: feeding microbes electricity to convert carbon dioxide and water to multicarbon extracellular organic compounds. MBio 1(2):e00103
11. Kumar A, Hsu LH-H, Kavanagh P, Barrière F, Lens PNL, Lapinsonnière L et al (2017) The ins and outs of microorganism–electrode electron transfer reactions. Nat Rev Chem 1(3). https://doi.org/10.1038/s41570-017-0024
12. Logan BE (2009) Exoelectrogenic bacteria that power microbial fuel cells. Nat Rev Microbiol 7(5):375–381. https://doi.org/10.1038/nrmicro2113
13. Chang IS et al (2007) Electrochemically active bacteria (EAB) and mediator-less microbial fuel cells. J Microbiol Biotechnol 16:163–177
14. Rittmann BE, Krajmalnik-Brown R, Halden RU (2008) Pre-genomic, genomic and postgenomic study of microbial communities involved in bioenergy. Nat Rev Microbiol 6:604–612
15. Lovley DR (2006) Bug juice: harvesting electricity with microorganisms. Nat Rev Microbiol 4:497–508
16. Zhu G et al (2016) Enhanced simultaneous nitrification/denitrification in microbial fuel cells. Process Biochem 51(1):80–88
17. Gorby YA et al (2006) Electrically conductive bacterial nanowires produced by Shewanella oneidensis strain MR-1 and other microorganisms. Proc Natl Acad Sci USA 103:11358–11363
18. Clauwaert P et al (2007) Open air biocathode enables effective electricity generation with microbial fuel cells. Environ Sci Technol 41:7564–7569
19. Rozendal RA, Jeremiasse AW, Hamelers HVM, Buisman CJN (2008) Hydrogen production with a microbial biocathode. Environ Sci Technol 42:629–634
20. Rabaey K, Boon N, Hofte M, Verstraete W (2005) Microbial phenazine production enhances electron transfer in biofuel cells. Environ Sci Technol 39:3401–3408
21. El-Naggar MY, Gorby YA, Xia W, Nealson KH (2008) The molecular density of states in bacterial nanowires. Biophys J 95:L10–L12

22. Brasseur G, Levican G, Bonnefoy V, Holmes D, Jedlicki E, Lemesle-Meunier D (2004) Apparent redundancy of electron transfer pathways via bc(1) complex and terminal oxidases in the extremophilic chemolithoautotrophic Acidithiobacillus ferrooxidans. Biochim Biophys Acta 1656:114–126
23. Schuchmann K, Müller V (2016) Energetics and application of heterotrophy in acetogenic bacteria. Appl Environ Microbiol 82(14):4056–4069
24. Kumar P, Chandrasekhar K, Kumari A, Sathiyamoorti E, Kim BS (2018) Electro-fermentation in aid of bioenergy and biopolymers. Energies 11(343):1–20
25. Tanushree A et al (2024) Whole-cell biocatalysis. In: Next-generation technology for green synthesis of pharmaceutical, chemicals, and biofuels whole-cell catalysts: sustainable green-chemical producing entities. CRC Press, Boca Raton
26. Zhao Z, Zhang Y, Wang L, Quan X (2015) Potential for direct interspecies electron transfer in an electric-anaerobic system to increase methane production from sludge digestion. Sci Rep 5:1–12
27. Bosire E, Blank LM, Rosenbaum M (2016) Strain- and substrate-dependent redox mediator and electricity production by Pseudomonas aeruginosa. Appl Environ Microbiol 82(16):5026–5038
28. Miran W, Jang J, Nawaz M, Shahzad A, Lee DS (2018) Sulfate-reducing mixed communities with the ability to generate bioelectricity and degrade textile diazo dye in microbial fuel cells. J Hazard Mater 352:70–79
29. Agostino V, Rosenbaum MA (2018) Sulfate-reducing ElectroAutotrophs and their applications in bioelectrochemical systems. Front Energy Res 6:55. https://doi.org/10.3389/fenrg.2018.00055
30. Ma J et al (2017) Photosynthetic microbial fuel cell for chlorella biomass production. Chemosphere 177:35–43
31. Saba B et al (2017a) Review on microbial fuel cells for energy generation and wastewater treatment. Environ Sci Technol 51(3):456–472
32. Lee MY et al (2015) Hydrogen production and internal resistance in forward osmosis membrane microbial electrolysis cells. Bioresour Technol 187:106–112
33. Kim HJ, Park HS, Hyun MS, Chang IS, Kim M, Kim BH (2002) A mediator-less microbial fuel cell using a metal reducing bacterium, Shewanella putrefaciens. Enzym Microb Technol 30(2):145–152. https://doi.org/10.1016/S0141-0229(01)00478-1
34. Reguera G, McCarthy KD, Mehta T, Nicoll JS, Tuominen MT, Lovley DR (2005) Extracellular electron transfer via microbial nanowires. Nature 435(7045):1098–1101. https://doi.org/10.1038/nature03661
35. Pham TH, Boon N, Aelterman P, Clauwaert P, De Schamphelaire L, Vanhaecke L et al (2008) Metabolites produced by pseudomonas sp enable a gram-positive bacterium to achieveextracellular electron transfer. Appl Microbiol Biotechnol 77(5):1119–1129. https://doi.org/10.1007/s00253-007-1248-6
36. Gregory KB, Lovley DR (2005) Remediation and recovery of uranium from contaminated subsurface environments with electrodes. Environ Sci Technol 39:8943–8947
37. Rabaey K, Rozendal RA (2010) Microbial electrosynthesis—revisiting the electrical route for microbial production. Nat Rev Microbiol 8:706–716
38. Bond DR, Lovley DR (2003) Electricity production by Geobacter sulfurreducens attached to electrodes. Appl Environ Microbiol 69:1548–1555
39. El-Naggar MY, Wanger G, Leung KM, Yuzvinsky TD, Southam G, Yang J, Lau WM, Nealson KH, Gorby YA (2010) Electrical transport along bacterial nanowires from Shewanella oneidensis MR-1. Proc Natl Acad Sci USA 107:18127–18131
40. Raghavulu SV, Goud RK, Sarma PN, Mohan SV (2011) Saccharomyces cerevisiae as anodic biocatalyst for power generation in biofuel cell: influence of redox condition and substrate load. Bioresour Technol 102:2751–2757
41. Schröder U, Nießen J, Scholz F (2003) A generation of microbial fuel cells with current outputs boosted by more than one order of magnitude. Angew Chem Int Ed 42:2880–2883
42. Castelle C, Guiral M, Malarte G, Ledgham F, Leroy G, Brugna M, Giudici-Orticoni M-T (2008) A new iron-oxidizing/O_2-reducing supercomplex spanning both inner and outer mem-

branes, isolated from the extreme acidophile Acidithiobacillus ferrooxidans. J Biol Chem 283:25803–25811
43. Thrash JC, Coates JD (2008) Review: direct and indirect electrical stimulation of microbial metabolism. Environ Sci Technol 42:3921–3931
44. Park HI, Dk K, Choi Y-J, Pak D (2005) Nitrate reduction using an electrode as direct electron donor in a biofilm-electrode reactor. Process Biochem 40:3383–3388
45. Aulenta F, Catervi A, Majone M, Panero S, Reale P, Rossetti S (2007) Electron transfer from a solid-state electrode assisted by methyl viologen sustains efficient microbial reductive dechlorination of TCE. Environ Sci Technol 41:2554–2559
46. Wang G, Guo Y, Cai J, Wen H, Mao Z, Zhang H et al (2019) Electricity production and the analysis of the anode microbial community in a constructed wetland-microbial fuel cell. RSC Adv 9:21460–21472
47. Rose ND, Regan JM (2015) Changes in phosphorylation of adenosine phosphate and redox state of nicotinamide-adenine dinucleotide (phosphate) in Geobacter sulfurreducens in response to electron acceptor and anode potential variation. Bioelectrochemistry 106:213–220. https://doi.org/10.1016/j.bioelechem.2015.03.003
48. Unden G, Bongaerts J (1997) Alternative respiratory pathways of Escherichia coli: energetics and transcriptional regulation in response to electron acceptors. Biochim Biophys Acta Bioenerg 1320:217–234. https://doi.org/10.1016/s0005-2728(97)00034-0
49. Liu F, Rotaru AE, Shrestha PM, Malvankar NS, Lovley DR (2012) Promoting direct interspecies electron transfer with activated carbon. Energy Environ Sci 5:8982–8989. https://doi.org/10.1016/j.biortech.2018.03.050
50. Smith JA, Nevin KP, Lovley DR (2015) Syntrophic growth via quinonemediated interspecies electron transfer. Front Microbiol 6:121. https://doi.org/10.3389/fmicb.2015.00121
51. Chen F, Feng XH, Liang JF, Xu H (2012) An oxidoreduction potential shift control strategy for high purity propionic acid production by Propionibacterium freudenreichii CCTCC M207015 with glycerol as sole carbon source. Bioprocess Biosyst Eng 36:1165–1176. https://doi.org/10.1007/s00449-012-0843-9
52. Emde R, Swain A, Schink B (1989) Anaerobic oxidation of glycerol by Escherichia coli in an amperometric poised-potential culture system. Appl Microbiol Biotechnol 32:170–175. https://doi.org/10.1007/bf00165883
53. Pandit AV, Mahadevan R (2011) In silico characterization of microbial electrosynthesis for metabolic engineering of biochemicals. Microb Cell Factories 10:1–14. https://doi.org/10.1186/1475-2859-10-76
54. Pham TH, Rabaey K, Aelterman P et al (2006) Microbial fuel cells in relation to conventional anaerobic digestion technology. Eng Life Sci 6:285–292
55. Logan BE (2005) Simultaneous wastewater treatment and biological electricity generation. Water Sci Technol 52:31–37
56. Sharma Y, Li B (2010) The variation of power generation with organic substrates in single-chamber microbial fuel cells (SCMFCs). Bioresour Technol 101:1844–1850
57. Lu L, Huggins T, Jin S, Zuo Y, Ren ZJ (2014) Environ Sci Technol 48:4021–4029
58. Watson VJ, Logan BE (2010) Power production in MFCs inoculated with *Shewanella oneidensis* MR-1 or mixed cultures. Biotechnol Bioeng 105(3):489–498
59. Ravi SN, Sankaranarayanan M (2023) Redox balanced co-production of propanediol and 3-hydroxypropionic acid from glycerol using novel recombinant Klebsiella quasipneumonia MSN12. J Inorg Organomet Polym Mater 33:5
60. Ravi SN, Sankaranarayanan M (2023) Enhanced synthesis of 3-hydroxypropionic acid by eliminating by-products using recombinant Escherichia coli as a whole cell biocatalyst. Top Catal 67:169–180
61. Rabaey K, Verstraete W (2005) Microbial fuel cells: novel biotechnology for energy generation. Trends Biotechnol 23(6):291–298

Electron Transfer Mechanisms in Bio-Electrochemical Systems

Bharat Kumar Harani, Khadija Quershi, Muhammad Suhaib Shaikh, Abdul Sattar Jatoi, and Nabisab Mujawar Mubarak

Abstract Electrochemical reactions facilitated by microorganisms enable bio-electrochemical systems (BESs) to produce biofuel, clean wastewater, and generate bioelectricity. The performance level of these systems heavily depends on the electron transport processes between microorganisms and electrodes. Three electron transfer mechanisms operate simultaneously in BESs: direct electron transfer (DET), mediated electron transfer (MET), and nanowire-based conduction. The transfer of electrons from Shewanella and *Geobacter* to electrodes occurs through direct electron transfer utilising conductive pili and cytochromes that reside in their outer membrane layers. Better power generation within MFCs heavily depends on this technique. The electron transfer method of MET functions through soluble redox mediators, such as flavins and quinones, which enable electron movement between microbial cells and electrode surfaces. Several bacteria transfer electrons through nanowires that conduct electrons from microbial networks to the electrode surface.

The advancement of BES technology requires an optimised understanding of electronic channels for higher efficiency levels in commercial applications.

B. K. Harani (✉)
Department of Petroleum and Gas Engineering, Dawood University of Engineering and Technology, Karachi, Pakistan
e-mail: bharat.kumar@duet.edu.pk

K. Quershi · M. S. Shaikh
Department of Chemical Engineering, Mehran University of Engineering and Technology, Jamshoro, Pakistan

A. S. Jatoi
Department of Chemical Engineering, Dawood University of Engineering and Technology, Karachi, Pakistan

N. M. Mubarak
Department of Chemical Engineering, College of Engineering, Imam Mohammad Ibn Saud Islamic University (IMSIU), Riyadh, Saudi Arabia

Department of Biosciences, Saveetha School of Engineering, Saveetha Institute of Medical and Technical Sciences, Chennai, India

A. Sattar Jatoi, N. M. Mubarak (eds.), *Bio-Electrochemical Conversion of Waste to Fuel*, Green Energy and Technology,
https://doi.org/10.1007/978-981-95-4310-6_7

Researchers will advance future work by creating new electrode materials in combination with enhanced extracellular electron transfer (EET) microbial strains to achieve improved conductivity and biocompatibility. These developments will yield innovations in renewable energy systems, environmental remediation procedures, and sustainable bioproduction methods.

Keywords Electron transfer · Mechanism · Bio-electrochemical system

1 Introduction

The progress in renewable energy, combined with environmental remediation and sustainable bioproduction methods, will stem from these scientific discoveries [1, 2]. The performance of these systems mostly depends on mechanisms that enable microorganisms to work with solid-state electrodes. Knowledge of these factors becomes essential to achieve maximum BES performance while enhancing practical applications [3].

Conditional system effectiveness arises from electron transfer techniques, which enable microorganisms to interact with solid-state electrodes [4]. Analysis of these factors serves the dual purpose of enhancing BES performance while facilitating more effective practical applications. Electroactive bacteria, including *Shewanella* and *Geobacter* transmit electrons to electrode surfaces through conductive pili and outer membrane cytochromes in DET [5]. The electron conductance in BESs occurs through three principal methods: direct electron transfer (DET), mediated electron transfer (MET), and nanowire-based transmission [6]. Shewanella and Geobacter's electroactive bacteria transfer electrons directly to electrode surfaces using conductive pili and outer membrane cytochromes[7]. MET enables electron transfer through redox mediators between cells and electrodes, but bacteria can produce these mediators naturally, or systems can accept external mediators[8]. Biosystems composed of bacteria also utilise conductive nanowires, allowing for extended electron transmission within biofilm structures [9].

The effectiveness and operational scale of BESs mainly depend on electron transfer channels. Scientists are currently researching the development of system designs and identifying optimal electrode materials while working to enhance microbial electron transfer genes [10]. Understanding BES systems will enable researchers to boost BES applications for bioprocessing, environmental remediation, and sustainable energy production.

2 Fundamentals of Electron Transfer in BES

The vital process of electron transfer enables microorganisms to conduct exchanges with electrodes within bio-electrochemical systems (BES) to trigger multiple bio-electrochemical reactions [11]. The basic electron transfer process enables

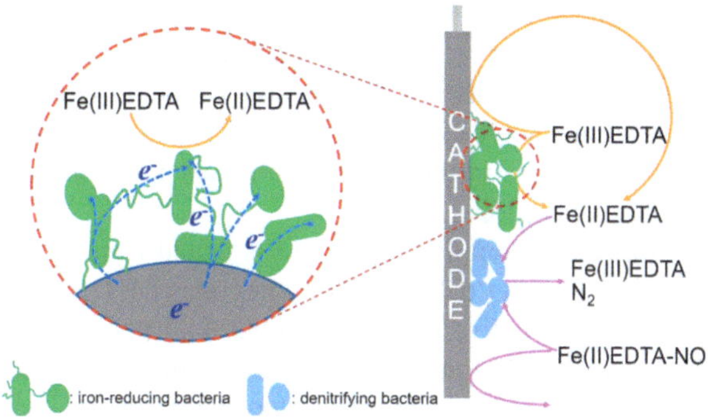

Fig. 1 Mechanism of electron transfer [13]

microorganisms to conduct electron exchange operations with electrodes in bio-electrochemical systems (BES). Cellular microorganisms utilise electron transfer reactions to exchange electrons with electrodes in bio-electrochemical systems (BES) for various bio-electrochemical operations [12]. The basic electron transfer process supports numerous bio-electrochemical reactions that microorganisms use to exchange electrons with biological electrodes in bio-electrochemical systems (BES). The direction of electron flow, in addition to the efficiency of transport, depends heavily on the redox potential differences between the donor (organic matter) and the acceptor (electrode). The redox potential between the donor organic matter and the acceptor electrode determines the flow of electrons along with their energy efficiency during transport processes. Maximum BES functionality along with optimised energy recovery and bioenergy and environmental remediation development rely on fundamental knowledge of these principles (Fig. 1).

3 Types of Electron Transfer Mechanisms in Bio-Electrochemical Systems

These three electron transfer mechanisms are direct electron transfer (DET), mediated electron transfer (MET), and conductive [14] biofilms/extracellular electron transfer (EET) in the bio-electrochemical (BES) [15]. In such systems, including bio-electrochemical systems (BES), electron transfer is mediated by three primary means: Direct Electron Transfer (DET), Mediated Electron Transfer (MET), and Conductive Biofilms/Extracellular Electron Transfer (EET). Electron transfer across the cell membrane is assisted by cytochromes, some outer membrane proteins, and conductive pili (nanowires). The efficiency of DET is dependent on pili conduction, biofilm structure, and the distance between the biofilm and the electrode surface [16]. The drawback of DET is that it requires bacteria and electrodes to be in direct contact with each other; however, it works very well for certain bacteria [17].

Oppositely, mediated electron transfer (MET) transfers electrons from microbes to electrodes or vice versa, using soluble redox mediators that may be generated by the microbes, such as flavins and quinones, or externally, such as ferricyanide or methylene blue [18]. The MET aids in electron transport efficiency for microorganisms that are not close to the electrode. The mediators must be stable, in a redox state, and at a concentration that allows them to diffuse through cells and the electrode for the MET to be effective. MET enables an increased range of electroactive microorganisms in the BES, which may produce toxicity, degrade mediators, and increase operational costs [19].

This more complex procedure is called Conductive Biofilms and Extracellular Electron Transfer (EET), which integrates DET and MET components. During this process, conductive biofilms of microbial populations form on the electrode surface, facilitating long-range electron transport opportunities. This is particularly important in systems utilising Geobacter species, which form highly conductive extracellular networks using conductive filaments and cytochromes. One key difference between EET and DET is that it allows deeper biofilms to participate in electron transfer, thereby enhancing current generation and system stability [20]. Some variables that determine EET efficacy include extracellular polymeric substances (EPS), which control the conductivity of electrons, biofilm thickness, and the composition of microbes. Understanding these electron transport channels is crucial to optimise BES performance and enhance its application in bioremediation, wastewater treatment, and bio-electrochemical energy generation [21].

3.1 Direct Electron Transfer (DET)

Some of the factors [presence of extracellular polymeric substances (EPS) controlling electron conductivity, presence of extracellular polymeric substances (EPS) controlling biofilm thickness, and microbial composition] that affect the effectiveness of EET are some [22]. We need to understand these electron transport channels to maximise BES performance and expand their use in bioremediation, wastewater treatment, and energy generation. The extracellular polymeric substances (EPS) that play a role in electron conductivity, biofilm thickness, and microbial composition are some factors responsible for the effectiveness of EET [23]. These electron transport channels maximise BES performance and extend the application into bioremediation, wastewater treatment, and energy generation. EET effectiveness is influenced by extracellular polymeric substances (EPS) that affect electron conductivity, biofilm thickness, and microbial composition. Consequently, understanding these electron transport channels is essential to maximise BES performance and further apply them in bioremediation, wastewater treatment, and energy production [24]. Factors such as electron conductivity, biofilm thickness, and microbial composition are all influenced by the presence of extracellular polymeric substances (EPS), which affect EET effectiveness [25]. They need knowledge of these electron transport channels to maximise BES performance and broaden their scope for bioremediation, wastewater treatment, and energy generation [26].

Among the factors affecting EET is the presence of EPS, which influences electron conductivity, biofilm thickness, and microbial composition. Since these electron transport channels are crucial for facilitating the highest BES performance and extending their application in bioremediation, wastewater treatment, and energy generation, knowledge of them is necessary. Several factors that influence EET effectiveness include extracellular polymeric substances (EPS), which impact electron conductivity and microbial composition, as well as EET thickness [27]. Knowledge of these electron transport channels is necessary to achieve maximum BES performance and expand its use as a bioremediation, wastewater treatment, and energy generation tool. Several factors influencing EET effectiveness include extracellular polymeric substances (EPS), biofilm thickness, changes in microbial composition, and biofilm electron conductivity. Knowledge of these electron transport channels is crucial for achieving maximum BES performance, and their application is beneficial in bioremediation, wastewater treatment, and energy generation [28].

Some factors influencing EET effectiveness include extracellular polymeric substances (EPS), which control electron conductivity, biofilm thickness, and microbial composition [29]. BES performance can be maximised to capitalise on these electron transport channels while applied in bioremediation, wastewater treatment, and energy generation [30]. Research has focused on enhancing biofilm adherence, genetic modification of bacteria with higher electron transfer properties, and modifying electrodes using nanomaterials to mitigate these constraints. BES can achieve improved performance and energy conversion efficiency when DET tuning is applied in applications such as microbial fuel cells (MFCs), microbial electrolysis cells (MECs), and bioremediation technologies [31].

3.2 *Mediated Electron Transfer (MET)*

The mechanism by which microorganisms can utilise soluble redox mediators to transfer electrons between a cell and an electrode in bio-electrochemical systems (BES) is known as "mediated electron transfer" (MET). MET enables electron transfer over longer distances than direct electron transfer (DET), in which electron transfer can occur only between the bacterium and the electrode in physical contact [32]. MET is advantageous for bio-electrochemical processes in which non-electrogenic or less electroactive bacteria can participate, as MET allows for electron transfer over longer distances than Direct Electron Transfer (DET), which requires physical contact between the bacteria and the electrode [33].

The MET process has three key steps: The mediator takes electrons from internal metabolic pathways (such as the electron transport chain) and a redox-active material created by, released by, or reacts with microbial cells (3). In the solution, the resultant reduced mediator diffuses to the electrode, where it is replenished for subsequent cycles and consists of electrons [34]. In this way, MET makes electron transfer more accessible to a greater diversity of metabolic species, as it does not require cell-to-electrode contact. As a result, MET is particularly useful in biofilm-limited or bacterium-free fluid scenarios[34].

Being an effective MEV Explorer has its advantages, but it also has its disadvantages. This process will be effective if mediators exhibit high stability, a high diffusion rate, and efficient redox cycling. Artificial mediators may degrade over time, reducing the system's function and causing microbial community toxicity [35]. Introducing synthetic mediators from the outside increases operating costs and may complicate system maintenance [21]. To overcome these challenges, researchers are focusing on bio-inspired aspects, such as the use of nanomaterials to enhance the kinetics of electron transfer, modifying bacteria to produce more flavins and quinones, and increasing the production of natural mediators by microorganisms [36]. Bio-electrochemical systems can optimise MET for biosensing, bioremediation, and energy-producing applications.

3.3 Conductive Biofilms and Extracellular Electron Transfer (EET)

Conductive biofilms are essential for facilitating the transport of electrons over long distances within a well-organised biofilm matrix, allowing microbial populations to transport electrons outside of the biofilm (extracellular electron transfer or EET) [37]. The transport of electrons over long distances within the well-organised biofilm matrix requires extracellular electron transfer (EET), and conductive biofilms are, therefore, necessary [38]. Conductive biofilms are needed for microbial populations to transport electrons over long distances within an organism's biofilm matrix via an extracellular electron transfer (EET) process [39]. The ability of biofilms to conduct electricity depends on their internal structure, microbial composition, and the presence of conductive appendages or redox-active compounds that facilitate charge transfer [40].

The biofilm's internal structure determines its capacity to conduct electricity, its composition of microbes, and the availability of conductive appendages or redox-active compounds that facilitate charge conduction [41]. These biofilms utilise multi-stage electron transfer processes, including outer membrane cytochromes, redox compounds secreted from the cell, and pili (nanowires). In *Geobacter* species, for example, pili exhibit a metallic-like conductivity that supports long-distance electron transport within biofilms [42]. Similarly, electrons are extracellularly transported by the network of cytochromes MtrC and OmcA, which are associated with flavins, within the *Shewanella* species. Like *Shewanella* species, the cytochromes that facilitate extracellular electron transport in Geobacter are coupled to a network of cytochromes, such as MtrC and OmcA, in combination with flavins. Two advanced techniques, namely electrochemical impedance spectroscopy (EIS) and conductive atomic force microscopy (c-AFM), have been applied to the study and optimisation of these conductive biofilms [28].

Thus, a critical area of research in increasing BES performance is to increase biofilm conductivity and stability. Some of the methods used in research include the use of artificial biofilm scaffolds to increase electron mobility, genetically modifying bacteria to overexpress conductive proteins, and modifying electrode surfaces with nanomaterials to enhance biofilm attachment [43]. Additionally, hybrid biofilms comprising graphene, carbon nanotubes, or metal nanoparticles have also demonstrated promising results in terms of conductivity improvement. Improving EET mechanisms will enhance power densities, improve electron recovery efficiency, and be beneficial in wastewater treatment, bioremediation, and bio-electrochemical energy generation in broader applications [44]. Additionally, evidence of improved conductivity in hybrid biofilms comprising nanomaterials, such as graphene, carbon nanotubes, and metal nanoparticles, is encouraging. Such improvements in EET mechanisms would enable bio-electrochemical systems to achieve higher power densities, enhance electron recovery efficiency, and extend their utilisation in wastewater treatment, bioremediation, and sustainable energy [45].

4 Mechanism of Liberating Electrons from Organic Matter by Microbes

Microorganisms perform metabolic oxidation processes in bio-electrochemical systems (BESs) that liberate electrons from organic substances. *Shewanella* and *Geobacter* are electroactive bacteria that utilise glycolysis and the tricarboxylic acid cycle (TCA cycle) to break down organic substrates, such as glucose, acetate, and lactic acid. Many metabolic activities involve the release or absorption of electrons by intracellular carriers, such as NADH and $FADH_2$ [45]. Ultimately, these high-energy electrons enter the microbial electron transport chain (ETC) and undergo a series of redox reactions that provide energy for microbial survival [46].

To perform extracellular electron transfer (EET), electrons must be extracted from organic materials and transported to the cell's outer membrane. There are three major methods for electron (E) transfer: direct electron (DET), mediated interfaces (MET), and conducting with the use of nanowires [47]. In DET, conductive pili can directly wire bacteria to electrodes or solid electron acceptors, including nanowires (bacterial pili) and outer membrane cytochromes. Unlike MET, however, soluble redox mediators such as flavins and quinones transfer electrons across the electrode and the microbial cell. In addition, some bacteria transfer electrons through conductive nanowires between populations of microbes to increase overall efficiency [48].

Microbes should be able to release and transport electrons for use in microbial fuel cells (MFCs), microbial electrolysis cells (MECs), and any other bioremediation systems. Using these techniques facilitates the breakdown of pollutants, allowing wastewater to be cleaned, power to be generated, and hydrogen to be produced. Genetic engineering, electrode surface enhancements, and improved system design have been the primary focus of BES technology development in enhancing electron transfer efficiency. These methods can break down contaminants into hydrogen, generate power, and produce clean wastewater. One of the primary goals in BES

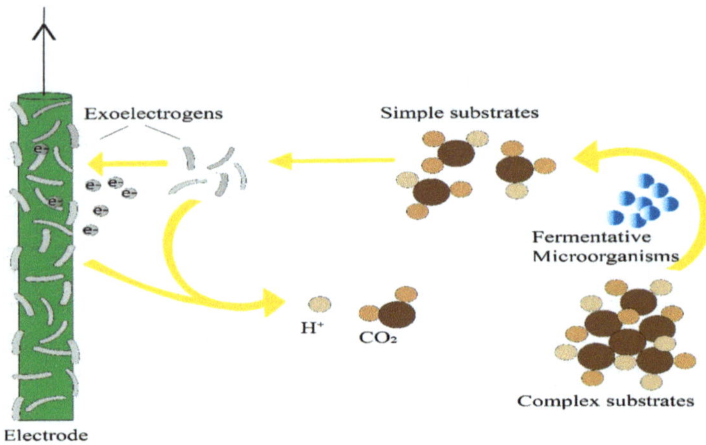

Fig. 2 Alternative electron transfer mechanisms [49]

technology development is to enhance electron transfer efficiency through improved system design, enhanced electrode surfaces (via surface modification and genetic engineering), and other advancements (Fig. 2).

5 Indirect Electrons Transported by Reduced Products

Mediated electron transfer (MET) is a process in which microorganisms in bioelectrochemical systems (BESs) transfer electrons indirectly through reduced metabolic products. Some bacteria produce redox-active compounds that act as electron shuttles, thereby bypassing the need for direct electron delivery to an electrode. The increased system efficiency of reduced products such as hydrogen (H_2), organic acids, quinones, and flavins enables electron transmission over longer distances.

Hydrogen is also a common electron carrier via indirect electron transport processes. Some microorganisms, such as fermentative bacteria, can break down organic molecules and produce hydrogen gas, which electroactive or hydrogenotrophic bacteria near the electrode surface can exploit. Hydrogen is a common intermediary electron carrier for indirect electron transport processes. However, some microorganisms, such as fermentative bacteria, can break down organic molecules to produce hydrogen gas that electroactive or hydrogenotrophic bacteria can exploit to obtain energy near the electrode surface [47].

Reduced products are particularly useful for applications such as bioremediation, microbial electrosynthesis, and microbial electrolysis cells (MECs) due to their potential for electron transport. Reduced products are advantageous for applications such as bioremediation, microbial electrosynthesis, and microbial electrolysis cells (MECs) because they can transfer electrons [50]. The use of reduced products for electron transport is highly preferred for bioremediation, microbial electrosynthesis, and microbial electrolysis cell (MEC) applications. The reduced products for electron transport are thus highly beneficial for bioremediation, microbial electrosynthesis, and microbial electrochemical cells (MECs).

6 Electrons Transported by Artificial Mediators

The reduced products are highly advantageous for bioremediation, microbial electrosynthesis, and microbial electrolysis cells (MECs). Unlike naturally occurring redox-active chemicals, artificial mediators are primarily used for electron shuttling with reduced DET limitations. In contrast to natural redox-active chemicals, artificial mediators are external electron shuttles that are more efficient than direct electron transfer (DET) [51].

Compared to naturally occurring redox-active chemicals, electron shuttle mediators are provided externally to facilitate electron transfer more effectively than direct electron transfer (DET). After being reduced by the mediator, the mediator diffuses across the extracellular medium, gives electrons to the electrode, and reoxidises. This cycle is so continuous that one can still transmit long-range electrons through electron transmission, even in microbial species lacking direct electron transmission channels. Artificial mediators decrease electron transfer resistance and enhance charge flow, significantly improving BES performance in microbial fuel cells (MFCs) and electrolysis cells (MECs).

Although they have advantages, artificial mediators also have disadvantages, including toxicity towards microbial communities, possible degradation over time, and increased operating costs. To address these problems, research is underway to develop more stable, biocompatible, and cheaper larger materials that mediate with high redox efficiency. To address these problems, research is underway to develop more stable, biocompatible, and cost-effective mediators with very high redox cycling efficiency.

7 Electrons Transport Through the Microorganism's Mediator

Many electroactive bacteria facilitate extracellular electron transfer (EET) through the use of naturally produced redox mediators [52]. The more these self-produced mediators increase electron transport efficiency, the better, especially for bacteria that do not embrace direct electron transfer (DET) pathways. Flavones, quinones, and phenazines are common microbial mediators for microbial fuel cells (MFCs) and other BES applications.

The first step in the stimulatory form of the microbial mediators is the intracellular oxidation of organic substrates to high-energy electrons and endogenous redox molecules. Oxidation of organic substrates by intracellular oxidation of its endogenous redox molecules is the initial stage of the electron transport system via microbial mediators. Shewanella species harness flavins to augment electron flow, while Pseudomonas species use phenazines for effective redox cycling.

For example, this includes the use of flavin in Shewanella species to enhance electron flow and the use of phenazine in Pseudomonas species for efficient redox cycling. Yet, one must regulate the supply of substrates, pH, and temperature to achieve maximal mediator production and efficiency. Future studies will continue to optimise BES through genetic engineering and metabolic pathway optimisation, aiming to create higher-yielding microbial mediator synthesis, thereby advancing BES applications in energy generation, wastewater treatment, and bioprocessing.

8 Factors Influencing Electron Transfer Efficiency

Among the main factors influencing the effectiveness of electron transfer in bioelectrochemical systems (BESs) are microbial characteristics, electrode characteristics, redox mediators, and operating conditions [53]. Among the main factors influencing the effectiveness of electron transfer in bio-electrochemical systems (BESs) are microbial characteristics, electrode characteristics, redox mediators, and operating conditions [54]. The formation of a persistent and conductive biofilm on the electrode surface further boosts electron transmission efficiency.

Electron transport performance is largely controlled by the electrode materials and their surface properties. Graphite, graphene, and carbon felt are widely utilised as carbon-based materials due to their high conductivity and biocompatibility [55]. Charge transmission and microbial adherence to the electrode increase by increasing electrode surface area and porosity. Improving electron mobility between microorganisms and electrodes can be achieved through surface modifications using conductive polymers or nanomaterials.

It includes operational parameters that can affect how good electrons travel, such as pH, temperature, and substrate availability. The ideal pH levels (typically 6.5–7.5) and moderate temperatures (25–40 °C) facilitate microbial activity. Microbial metabolism and power output are affected by the presence of suitable electron acceptors and the availability of organic substrates. Microbial metabolism and power output are dependent on the presence of suitable electron acceptors and organic substrates.

Therefore, by optimising these factors, researchers can design BESs to enhance the efficiency and scalability of these systems for use in wastewater treatment, biofuel production, renewable energy, and other applications. Future improvements in system optimisation, electrode design, and microbial engineering are likely to enhance electron transport mechanisms. They will increase the feasibility of using BESs in commercial and industrial contexts.

9 Applications of Electron Transfer in Bio-Electrochemical Systems

Electron transfer occurs in bio-electrochemical systems (BESs) to generate power, biomining, or remediate environmental waste as part of bioprocessing. One of the best-known applications of microbial fuel cells is the use of electroactive microorganisms that oxidise organic molecules from wastewater or other substrates to produce energy. This technique offers an environmentally beneficial approach to managing wastewater and generating bioelectricity, making it a viable and appealing option for local energy solutions. Like MECs, similar electron transport concepts are used in microbial electrolysis cells (MECs) that generate usable biofuel through voltage-powered external voltage, including hydrogen, alcohol (ethanol), and methane.

Electron transfer is also utilised in another significant application of BESs: bioremediation, which involves the use of electroactive microorganisms to break down environmental pollutants. The redox reactions at the electrode interface have removed heavy metals, nitrates, and persistent organic pollutants from soil and wastewater using BESs. At the same time, this technique also facilitates positive, ecologically responsible waste management by accelerating the breakdown of toxic materials, generating electrical energy, or producing useful chemical compounds.

In addition to its energy and environmental applications, bioelectrosynthesis is a process in which microorganisms utilise electrodes as electron donors or acceptors to produce useful biochemicals. Therefore, electron transport is also needed in BESs. This technique is sustainable for synthesising organic acids, alcohols, and biopolymers from carbon dioxide or organic waste streams. Similarly, BESs are being researched for biosensing applications, including the detection of contaminants, diseases, and metabolic changes in soil and water through microbial electron transfer.

Advancements in system optimisation, electrode materials, and microbial engineering have broadened the applications for BES. Scaling up BES technology to broad applications in energy generation, wastewater treatment, and other sustainable biomanufacturing requires improving electron transfer efficiency.

10 Challenges and Future Perspectives of Electron Transfer in Bio-Electrochemical Systems

Despite significant advancements, several challenges still hinder the widespread application of bio-electrochemical systems (BESs). The low electron transfer efficiency is one of the primary challenges posed by biofilm restrictions, the adsorption of foulants on the electrode, and high intrinsic resistance. The most significant advancements have been made in overcoming such challenges; however, there are still limitations on the broad application of bio-electrochemical systems (BESs).

One major barrier is the low electron transfer efficiency from the electrode, which is hindered by electrode fouling, biofilm limitations, and high internal resistance. Economic obstacles to the commercialisation of BES also include the high cost and limited lifespan of electrode materials, particularly those that require modification to enhance conductivity.

Optimisation of operational conditions and system design is another essential job. Many BESs operate at low power densities; therefore, improvements in reactor configuration, microbial consortia engineering, and electrode architecture are needed. Due to competition with non-electroactive bacteria, maintaining stable microbial communities is challenging, resulting in lower efficiency. In addition, scaling up BESs for industrial applications presents engineering challenges, including mass transfer limitations, increased internal resistance, and maintaining system stability under actual operating conditions.

In future research on BESs, efforts to improve electronic transport channels will be made by employing microbial engineering and advanced nanomaterials. The alteration of genetic changes in electroactive bacteria may enhance metabolic efficiency, biofilm development, and EET pathways. New nanostructured, biofunctionalized electrodes have been developed to provide novel enhancements in long-term stability, conductivity, and surface area. BESs can also be integrated with renewable energy sources and hybrid technologies, such as bio-hybrid fuel cells and bioreactors, thereby expanding their potential for generating energy and cleaning the environment.

With further advancements in reactor engineering, materials science, and synthetic biology, BESs have the potential to become a significant technology in sustainable energy and waste management. To move from current limitations and advance towards a cleaner and more sustainable future, BESs will be transitioned from laboratory-scale experiments to commercial-scale applications with interdisciplinary research and innovation.

11 Conclusion

Bio-electrochemical systems (BESs) are promising for wastewater treatment and bio-electrochemical processes (bioremediation, biofuel production, and sustainable energy generation) and gas, cells, or water treatment. Therefore, the electron transfer techniques that mostly affect the efficiency of these devices are direct electron transfer (DET), mediated electron transfer (MET), and conduction based on nanowires. Bio-electrochemical systems (BESs) are promising in Bio-electrochemical, biofuel production, and distributed sustainable energy generation. Electron transfer techniques (DET, mediated electron transfer, or MET) and conduction by nanowires primarily impact the efficiency of these devices.

BESs continue to improve performance through advances in microbial engineering, nanomaterials, and reactor design. Functionalized electrode materials, improved system designs, and genetic changes in the electroactive bacteria can help improve electron transport and overall efficiency. Furthermore, when BESs are coupled to hybrid renewable energy systems, they may be more utility in clean energy and environmental sustainability.

However, the present realities do not prevent the BES technology from becoming the future. As more multidisciplinary research and technological development are developed, BESs will become a vital tool in addressing global energy and environmental issues. BESs may be preferable to waste management and sustainable energy generation by improving scalability and simplifying electron transfer techniques.

References

1. Jatoi AS, Akhter F, Mazari SA, Sabzoi N, Aziz S, Soomro SA, Mubarak NM, Baloch H, Memon AQ, Ahmed S (2021) Advanced microbial fuel cell for waste water treatment—a review. Environ Sci Pollut Res 28:5005–5019
2. Jatoi AS, Unar IN, Aziz S, Soomro SA, Abro M, Khoso JA, Siddique M (2018) Microbial fuel cell for bio-electricity generation via bio-electrochemical conversion of fermentation industry sludge. A J Life Sci 2(1):42–53
3. Gupta S, Patro A, Mittal Y, Dwivedi S, Saket P, Panja R, Saeed T, Martínez F, Yadav AK (2023) The race between classical microbial fuel cells, sediment-microbial fuel cells, plant-microbial fuel cells, and constructed wetlands-microbial fuel cells: applications and technology readiness level. Sci Total Environ 879:162757
4. Boas JV, Oliveira VB, Simões M, Pinto AM (2022) Review on microbial fuel cells applications, developments and costs. J Environ Manag 307:114525
5. Obileke K, Onyeaka H, Meyer EL, Nwokolo N (2021) Microbial fuel cells, a renewable energy technology for bio-electricity generation: a mini-review. Electrochem Commun 125:107003
6. Maddalwar S, Nayak KK, Kumar M, Singh L (2021) Plant microbial fuel cell: opportunities, challenges, and prospects. Bioresour Technol 341:125772
7. Roy H, Rahman TU, Tasnim N, Arju J, Rafid MM, Islam MR, Pervez MN, Cai Y, Naddeo V, Islam MS (2023) Microbial fuel cell construction features and application for sustainable wastewater treatment. Membranes 13(5):490
8. Gul H, Raza W, Lee J, Azam M, Ashraf M, Kim K-H (2021) Progress in microbial fuel cell technology for wastewater treatment and energy harvesting. Chemosphere 281:130828
9. Arun J, SundarRajan P, Pavithra KG, Priyadharsini P, Shyam S, Goutham R, Le QH, Pugazhendhi A (2024) New insights into microbial electrolysis cells (MEC) and microbial fuel cells (MFC) for simultaneous wastewater treatment and green fuel (hydrogen) generation. Fuel 355:129530
10. Shrivastava A, Sharma RK (2022) Lignocellulosic biomass based microbial fuel cells: performance and applications. J Clean Prod 361:132269
11. Ren W, Cheng C, Shao P, Luo X, Zhang H, Wang S, Duan X (2021) Origins of electron-transfer regime in persulfate-based nonradical oxidation processes. Environ Sci Technol 56(1):78–97
12. Baek G, Kim D, Choi Y, Cha J, Lee C (2023) Challenges in engineering direct interspecies electron transfer for enhanced methanogenesis. Renew Sust Energ Rev 183:113503

13. Zhao J, Zhang C, Sun C, Li W, Zhang S, Li S, Zhang D (2018) Electron transfer mechanism of biocathode in a bioelectrochemical system coupled with chemical absorption for no removal. Bioresour Technol 254:16–22
14. Kižys K, Zinovičius A, Jakštys B, Bružaitė I, Balčiūnas E, Petrulevičienė M, Ramanavičius A, Morkvėnaitė-Vilkončienė I (2023) Microbial biofuel cells: fundamental principles, development and recent obstacles. Biosensors 13(2):221
15. Danhassan UA, Lin H, Lawan I, Zhang X, Ali MH, Muhammad AI, Sheng K (2023) Critical insight into sediment microbial fuel cell: fundamentals, challenges, and perspectives as a barrier to black-odor water formation. J Environ Chem Eng 11(1):109098
16. Selvasembian R, Mal J, Rani R, Sinha R, Agrahari R, Joshua I, Santhiagu A, Pradhan N (2022) Recent progress in microbial fuel cells for industrial effluent treatment and energy generation: fundamentals to scale-up application and challenges. Bioresour Technol 346:126462
17. Hamedani EA, Abasalt A, Talebi S (2024) Application of microbial fuel cells in wastewater treatment and green energy production: a comprehensive review of technology fundamentals and challenges. Fuel 370:131855
18. Adarakatti PS (2025) Fundamentals of electrochemistry. In: Agricultural electrochemistry. ACS Publications, Washington, pp 17–46
19. Naaz T, Kumar A, Vempaty A, Singhal N, Pandit S, Gautam P, Jung SP (2023) Recent advances in biological approaches towards anode biofilm engineering for improvement of extracellular electron transfer in microbial fuel cells. Environ Eng Res 28(5):220666
20. Koók L, Nemestóthy N, Bélafi-Bakó K, Bakonyi P (2021) The influential role of external electrical load in microbial fuel cells and related improvement strategies: a review. Bioelectrochemistry 140:107749
21. Meng L, Feng M, Sun J, Wang R, Qu F, Yang C, Guo W (2022) High-performance freestanding microbial fuel cell anode derived from Chinese date for enhanced electron transfer rates. Bioresour Technol 353:127151
22. Pandya RS, Kaur T, Bhattacharya R, Bose D, Saraf D (2024) Harnessing microorganisms for bioenergy with microbial fuel cells: powering the future. Water-Energy Nexus 7:1–12
23. Liu Y, Wang J, Sun Y, Li H, Zhai Z, Guo S, Ren T, Li C (2022a) Nitrogen-doped carbon nanofibers anchoring Fe nanoparticles as biocompatible anode for boosting extracellular electron transfer in microbial fuel cells. J Power Sources 544:231890
24. Jiang J, Wang H, Zhang S, Li S, Zeng W, Li F (2021) The influence of external resistance on the performance of microbial fuel cell and the removal of sulfamethoxazole wastewater. Bioresour Technol 336:125308
25. Etghani SA, Hosseini M, Mousavi MR, Ansari E, Sanaee Z, Mohajerzadeh S (2025) Facile and scalable synthesis of vertically oriented TiS2 Nanosheets for green energy harvesting in microfluidic microbial fuel cells. Energy Fuel 39(10):4834–4848
26. Chu N, Zhang L, Hao W, Liang Q, Jiang Y, Zeng RJ (2021) Rechargeable microbial fuel cell based on bidirectional extracellular electron transfer. Bioresour Technol 329:124887
27. Fang C, Li J, Feng Z, Li X, Cheng M, Qiao Y, Hu W (2022) Spatiotemporal mapping of extracellular electron transfer flux in a microbial fuel cell using an oblique incident reflectivity difference technique. Anal Chem 94(30):10841–10849
28. Ida TK, Mandal B (2023) Microbial fuel cell design, application and performance: a review. Mater Today Proc 76:88–94
29. Yaqoob AA, Ibrahim MNM, Umar K (2021) Electrode material as anode for improving the electrochemical performance of microbial fuel cells. In: Energy storage battery systems-fundamentals and applications. IntechOpen, London
30. Fadzli FS, Bhawani SA, Adam Mohammad RE (2021) Microbial fuel cell: recent developments in organic substrate use and bacterial electrode interaction. J Chem 2021(1):4570388
31. Wang H, Long X, Sun Y, Wang D, Wang Z, Meng H, Jiang C, Dong W, Lu N (2022) Electrochemical impedance spectroscopy applied to microbial fuel cells: a review. Front Microbiol 13:973501

32. Zheng X, Hou S, Amanze C, Zeng Z, Zeng W (2022) Enhancing microbial fuel cell performance using anode modified with Fe3O4 nanoparticles. Bioprocess Biosyst Eng 45(5):877–890
33. Siddiqui S, Bhatnagar P, Dhingra S, Upadhyay U, Sreedhar I (2023) Wastewater treatment and energy production by microbial fuel cells. Biomass Convers Biorefinery 13(5):3569–3592
34. Zhou E, Lekbach Y, Gu T, Xu D (2022) Bioenergetics and extracellular electron transfer in microbial fuel cells and microbial corrosion. Curr Opin Electrochem 31:100830
35. Mahto KU, Das S (2025) Electroactive biofilm communities in microbial fuel cells for the synergistic treatment of wastewater and bioelectricity generation. Crit Rev Biotechnol 45(2):434–453
36. Palanivel PD, Naina Mohamed S, Mohanakrishna G, Ravinuthala S, Saravanan S (2025) An insight on energy production using microbial fuel cell: a microbiological perspective and electron transfer improvement. J Chem Technol Biotechnol 100(4):672–687
37. Wang R, Li H, Sun J, Zhang L, Jiao J, Wang Q, Liu S (2021) Nanomaterials facilitating microbial extracellular electron transfer at interfaces. Adv Mater 33(6):2004051
38. Taşkan B, Taşkan E (2021) Inhibition of AHL-mediated quorum sensing to control biofilm thickness in microbial fuel cell by using Rhodococcus sp. BH4. Chemosphere 285:131538
39. Song YE, Mohamed A, Kim C, Kim M, Li S, Sundstrom E, Beyenal H, Kim JR (2022) Biofilm matrix and artificial mediator for efficient electron transport in CO2 microbial electrosynthesis. Chem Eng J 427:131885
40. Andriukonis E, Celiesiute-Germaniene R, Ramanavicius S, Viter R, Ramanavicius A (2021) From microorganism-based amperometric biosensors towards microbial fuel cells. Sensors 21(7):2442
41. Li Y, Liu J, Chen X, Yuan X, Li N, He W, Feng Y (2021) Tailoring spatial structure of electroactive biofilm for enhanced activity and direct electron transfer on iron phthalocyanine modified anode in microbial fuel cells. Biosens Bioelectron 191:113410
42. Shirkosh M, Hojjat Y, Mardanpour MM (2022) Boosting microfluidic microbial fuel cells performance via investigating electron transfer mechanisms, metal-based electrodes, and magnetic field effect. Sci Rep 12(1):7417
43. Verma J, Kumar D, Singh N, Katti SS, Shah YT (2021) Electricigens and microbial fuel cells for bioremediation and bioenergy production: a review. Environ Chem Lett 19(3):2091–2126
44. Gu T, Wang D, Lekbach Y, Xu D (2021) Extracellular electron transfer in microbial biocorrosion. Curr Opin Electrochem 29:100763
45. Gunaseelan K, Jadhav DA, Gajalakshmi S, Pant D (2021) Blending of microbial inocula: an effective strategy for performance enhancement of clayware biophotovoltaics microbial fuel cells. Bioresour Technol 323:124564
46. Leung DHL, Lim YS, Uma K, Pan G-T, Lin J-H, Chong S, Yang TC-K (2021) Engineering S. oneidensis for performance improvement of microbial fuel cell—a mini review. Appl Biochem Biotechnol 193(4):1170–1186
47. Krishnan SK, Kandasamy S, Subbiah K (2021) Chapter 32 - fabrication of microbial fuel cells with nanoelectrodes for enhanced bioenergy production. In: Kumar RP, Bharathiraja B (eds) Nanomaterials. Academic, New York, pp 677–687
48. Moradian JM, Yang F-Q, Xu N, Wang J-Y, Wang J-X, Sha C, Ali A, Yong Y-C (2022) Enhancement of bioelectricity and hydrogen production from xylose by a nanofiber polyaniline modified anode with yeast microbial fuel cell. Fuel 326:125056
49. Thengumthottathil V, Ponnusamy K, Naina Mohamed S (2024) Bioelectrochemical systems: exploring microbial communities, interactions, and electron transfer. Biochem Eng J 211:109442
50. Agrahari R, Bayar B, Abubackar HN, Giri BS, Rene ER, Rani R (2022) Advances in the development of electrode materials for improving the reactor kinetics in microbial fuel cells. Chemosphere 290:133184
51. Baziną N, Ahmed TG, Almdaaf M, Jibia S, Sarker M (2023) Power generation from wastewater using microbial fuel cells: a review. J Biotechnol 374:17–30

52. Hidayat ARP, Widyanto AR, Asranudin A, Ediati R, Sulistiono DO, Putro HS, Sugiarso D, Prasetyoko D, Purnomo AS, Bahruji H, Ali BTI, Caralin IS (2022) Recent development of double chamber microbial fuel cell for hexavalent chromium waste removal. J Environ Chem Eng 10(3):107505
53. Liu Y, Guo S, Wang J, Li C (2022b) Fundamental development and research of cathodic compartment in microbial fuel cells: a review. J Environ Chem Eng 10(3):107918
54. Dutta S, Patil R, Dey T (2022) Electron transfer-driven single and multi-enzyme biofuel cells for self-powering and energy bioscience. Nano Energy 96:107074
55. Nguyen HD, Babel S (2022) Insights on microbial fuel cells for sustainable biological nitrogen removal from wastewater: a review. Environ Res 204:112095

Innovative Energy Solutions: Merging Bio-Electrochemical Systems with Renewables

K. Siva Kumar, K. Sunil, and Albert Selvakumar

Abstract At a time when the world is urgently seeking cleaner, more innovative ways to produce and manage energy, bio-electrochemical systems (BES) offer a quiet yet powerful solution. They don't just generate electricity or hydrogen-they do it while treating waste, recovering nutrients, and working in harmony with natural processes. This chapter is a journey through what makes BES so compelling: its biological foundation, its technical elegance, and, most importantly, its potential to reshape our understanding of energy. We explore how these systems integrate with renewable technologies, such as solar, wind, and biomass, to create small-scale, circular, and resilient energy solutions. Along the way, we examine recent breakthroughs, real-world case studies, and the often overlooked policy and infrastructure challenges that hinder broader adoption. But this isn't just about the science—it's about what's possible when we take seriously the idea that waste can be a resource, that microbes can be engineers, and that innovation doesn't always have to be loud to be revolutionary. This chapter presents both a technical roadmap and a hopeful vision for researchers, practitioners, and anyone interested in the future of energy.

Keywords Energy · Electrochemical · Renewable energy · Wind energy

K. S. Kumar (✉)
Department of Petroleum Engineering, Godavari Global University, Rajahmundry, Andhra Pradesh, India
e-mail: ksivakumar.pt@giet.ac.in

K. Sunil
Department of Freshman Engineering, Sanketika Vidya Parishad Engineering, Vishakhapatnam, Andhra Pradesh, India

A. Selvakumar
Department of Water Technology, Al Khabourah Vocational College, Al Khabourah, North Al Batinah, Oman

A. Sattar Jatoi, N. M. Mubarak (eds.), *Bio-Electrochemical Conversion of Waste to Fuel*, Green Energy and Technology,
https://doi.org/10.1007/978-981-95-4310-6_8

1 Introduction

The sudden and global rise in demand for eco-friendly energy features advanced technologies such as bio-electrochemical systems (BES) integrated with renewable energy sources. Such systems have emerged as the most promising option due to the unsustainability of traditional fossil fuel systems in creating a low-carbon economy. Research on alternative energy sources has intensified in parallel with the growing environmental concerns caused by traditional energy systems. AI-powered systems are becoming astonishingly powerful, leading to paradigm shifts in energy innovation. The integration of smart transportation alongside electric vehicles is relatively straightforward compared to the rapid 56.4% annual growth in research relating AI to electric vehicles, as showcased in studies [1]. This, alongside the automation of energy generation, storage, and consumption systems, as well as the simplification of equipped BES-renewable systems, showcases a clear trend concerning power systems. The further evolution of wearable technology, such as smartwatches and the devices mentioned above, demands more self-sufficient and longer-lasting energy sources. Considerable advances in the development of self-reliant and portable energy storage and harvesting devices enabled the invention of self-powered healthcare wearables and powered devices [2]. These devices perfectly encapsulate how users are progressively shifting their approach toward energy systems and power generation, focusing on modular designs that can be incorporated into building energy systems (BESs).

Innovation in materials is important for making these technologies more sustainable. Polysaccharides from biomass have substituted traditional materials used in energy storage systems as a more environmentally friendly and electrochemically active constituent. Their structural differences enable them to act as functional and structural components in the most revolutionary batteries and capacitors, making nature-friendly energy design possible [3]. From a wider perspective, the constructional elements of the system should integrate with sustainable energy systems. Earth-based materials have received increasing attention due to their low embodied energy, recyclability, and contribution to providing thermally efficient and resilient urban infrastructure. A combination of natural building materials and digital systems is augmented by the principle of providing a new paradigm of urban energy resilience [4].

This chapter reviews the integration of BES with renewables as a systemic innovation—technological, ecological, and social—with BES and renewable energy at its core. It details the background, existing synergies, technological innovations, case studies, and foresight. By integrating literature from various fields, this paper aims to provide insight into how bio-electrochemical systems can be coupled with renewable energy sources to meet critical energy and environmental needs.

2 Principles of Bio-Electrochemical Systems (BES)

Bio-electrochemical systems (BES) represent a new class of technologies that utilize microorganisms as catalysts for the oxidation or reduction of organic matter. These systems encompass microbiology, electrochemistry, and materials science to devise environmentally friendly methods for energy recovery, wastewater treatment, and bioproduction. Such systems are integrated with fields such as microbiology, electrochemistry, and materials engineering to find eco-friendly methods for energy recovery, wastewater biotreatment, and bioproduct synthesis.

BES centers on bioelectrodes, which are conductive surfaces colonized by electroactive microbes and foamed carbon materials. Microorganisms, such as Geobacter and Shewanella species, enable direct or mediated bioelectrocatalysis, facilitating electron transfer between the electrode and their metabolic processes. Due to advances in organophosphorus chemistry, recently developed highly active and biologically interesting molecular systems are organophosphorus compounds, especially azaheterocyclic phosphonates. These compounds are stable and functionally versatile, thus possessing considerable bio-electrochemical activity [5]. Besides these, biological stress tolerance is another important characteristic of BES's performance, particularly in cases of environmental pollution. Bioremediation of plants and attempts to genetically mitigate metal and metalloid stress in plants have provided insights into enhancing the robustness of microbes in bio-electrochemical processes [6]. This is essential knowledge for creating a BES that will be operational in contaminated or high-salinity environments. The system's efficiency of electron transfer and stability are greatly affected by the design of the electrodes.

Developing renewable energy sources, like hydrogen production through solar-driven biomass conversion, brings new capabilities to BES. Concerning other technologies that combine biomass-derived photoelectrochemical systems and BES, these demonstrate the ability of light-driven microbial or catalytic processes to promote sustainable and decentralized hydrogen generation. Boddula et al. demonstrate how biomass systems can also contribute to meeting energy conversion targets while achieving low greenhouse gas emissions [7].

The use of silicon anodes in BES and hybrid systems is promising due to their high specific capacity and conductivity. Nonetheless, capacity fading and structural instability pose challenges. Other studies have been conducted so far, and these challenges have been addressed through material engineering and surface modification aimed at silicon-based components in BES and hybrid systems [8]. All these efforts deepen the understanding of and interplay between different aspects of BES, forming the foundation for its effective use. It also marks the significant contribution of non-traditional disciplines to the aforementioned fields, including interdisciplinary approaches combining synthetic chemistry, plant and microbial biology, materials science, and renewable energy engineering. The following section describes the interplay of these systems with established renewable energy technologies.

3 Overview of Renewable Energy Technologies

Recently, there has been significant progress in developing renewable energy technologies, driven by the global shift toward a decarbonized energy future. The use of solar, wind, hydro, biomass, and geothermal power are being adopted as alternatives to fossil fuels for energy generation. However, the lack of reliable storage methods and the inconsistent quality of several renewable energy sources remain significant obstacles. These issues could be addressed by integrating other energy systems with bio-electrochemical platforms, thereby enhancing the system's reliability and sustainability.

3.1 Comparison of Renewable Energy Technologies

Catalytic water splitting is a cornerstone of renewable energy methods, especially for hydrogen production. The effectiveness of bubble generation and detachment at catalytic interfaces plays an important role in determining the reaction kinetics of electrochemical water splitting. Zhang et al. claim that advanced bubble regulation properties can exhibit a higher catalytic effect, thereby enhancing the overall system performance [9]. Table 1 gives comparison of key renewable energy technologies.

3.2 Integration Potential with BES

The oxygen reduction reaction (ORR) plays a crucial role in various electrochemical energy systems, including fuel and microbial fuel cells. Biomass-based carbon materials have made significant advancements as inexpensive and electrocatalytically active substitutes for noble metals in the oxygen reduction reaction (ORR) process [10]. Table 2 shows the potential for integration of renewable technologies with BES.

Table 1 Comparison of key renewable energy technologies

Technology	Efficiency	Scalability	Intermittency	Env. impact
Solar PV	Medium (1522%)	High	High	Low
Wind energy	High (30%–45%)	Medium	High	Low
Hydro power	Very High (7090%)	Site dependent	Low	Medium
Biomass	Medium (2030%)	High	Low	Medium
Geothermal	High (40%–50%)	Site dependent	Low	Low

Table 2 Potential for integration of renewable technologies with BES

Renewable source	BES integration use case	Challenges
Solar PV	Powering MFCs, photo-assisted electrosynthesis	Variability, storage
Wind	Energy for MEC-driven hydrogen production	Fluctuation management
Biomass	Substrate source for BES	Pretreatment, feedstock logistics
Geothermal	Temperature-controlled bioreactor operation	Regional limitations
Hydro	Co-location for water treatment MFCs	Infrastructure integration

Innovations in coordination chemistry also enable sustainable chemical processes. Osman et al. explain low-energy catalytic reactions supported by coordination complexes, which can be leveraged in integrated waste-to-energy recovery systems [11]. Energy storage remains crucial for the stability of renewable energy systems. Njema et al. discuss progress in lithium-ion and solid-state battery technologies that may facilitate matching intermittent energy input with sustained, continuous battery energy storage (BES) operation [12].

3.3 Conceptual Integration Model

Figure 1 shows the conceptual flow of renewable-BES integration.

4 Synergy Between BES and Renewables

A strong strategy for creating decentralized, clean, and robust energy infrastructures is the combination of bio-electrochemical systems (BESs) and renewable energy technologies. BES can simultaneously treat waste streams and produce electricity, hydrogen, or value-added chemicals by utilizing microbial metabolism with renewable energy sources. The complementary functions of BES and renewables-BES add biochemical flexibility and closed-loop resource recovery, while renewables supply clean energy, creating the synergy.

4.1 Integration Use Cases

In the industrial context, integrating renewable energy with building energy systems (BESs) can transform manufacturing operations. Usman et al. highlight how manufacturing sectors can benefit from renewable-powered systems by achieving greater energy efficiency, lower emissions, and improved cost-effectiveness [13]. BES can

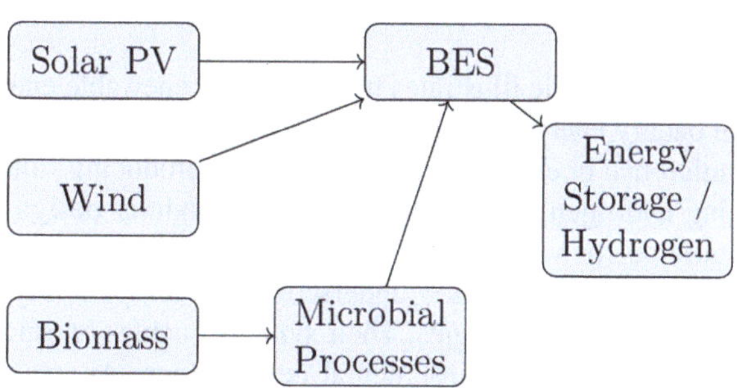

Fig. 1 Conceptual flow of renewable-BES integration

Table 3 Examples of renewable-BES synergy in different sectors

Sector	Renewable integration	BES contribution
Manufacturing	Rooftop solar or biomass CHP systems	Wastewater treatment and on-site energy recovery
Transportation	Wind/solar-powered electrolysis for hydrogen fuel	Microbial electrolysis for green hydrogen production
Carbon capture	Photovoltaics powering DAC or electrochemical units	Microbial CO_2 reduction into methane/ acetate
Hydrogen economy	Distributed solar or wind farms	Decentralized hydrogen generation from biomass or organics

complement these strategies by enabling on-site waste-to-energy recovery and wastewater treatment, contributing to circular manufacturing processes. BES-based hydrogen production can support the decarbonization of fuel-cell electric vehicles in the transportation sector. Reddy et al. compare alternative transportation systems, including biofuel, electric, hydrogen, and solar-powered vehicles. Their study suggests that hydrogen systems, particularly when supported by biomass-based hydrogen generation, have a strong potential for long-term sustainability [14]. BES integrated with solar or wind energy can drive microbial electrolysis for hydrogen production, reinforcing this vision.

Carbon capture and conversion are other crucial applications where BES-renewable integration can offer added value. Mohammed et al. review the emerging landscape of carbon capture technologies, particularly in the context of hydrogen energy systems [15]. BES can assist in carbon fixation through microbial pathways and, when coupled with renewable inputs, provide a sustainable route for transforming CO_2 into bio-based fuels or chemicals.

Developing a hydrogen supply chain based on renewable and bio-electrochemical technologies is gaining momentum. Li et al. assess the challenges of building a future hydrogen economy, emphasizing the need for scalable production methods and stable supply infrastructures [16]. BES offers a decentralized and modular approach to hydrogen generation, making it an ideal candidate for distributed hydrogen systems, especially when powered by renewable energy (Table 3).

4.2 Conceptual Model of BES-Renewable Integration

This schematic illustrates how multiple renewable energy sources can contribute to a battery energy storage system (BESS) system. The BES processes inputs through microbial or electrochemical pathways, producing valuable forms of energy, including hydrogen and bioelectricity. The modular design enables these systems to be tailored for various applications, including industrial, municipal, and agricultural uses. These examples underscore the compatibility between BES and various renewable technologies. Their synergy enables enhanced system resilience, energy autonomy, and environmental performance. As we progress toward smarter and

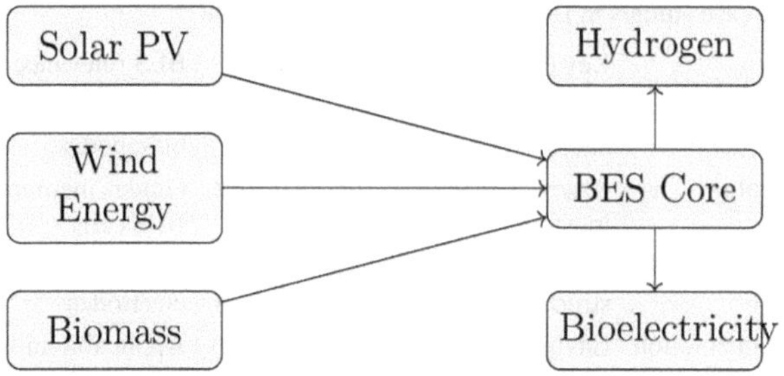

Fig. 2 Synergistic integration of BES with renewable energy systems

more sustainable infrastructures, BES-renewable hybrid systems represent a vital frontier in innovation (Fig. 2).

5 Case Studies and Recent Research

Bio-electrochemical systems are gaining new life as they integrate with modern materials and renewable energy setups. Recent studies indicate that they're more feasible and efficient than previously believed. New twists in microalgae cultivation, updated membranes, quirky 3D nanostructures, naturally sourced electrolytes, and even electrodes built from biomass continue to push performance and sustainability boundaries further, sometimes in ways that feel as serendipitous as they are planned.

5.1 Summary of Recent Research Directions

Tocca et al. [17] found that acetate reclaimed from a mix of industrial streams can kickstart microalgal growth under both mixotrophic and heterotrophic conditions (see [17]). Interestingly, this method not only increases overall biomass but also lays the groundwork for crafting algal bioelectrodes in microbial fuel cells (MFCs), offering a dual benefit in bioenergy research. See Table 4 for few such selected case studies.

Membranes do a lot of heavy lifting in sustainable energy systems—they quietly separate anodic and cathodic zones in BES while allowing ions to move freely, a trick that becomes especially vital in microbial electrolysis cells. In a broad review, Osman et al. noted that these barriers are generally prized for their efficiency, selectivity, and solid chemical stability [18].Graphene-based materials have emerged as strong candidates for enhancing BES performance. Uddin et al. synthesized a novel 3D graphene with exceptional porosity and surface area, making it an ideal material for supercapacitor applications [19]. Such nanostructures offer enhanced electron mobility and microbial adhesion, which are key for effective biofilm-electrode

Table 4 Selected case studies in BES and renewable integration

Study focus	Key contribution	BES relevance
Microalgae cultivation using acetate [17]	Enhanced biomass growth via alternative carbon source	Boosts the productivity of algal bio-anodes
Membrane technology review [18]	Assesses membrane performance in energy applications	Guides membrane selection for BES cells
3D graphene synthesis [19]	High-surface-area carbon structures for supercapacitors	Enables high-capacitance BES electrodes
Natural solvents in zinc-ion batteries [20]	Environment-friendly electrolytes with high stability	Applicable in sustainable BES electrolyte design
Sulfur confinement in jute-derived carbon [21]	Improved energy density in supercapacitors	Promotes the use of agricultural waste in BES materials

interfaces. Yaman and Yüksel explored the use of natural solvent-based electrolytes in zinc-ion batteries, showcasing their potential for sustainability and electrochemical efficiency [20]. Similar electrolytes could be adapted for BES, where green chemistry is a growing priority. Jute-derived activated carbon was studied by Shah et al. for sulfur confinement in supercapacitor electrodes [21]. Their work highlights a promising route for incorporating agricultural waste into high-performance energy storage, aligning well with the BES objective of utilizing circular resources.

5.2 *Performance Attributes Highlighted in Recent Studies*

Researchers today are exploring new electrode materials and various biological feedstocks that often blend insights from biology with materials science in a way that is rather unexpected. BES is steadily being positioned as a central role in mapping the future of next-generation energy systems. See Fig 3.

6 Technological Innovations

Bio-electrochemical systems (BES) have undergone a quiet yet steady transformation over the past few years, driven largely by innovations that, while sometimes subtle, have profoundly impacted our thinking about sustainable energy. Whether it's the chemistry of carbon reduction or the clever repurposing of waste, researchers are beginning to stretch the possibilities of these systems in impressive ways.

Take, for instance, the reduction of carbon dioxide. Mukhopadhyay and colleagues introduced an interesting idea, designing a layered structure that essentially creates a microreservoir for CO_2 right at the reaction site [22]. This isn't just about improving reaction rates; it's also about enhancing overall performance. It's about rethinking how we approach carbon utilization altogether, particularly in microbial

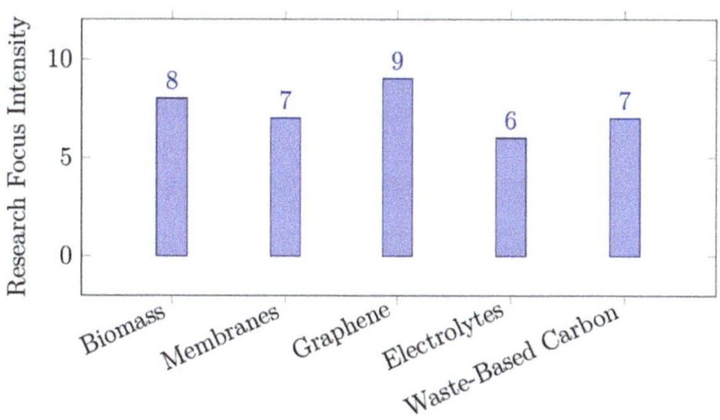

Fig. 3 Key attributes enhanced in recent BES-related research

electrolysis setups where every bit of efficiency counts. Then there's the issue of food waste. We often think of leftovers as a disposal problem, but Ramandani et al. remind us that food waste is also an untapped source of energy [23]. Their review sheds light on several practical methods for turning this waste into power, which aligns remarkably well with BES, which can digest organic material and convert it into electricity. It's a simple concept, but the environmental payoff could be significant. Energy generation doesn't stop at organic waste. In another study, Baldinelli and his team investigated the combination of biogas systems with solid oxide fuel cells (SOFCs) [24]. While the technical details are complex, the takeaway is clear. By merging SOFCs with BES, we may be able to create more stable and efficient systems that handle both gas conversion and electricity generation, particularly in decentralized settings. Finally, Lee et al. explored the broader picture by focusing on waste biorefineries [25]. Their work makes the case for integrating BES not as standalone technologies but as functional modules within larger processing networks, handling side tasks such as nutrient recovery or low-grade waste treatment. Essentially, BES becomes part of a larger puzzle rather than a separate solution. Together, these developments suggest that BES is finally shedding its experimental label. It's no longer about what BES can do in isolation but what it can do as part of a smarter, cleaner energy ecosystem.

7 Environmental and Economic Impact

When we discuss the future of energy, it's no longer just about generating more; it's about being smarter. Though still emerging, bio-electrochemical systems (BES) have quietly carved out a compelling case as clean technologies and tools that blend environmental benefits with economic potential. While engineering is impressive, its real-world impact on ecosystems and economies may prove most important in the years ahead.

Table 5 Environmental contributions of bio-electrochemical systems

Impact area	Contribution of BES
Wastewater treatment	Simultaneous treatment and energy recovery via microbial fuel cells (MFCs)
Carbon reduction	Microbial CO_2 fixation and electrochemical reduction into methane or acetate
Organic waste utilization	Conversion of food and agricultural waste into electricity or hydrogen
Heavy metal recovery	Selective electrodeposition from industrial effluents using BES
Decentralized sanitation	Off-grid sanitation with energy-positive systems in developing regions

7.1 How BES Helps the Environment Quietly and Effectively

BES systems may not dominate headlines, but they're already making a difference in areas such as wastewater treatment, waste recovery, and emissions reduction. These aren't just theoretical applications—microbial fuel cells treat wastewater while producing electricity. Microbial electrolysis cells, meanwhile, can generate hydrogen while cleaning organic waste. Their strength lies in doing many things at once: cleaning, converting, recovering, and developing (Table 5).

Hydrogen production is a good example of where BES could shift the balance. As we seek alternatives to fossil fuels, hydrogen continues to stand out; however, producing it cleanly and storing it efficiently remains a puzzle that's far from solved. Alasali et al. examine hydrogen systems and materials, pointing out the ongoing trade-offs between cost, stability, and efficiency [26]. BES brings something new: it can produce hydrogen from waste using low-energy biological processes rather than high-temperature or high-pressure setups. It's a quiet, innovation-less, flashy, but grounded in practicality. Catalyst development is another space where BES quietly benefits from broader research. Dai et al. worked on improving methanol oxidation catalysts, which are important in fuel cells, and their insights into catalyst design can also be applied to BES [27]. After all, BES relies on efficient electron transfer, and the better the materials are, the more output you get from the same biological input. Then there's the matter of carbon. While capturing and storing CO_2 It is often discussed at the macro level. Cruz-Navarro and his team explored how copper-based metal-organic frameworks (Cu-MOFs) could reduce CO_2 into methane electrochemically [28]. The link to BES is subtle but important: with the right materials, BES could 1 day help transform greenhouse gases into usable energy, not just capture them.

7.2 Where Economics Fits into the Picture

Of course, environmental impact is only half the story. The other half is whether BES makes economic sense. Currently, the setup costs remain high, and reliability over extended periods is a concern. But the picture is improving. As feedstock

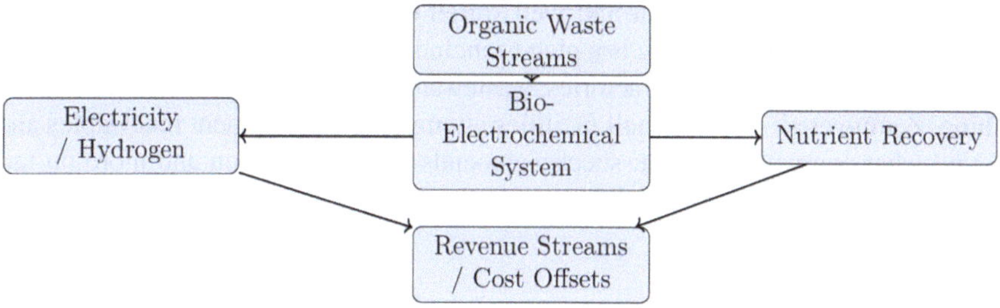

Fig. 4 Economic value flows enabled by BES integration

availability increases, particularly from food and agricultural waste, and as materials become more affordable and robust, the financial side of BES is starting to look more promising.

To understand the potential, consider the following conceptual model of how BES can turn waste into value (Fig. 4).

This is where BES begins to shine—not as a one-size-fits-all technology but as a flexible tool for specific contexts: rural communities with food waste issues, urban settings with wastewater needs, and industries seeking to reduce chemical runoff. Each of these environments offers a different value, and BES is adaptable enough to meet them where they are. The environmental argument for BES is already strong, built on real use cases and tested outcomes. The economics are coming along, and as technologies evolve and costs drop, the case for integrating BES into mainstream energy and waste systems grows stronger by the day.

8 Policy and Implementation Challenges

One of the biggest ironies in energy innovation is this: the more interdisciplinary a solution is, the harder it is to implement. Bio-electrochemical systems (BES) are a textbook case. They're part biology, engineering, chemistry, and waste management, yet they don't fully "belong" anywhere. That's not just a philosophical dilemma; it's a policy one. When no one sector claims ownership, things slip through the cracks. Marcos and colleagues noted this when discussing the decarbonization of European industry [29]. Their argument was clear: having the tech isn't enough to rethink how our systems handle energy, waste, and heat together. BES could be a perfect fit in that reimagined model. However, for now, they tend to live on the margins, underfunded and underprioritized, despite ticking all the right sustainability boxes. There's also a practical issue that's less about red tape and more about infrastructure. Hybrid renewable systems, such as solar or wind paired with batteries, are making progress partly because policy support is clearer. However, as Sabhi et al. point out, without flexible energy management strategies and smart power electronics, even the best technology can become bottlenecked

[30]. BES, inherently modular and well-suited to decentralization, could thrive in these systems. What's missing is a plan to include them. And it's not just big industry that's in play. Small-scale factories, wastewater plants, farms—where BES could shine. Zemīte et al. studied small facilities attempting to implement renewables and found what one would expect: success depends less on ambition and more on tailored, realistic planning [31]. BES needs the exact solutions grounded in place, budget, and use case.

8.1 A Quick Reality Check and a Way Forward

None of these problems is insurmountable. But ignoring them slows everything down. In many ways, BES doesn't need massive investment or breakthrough discoveries. They need space to prove themselves: a few testbeds, some recognition in funding schemes, and a voice in long-term planning conversations. That's how other clean technologies got off the ground, and it's not too much to ask. Table 6 gives a list of some of the challenges and potential responses.

8.2 How It Could Work: If We Let It

This kind of flow-through policy sparks real-world adoption, which isn't speculative. We've seen it happen with solar, wind, and electric vehicles. The model is there. BES needs to be written into the script. It comes down to this: BES is not some futuristic fantasy. They're here. They work. But they need a little help to scale. And if policymakers are willing to look beyond the usual suspects and support ideas that live in between categories, they might just find that BES aren't a fringe innovation—they're part of the foundation we've been searching for (see Fig. 5).

Table 6 Policy and implementation challenges for BES and suggested responses

Challenge	Potential response
Blurred lines between waste and energy sectors. Upfront costs are still too high for small players. Too few trained technicians Low visibility in national energy strategies Rigid infrastructure not built for modular tech	Create integrated, cross-sector policy guidelines for BES Support with subsidies, grants, and shared pilot infrastructure Build a workforce through targeted training and educational outreach Add BES to regional sustainability pilots and innovation funds Promote BES in microgrid, off-grid, and decentralized projects

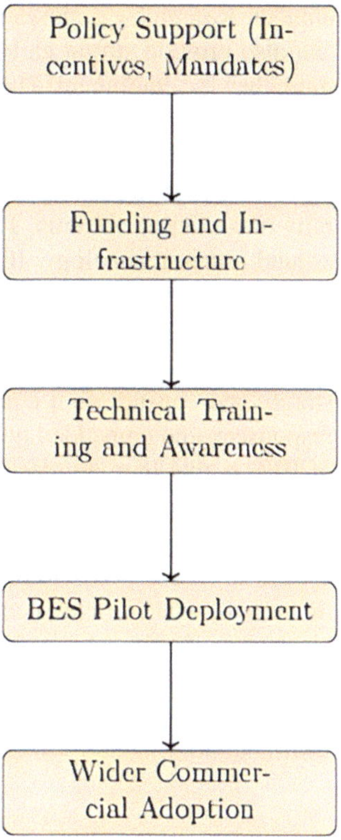

Fig. 5 From policy to practice: driving BES adoption

9 Future Directions

If there's one thing we've learned from the slow, uneven adoption of emerging technologies, it's this: potential alone isn't enough. It doesn't matter how elegant the science is, how efficient the outputs are, or how compelling the sustainability story sounds. If the pieces don't come together in the real world, the whole thing remains theoretical.

Bio-electrochemical systems (BES) are at a familiar crossroads. On paper, they almost sound too good to be true. They can clean waste, generate energy, recover nutrients, and operate with microbes and sunlight. The real question, however, is not what BES can do. It is whether we are willing to support them long enough for them to prove their value in the real world. The way forward is not a single breakthrough. It is a sequence of deliberate steps. Yes, the technology must keep improving. Materials still need to become cheaper, tougher, and easier to maintain. Systems also need better efficiency and simpler designs. But the biggest barrier today is often not the lab performance. It is the slow work of integration—into policy, into energy planning, into education and training, and into funding systems that reward long-term impact. In the short term, BES are most likely to succeed in niche applications. Rural water treatment is one clear example. Decentralized hydrogen production is another. Research campuses that are serious about living their sustainability goals can also provide strong early use cases. These locations matter because they test more than the equipment. They also test the surrounding ecosystem, including regulation, workforce readiness, supply chains, and maintenance support. In the medium term, the picture can become more promising. As infrastructure shifts toward flexibility—through modular grids and circular resource flows—BES may fit more naturally into existing systems. The goal is to avoid forcing the whole network to bend around a new technology. If this integration is done well, people may eventually stop asking, "Why BES?" and start asking, "Why did we not adopt this earlier?" Over the long term, BES could reshape how we think about energy. That shift is subtle but important. Energy would no longer be seen only as something extracted, stored, and consumed. It could also be understood as a living process—adaptive, responsive, and embedded in local environments. One of the most valuable features of BES is that they are designed to connect things. They can bridge biology and engineering in a practical way. They can link waste streams with energy recovery. They can also connect local problems with global sustainability goals. This is not always an easy message to sell in a world that expects quick fixes and dramatic solutions. Still, what we may need now is not another miracle. We may need technologies that work with what we already have, especially in places that are often ignored.

That's the future BES points toward. Not a shiny techno-utopia, but a quieter, smarter, more grounded way forward. And if we're paying attention, it's a future worth building.

10 Conclusion

Sometimes, the technologies that change everything don't arrive with a bang. They don't always have flashy headlines or billion-dollar IPOs. They appear quietly, in lab benches and field trials, in conversations between disciplines, in the corners of the world where problems are urgent and resources are scarce. Bio-electrochemical systems feel like one of those technologies. They aren't perfect. Not yet. They won't solve the climate crisis on their own, and they won't replace every part of our energy infrastructure. But maybe that's the point. BES aren't trying to do everything—they're trying to do a few things well. Recover energy from waste. Treat water while generating power. Close the loop in places where the loop has been broken for a long time. Throughout this chapter, we've examined what BES is capable of—not just chemistry or engineering but also vision. We've seen how they can fit into a more decentralized, circular, and human-centered future. A future where energy is produced locally, where waste isn't something to be hidden but something to be transformed, and where innovation doesn't just mean "new" but "right-sized," "thoughtful," and "regenerative." But none of that happens by accident. It will take work to bring BES out of the margins and into the mainstream. It will take flexible policies, patient funding, and many people who believe that small, local solutions matter as much as big, global ones. The good news is that the pieces are already in place. The research is strong. The early applications are working. The ideas are ready. What's needed now is belief and a little courage to back technologies that don't fit neatly into old categories.

Ultimately, BES reminds us of something we often forget: progress isn't always loud. Sometimes, it hums softly in a reactor powered by microbes and leftovers, quietly creating energy where no one thought to look.

References

1. Domenteanu A, Cotfas L-A, Diaconu P, Tudor G-A, Delcea C (2025) Ai on wheels: bibliometric approach to mapping of research on machine learning and deep learning in electric vehicles. Electronics 14:378–378
2. Zhang Q, Das S, Zheng L, Wan J (2025) Advances in wearable energy storage and harvesting systems. Med-X 3:3
3. Thomas SM, Gómez-Romero P, González-Gil RM (2025) Polysaccharides: The sustainable foreground in energy storage systems. Polysaccharides 6:5–5
4. Bühler MM, Hollenbach P, Köhler L, Armstrong R (2024) Unlocking resilience and sustainability with earth-based materials: a principled framework for urban transformation. Front Built Environ 10:1385116
5. Mayorquín-Torres MC, Simoens A, Bonneure E, Stevens CV (2024) Synthetic methods for azaheterocyclic phosphonates and their biological activity: an update 2004-2024. Chem Rev 124:7907–7975
6. Almotairy HM (2024) Mitigating metal/metalloid stress in crops: strategies for sustainable agricultural resilience. In: IntechOpen eBooks. IntechOpen, London
7. Boddula R, Lee Y-Y, Masimukku S, Chang-Chien G-P, Pothu R, Srivastava RK, Sarangi PK et al (2024) Sustainable hydrogen production: solar-powered biomass conversion explored through (photo)electrochemical advancements. Process Saf Environ Prot 186:1149–1168
8. Khan M, Yan S, Ali M, Mahmood F, Yang Z, Li G, Liu J et al (2024) Innovative solutions for high-performance silicon anodes in lithium-ion batteries: overcoming challenges and real-world applications. NanoMicro Lett 16. https://doi.org/10.1007/s40820-024-01388-3
9. Zhang Z, Chen G, Wang K, Haoxuan Y, Qiu J, Wang S, Wang L et al (2024) Bubbles management for enhanced catalytic water splitting performance. Catalysts 14:254–254. https://www.mdpi.com/2073-4344/14/4/254
10. Liu S, Wang A, Liu Y, Zhou W, Wen H, Zhang H, Sun K et al (2024) Catalytically active carbon for oxygen reduction reaction in energy conversion: recent advances and future perspectives. Adv Sci 11. https://doi.org/10.1002/advs.202308040
11. Osman AI, Ayati A, Krivoshapkin PV, Tanhaei B, Farghali M, Yap P-S, Abdelhaleem A (2024) Coordination-driven innovations in low-energy catalytic processes: advancing sustainability in chemical production. Coord Chem Rev 514:215900–215900. https://www.sciencedirect.com/science/article/pii/S0010854524002467
12. Njema GG, Ouma RBO, Kibet JK (2024) A review on the recent advances in battery development and energy storage technologies. J Renew Energy 2024:1–35. https://doi.org/10.1155/2024/2329261
13. Usman FO, Ani EC, Ebirim W, Montero DJP, Olu-lawal KA, NinduwezuorEhiobu N (2024) Integrating renewable energy solutions in the manufacturing industry: challenges and opportunities: a review. Eng Sci Technol J 5:674–703. https://fepbl.com/index.php/estj/article/view/865
14. Reddy VJ, Hariram NP, Maity R, Ghazali MF, Sudhakar K (2024) Sustainable vehicles for decarbonizing the transport sector: a comparison of biofuel, electric, fuel cell and solar-powered vehicles. World Electr Vehicle J 15:93–93. https://www.mdpi.com/2032-6653/15/3/93

15. Mohammed S, Eljack F, Al-Sobhi SA, Kazi M-K (2024) A systematic review: the role of emerging carbon capture and conversion technologies for energy transition to clean hydrogen. J Clean Prod 447:141506141506. https://qspace.qu.edu.qa/handle/10576/57855
16. Li F, Liu D, Sun K, Yang S, Peng F-Z, Zhang K, Guo G et al (2024) Towards a future hydrogen supply chain: a review of technologies and challenges. Sustainability 16:1890–1890. https://www.mdpi.com/2071-1050/16/5/1890
17. Tocca GP, Agostino V, Menin B, Tommasi T, Fino D, Di Caprio F (2024) Mixotrophic and heterotrophic growth of microalgae using acetate from different production processes. Rev Environ Sci Biotechnol 23:93–132. https://doi.org/10.1007/s11157-024-09682-7
18. Osman AI, Chen Z, Elgarahy AM, Farghali M, Israa MA, Mohamed AK, Priya HB, Hawash et al (2024) Membrane technology for energy saving: principles, techniques, applications, challenges, and prospects. Adv Energy Sustain Res 5. https://doi.org/10.1002/aesr.202400011
19. Nizam Uddin M, Aliahsan Bappy M, Fozle Rab M, Znidi F, Morsy M (2024) Recent progress on synthesis of 3D graphene, properties, and emerging applications. IntechOpen, London
20. Yaman G, Yüksel R (2025) Toward green and sustainable zinc-ion batteries: the potential of natural solvent-based electrolytes. Small. https://doi.org/10.1002/smll.202411478
21. Shah SS, Abdul Aziz M, Oyama M (2024) Advancing electrochemical energy storage: the synergistic impact of sulfur nano-confinement in hierarchically porous jute-derived activated carbon for enhanced supercapacitor performance. Meeting abstracts (electrochemical society. CD-ROM), MA2024-01:11–11. https://www.researchgate.net/publication/380101010_Advancing_Electrochemical_Energy_Storage_The_Synergistic_Impact_of_Sulfur_Nano-Confinement_in_Hierarchically_Porous_Jute-Derived_Activated_Carbon_for_Enhanced_Supercapacitor_Performance
22. Mukhopadhyay S, Naeem MS, Shiva Shanker G, Ghatak A, Kottaichamy AR, Shimoni R, Avram L et al (2024) Local co2 reservoir layer promotes rapid and selective electrochemical co2 reduction. Nat Commun 15. https://www.nature.com/articles/s41467-024-47498-9
23. Ramandani AA, Rachmadona N, Munawaroh HSH, Lan JC-W, Kataria N, Khoo KS (2025) Sustaining food waste for energy conversion: a mini review. Green Energy Fuel Res. https://www.semanticscholar.org/paper/585ba218ae707394361f2e7543e1beeb6b45b931
24. Arianna Baldinelli U, Desideri FF, Cinti G (2024) Biogas-topower systems based on solid oxide fuel cells: thermodynamic analysis of stack integration strategies. Energies. https://www.semanticscholar.org/paper/565824f9626cb62d921fa7a438bbc614600ab388
25. Lee Y-Y, Srinivaas M, Li IC, Keharika K, Pothu R, Boddula R, Al-Qahtani N et al (2024) Enhancing sustainable energy through cutting-edge waste biorefinery technologies. Reactions. https://www.semanticscholar.org/paper/7d2175794e56951e7005cf6f68fa2131b9f12ce1
26. Alasali F, Abuashour MI, Hammad W, Almomani D, Obeidat AM, Holderbaum W (2024) A review of hydrogen production and storage materials for efficient integrated hydrogen energy systems. Energy Sci G3 Eng 12:1934–1968. https://doi.org/10.1002/ese3.1723
27. Dai S, Li M, Li H, Shi Y, Zhang H, Wang D, Xiang K et al (2025) Advancements in electrocatalytic methanol oxidation: catalyst design, reaction mechanisms, and renewable energy applications. ChemSusChem. https://www.semanticscholar.org/paper/ac0bcd2699b089654a7cdd2301095dcb905b3e80
28. Cruz-Navarro JA, Hernández-García F, Sanchez-Mora AT, Moreno-Narváez ME, Reyes-Márquez V, Peralta RC, Morales-Morales D (2024) Copper-based metal-organic frameworks applied as electrocatalysts for the electroreduction of carbon dioxide (co2er) to methane: a review. Methane. https://www.semanticscholar.org/paper/29ff08d7870cff68953686a51a047cf0541e6cc0
29. Marcos JD, Iman Golpour R, Barbero, and Antonio Rovira. (2024) Decarbonizing european industry: a novel technology to heat supply using waste and renewable energy. Appl Sci. https://www.semanticscholar.org/paper/3aed29419b5fcc8e702ca0fa67a1e5cee748f739
30. Sabhi K, Talea M, Bahri H, Dani S (2025) Integrating dual active bridge dc-dc converters: a novel energy management approach for hybrid renewable energy systems. Electr Eng E. https://www.semanticscholar.org/paper/607f33e02b6aafc7497c324ba668832b9f55bf5b
31. Laila Zemīte J, Kozadajevs L, Jansons IB, Dzelzītis E, Palkova K (2024) Integrating renewable energy solutions in small-scale industrial facilities. Energies. https://www.semanticscholar.org/paper/ec9137dcb35341665c18d02083be26a41a00fce2

Microbial Fuel Cell for Wastewater Treatment and Bioenergy Generation

Abdul Sattar Jatoi, Muhammad Azam Usto, Nabisab Mujawar Mubarak, Arshad Iqbal, Abdul Karim Shah, Abdul Basit Qazi, and Shoaib Ahmed

Abstract Through innovative bioelectrochemical technology, MFCs provide water treatment solutions that generate reusable bioenergy. The conversion of organic pollutants in wastewater by microorganisms through microbial fuel cells (MFCs) produces electrical energy, providing a novel method for wastewater management and the generation of green energy. MFC fundamentals, design components, and operational principles are investigated by examining power density in conjunction with coulombic efficiency and chemical oxygen demand (COD) removal. According to this study, the performance of MFCs depends on the electrode materials, microbial communities, and optimisation strategies for the system. Before technological development can resolve these barriers, the chapter focuses on commercial MFC scale-up limitations, including reduced power output, electrode degradation problems, and system maintenance issues. This work evaluates three emerging research directions in MFC studies: combining different MFC systems, improving microbial communities and their associated optimization approaches, and self-powered biosensing technologies. This research evaluates how MFCs can support sustainable wastewater cleanup while providing decentralized power generation capabilities for isolated locations and discusses their market potential in building a circular economy framework. Future developments in MFC technology hold promise through active materials research combined with improved system designs and microbial modifications, which may become essential for addressing global power and environmental issues.

A. S. Jatoi (✉) · M. A. Usto · A. Iqbal · A. K. Shah · A. B. Qazi · S. Ahmed
Department of Chemical Engineering, Dawood University of Engineering and Technology, Karachi, Pakistan
e-mail: abdul.sattar@duet.edu.pk

N. M. Mubarak
Department of Chemical Engineering, College of Engineering, Imam Mohammad Ibn Saud Islamic University (IMSIU), Riyadh, Saudi Arabia

Department of Biosciences, Saveetha School of Engineering, Saveetha Institute of Medical and Technical Sciences, Chennai, India

A. Sattar Jatoi, N. M. Mubarak (eds.), *Bio-Electrochemical Conversion of Waste to Fuel*, Green Energy and Technology,
https://doi.org/10.1007/978-981-95-4310-6_9

Keywords Microbial fuel cell · Wastewater treatment · Bioenergy production

1 Introduction

The worldwide surge in population growth coupled with developed urban areas and rising industry has led wastewater production to increase dramatically, thus challenging existing wastewater treatment installations because they use large amounts of energy and cost money while often failing to eliminate new pollutants, including pharmaceutical residues as well as endocrine-disrupting chemicals together with antibiotic-resistant strains of bacteria [1]. The activated sludge wastewater treatment method consumes substantial amounts of energy from non-renewable sources, resulting in carbon emissions that contribute to climate change [2]. The energy content of wastewater organic matter remains largely unrecognized despite wastewater treatment consuming additional energy for wastewater management [3]. Wastewater treatment requires innovative green technologies that serve both wastewater treatment functions and energy recovery tasks with environmental responsibility. Research has focused on microbial fuel cells (MFCs) because these devices perform dual wastewater treatment functions while generating power without external power sources [4]. Electroactive microorganisms, known as exoelectrogens, function within MFCs by biologically oxidizing organic compounds in wastewater. This results in electron transfer toward the anode, generating external electrical energy between the linked cathode and anode terminals [4]. Over a century ago, scientists discovered that microorganisms could produce electricity; however, fundamental advancements in power technology have occurred over the last two decades due to better insights into microorganism electron transfer mechanisms, electrode materials, and reactor designs [5]. Microbial fuel cells (MFCs) offer promising implementation features, including low power requirements, decentralized application options, minimal waste production, and the ability to enable wastewater treatment operations to recover energy [6]. MFCs show great potential, yet they must overcome multiple technical, economic, and scale-up challenges to achieve widespread commercial adoption [7]. This chapter aims to provide a comprehensive exploration of microbial fuel cells for wastewater treatment and bioenergy generation, beginning with the fundamental principles of MFC operation, delving into the types of microbial communities and electron transfer pathways involved, discussing various MFC configurations and performance optimization strategies, highlighting recent innovations in materials development, analyzing the economic and environmental aspects of scaling up MFC technology, and finally examining future directions and research opportunities [8]. Through the combination of microbiology principles with electrochemistry principles along with materials science and environmental engineering methods, this chapter demonstrates how MFCs will revolutionize wastewater sustainability and renewable energy creation while building the path toward a cleaner future (Fig. 1).

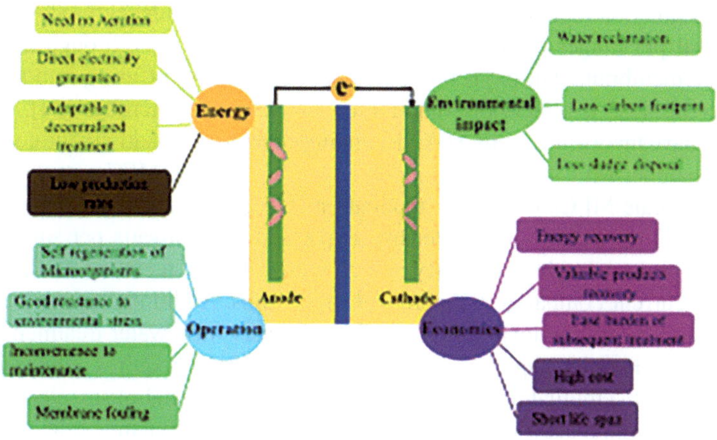

Fig. 1 Overview of microbial fuel cell and their application [1]

2 Principles of Microbial Fuel Cells

Microbial fuel cells (MFCs) utilize bioelectrochemical processes to generate electricity from microbial metabolism and purify wastewater streams through the same mechanism, thereby addressing global energy shortages and environmental degradation issues. The main function of MFCs relies on exoelectrogen microorganisms, which release electrons outside the cell to produce an electric current through external electrodes. A standard MFC design incorporates two key compartments: the anaerobic anode chamber and the aerobic cathode chamber, separated by proton exchange membrane (PEM) technology or by operating without membranes [9]. The bacterial process in the anode chamber breaks down organic wastewater substances into electrons along with protons and carbon dioxide molecules. The electric current flow is designed to harness protons, which pass through the membrane to the cathode chamber, while electrons conduct power outside, producing a usable electric current. The electrons produced at the cathode react with protons and oxygen as electron acceptors to complete the redox process, resulting in the formation of water as the reaction outcome [10]. The functioning efficiency of MFCs depends on three foundational pathways responsible for bacterial electron transfer to the anode, including conductive pili-mediated and redox-active compound-mediated transfer, as well as outer membrane-bound cytochromes-assisted contact between cells and the anode surface [11]. The three available electron transfer pathways between bacterial cells and anodes present multiple performance benefits and system design limitations. The MFC performance and stability depend heavily on four key factors: electrode materials, reactor design, microorganism selection, and operational condition management. The ideal anode material must exhibit high electrical conductivity while being biocompatible enough to support microbial attachment and the development of biofilms. It should offer a vast contact surface area for microbial exposure [12]. The cathode requires specific components to perform its reduction function rapidly, as the oxygen reduction reaction develops slowly within the neutral pH ranges of wastewater [13]. Protons must pass through the proton exchange membrane, but oxygen must be prevented from diffusing into the anode chamber because it affects the functioning of anaerobic microbes [14]. Years of research have involved various MFC design improvements to address the constraints of two-chamber MFCs through single-chamber air-cathode MFCs, stacked MFCs, and micro-scale MFCs for sensor applications [15]. The critical measurements for MFC operations include power density, expressed in milliwatts per square meter (mW/m^2), and coulombic efficiency, which represents the recovery of electrons as electrical current compared to theoretical maximums. Additionally, chemical oxygen demand (COD) removal efficiency demonstrates wastewater treatment performance [16]. Funding optimal MFC system operation requires comprehensive knowledge of the mutual relationships among microbial metabolism, electrochemical processes, and reactor design features [17]. The thermodynamic rules for microbial electron transport along with substrate degradation velocity and system resistance levels and mass transfer restrictions determine MFC efficiency as well as scalability. MFCs have gained considerable appeal for sustainable innovation because they harvest power from waste materials while carrying out wastewater processing even though their electrical output remains below traditional chemical fuel cells. Living organisms employed as electrochemical cell catalysts create a union that connects microbiology and engineering and establishes the framework for future biohybrid technologies which will tackle important environmental and energy issues in the upcoming years.

3 Microbial Communities and Electron Transfer

Microbial fuel cells operate using two essential microbial characteristics: the oxidation of organic compounds and solid-state electrode electron transfer, which form the basis of power generation [18]. Exoelectrogens and electroactive bacteria function as specialized microorganisms that utilize their biofilm-growing capability at the anode surface for extracellular electron transfer (EET) to convert the organic chemical energy in wastewater into electrical power. Bioelectrochemical starters in MFCs comprise five genera: *Geobacter*, *Shewanella*, *Pseudomonas*, *Clostridium*, and *Desulfuromonas* [18]. Two different approaches of bacterial electron transfer operate in *Geobacter sulfurreducens*, which uses conductive pili to shuttle electrons throughout the cell to the anode surface, and *Shewanella oneidensis*, which creates soluble redox mediators using flavins to transfer electrons to the anode surface. The biofilm development status and stability on the anode determine the efficiency of EET, which directly affects MFC performance [19]. The structured biofilm infrastructure enables various microbial species to work together, creating syntrophic relationships that allow electrochemically non-capable microbes to break down complex organic substances into simpler molecules, which serve as electron donors for exoelectrogenic species [20]. Biofilms arrange microbial cells into spatial patterns that establish nutritional, electron acceptor, and metabolic byproduct gradients that affect microbial metabolic activities and electron transfer processes [21].

Microbial community composition and microbial activity undergo significant changes due to various environmental factors, including pH, temperature, substrate concentration, and the presence of toxic compounds. An anaerobic atmosphere must prevail at the anode because any oxygen intrusion would disrupt the electrogenic anaerobic metabolism, resulting in decreased electron recovery. Molecular biology and various omics technologies, including metagenomics, metatranscriptomics, and proteomics, have provided extensive knowledge about MFC bacterial communities by revealing their basic population elements, as well as their metabolic pathways, operational condition adaptations, and gene expression behaviors. Electroactive bacteria exhibit extensive metabolic adaptabilities, enabling them to survive under fluctuating substrate environments and other environmental changes, thereby enhancing the operational stability of electricity-producing bioelectrochemical systems [22]. Researchers utilize naturally occurring microbial groupings and synthetic, engineered microbial communities to enhance electron generation speed and substrate flexibility while improving operational stability [23]. Two methods for improving EET capabilities involve modifying natural strains to enhance electron transfer abilities and co-growing microbial species that complement each other during degradation and electron production processes. Scientific studies now demonstrate how quorum sensing influences biofilm development and EET mechanisms, suggesting that controlling biofilm behavior could be a new method to enhance MFC function [24]. The complex relationships between operational environment components and microbial species, as well as their electron transfer activities, become crucial for optimizing MFCs, particularly when treating dynamic wastewater streams [25]. Advancements in the study of exoelectrogenic species have occurred. However, numerous questions persist regarding EET biology, microbial community dynamics, and ideal strategies to control microbial communities for enhanced energy recovery and wastewater processing effectiveness (Fig. 2). Complete exploitation of microbial fuel cell technology requires a combination of microbiology, electrochemistry, systems biology, and environmental engineering knowledge to address existing challenges [27].

4 Types and Configurations of Microbial Fuel Cells

The design specifications of microbial fuel cells (MFCs) determine their performance, efficiency, and scalability for wastewater treatment and bioenergy production, prompting researchers to study various MFC models that match different operational needs and challenges [28]. Researchers widely study the two-chamber MFC system because of its essential design, which separates the anode and cathode compartments through a proton exchange membrane (PEM). This membrane allows proton flow but blocks oxygen diffusion from the cathode toward the anode chamber. Under this setup, electroactive microorganisms receive their anaerobic

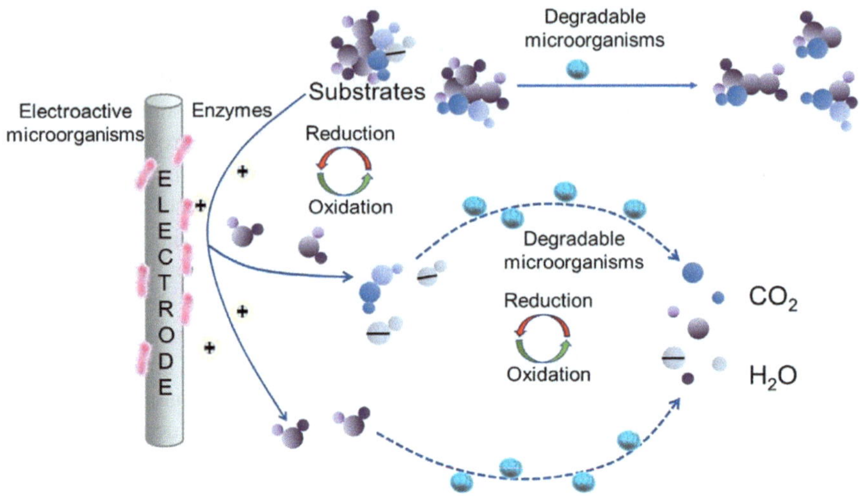

Fig. 2 Microbial communities and electron transfer mechanism [26]

environment in the anode chamber, while the cathode chamber requires oxygen for reduction reactions [29]. The high internal resistance, membrane fouling, and operational complexity restrict the practical applications of two-chamber MFCs during laboratory investigations. Single-chamber MFCs overcome these limitations by exposing the cathode to open air, thereby decreasing installation costs and reducing internal resistors [30]. The combination of basic design elements, space efficiency, and enhanced oxygen transport capabilities underpins Air-cathode MFCs as an effective system for wastewater applications, as they produce greater power outputs than typical two-chamber implementations [31]. Membrane-less MFC systems represent an alternative approach that relies on physical separation methods or electrode placement to establish suitable environmental parameters rather than relying on traditional membrane use and associated expenses. Several engineering approaches have been developed to maximize stacked MFC production capabilities through serial or parallel connections of individual cells. Stacked systems present both scalability possibilities and operational obstacles, including voltage reversal, uneven flow distribution, and system complexity problems that need proper design optimization techniques and flow distribution methods to adequately control these challenges. Scientists have employed microfabrication approaches to build microscale MFCs also known as microfluidic MFCs which provide precise management over electrode distance and flow rate and chemical gradients. The small size of these systems delivers effective self-impowered sensor technologies for environmental testing and pathogen detection and biochemical investigations but presents barriers for wastewater treatment of significant volumes. Scientists developed sediment MFCs (SMFCs) and benthic MFCs (BMFCs) by placing the anode into organic-rich sediments underneath water and using the cathode above them in oxygenated water. The natural geochemical differences located within sediments power such devices leading them to generate electrical energy that supports remote sensors and data loggers while offering sustainable power solutions for distant marine and riverine sites. Research has developed hybrid systems of MFC technology alongside wastewater treatment technologies, including constructed wetlands alongside anaerobic digestion and microbial electrolysis cells for enhancing treatment performance and energy production. Researchers are exploring the potential of combining microbial fuel cells (MFCs) with algae-based photosynthetic organisms to develop biophotovoltaic systems that integrate cathode oxygen production and biomass production (Fig. 3). The design configuration for MFC systems determines treatment capabilities and operating constraints because they relate to application requirements, wastewater qualities, operational scenarios, and economic feasibility criteria. The optimization of fundamental design variables, such as electrode component selection and chamber measurements alongside flow pattern optimization, along with substrate addition speed control and electrode distance, provides maximum power output in addition to effective wastewater treatment capabilities. Long-term system stability and performance heavily depend on the active or passive use of cathode aeration methods, external resistances, and modular or integrated architectural styles. Controlling MFC systems through optimal decisions between affordability and longevity while ensuring reliable operation and maximum power output represents an intense engineering task. Upcoming MFC advancement endeavors will focus on developing adaptable, modular systems with scalability capabilities that respond effectively to variations in wastewater content, volume, and power requirements, thereby reducing operational complexity and maintenance needs [32]. The continuous advancement of MFC designs through interdisciplinary research between microbiology, materials

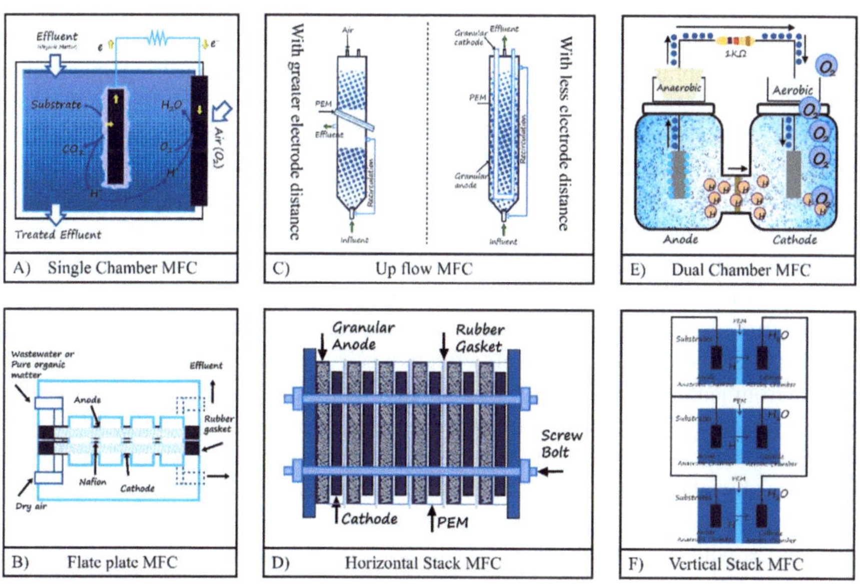

Fig. 3 Types and configuration of microbial fuel cells [4]

science, fluid dynamics, and electrochemical engineering sets the stage for maximizing the sustainable functionality of microbial fuel cells as environmental cleanup units and renewable energy generators [33].

5 Wastewater as a Substrate for Microbial Fuel Cells

The design specifications of microbial fuel cells (MFCs) determine their performance, efficiency, and scalability for wastewater treatment and bioenergy production, prompting researchers to study various MFC models that match different operational needs and challenges [34]. Researchers widely study the two-chamber MFC system due to its essential design, which separates the anode and cathode compartments through a proton exchange membrane (PEM). This membrane allows proton flow but blocks oxygen diffusion from the cathode toward the anode chamber. Under this setup, electroactive microorganisms receive their anaerobic environment in the anode chamber, while the cathode chamber requires oxygen for reduction reactions [35]. The high internal resistance, membrane fouling, and operational complexity restrict the practical applications of two-chamber MFCs during laboratory investigations. Single-chamber MFCs overcome these limitations by exposing the cathode to open air, thereby decreasing installation costs and reducing internal resistors [22]. The combination of basic design elements, space efficiency, and enhanced oxygen transport capabilities underpins Air-cathode MFCs as an effective system for wastewater applications, as they produce greater power outputs than typical two-chamber implementations [36]. Membrane-less MFC systems represent an alternative approach that relies on physical separation methods or electrode placement to establish suitable environmental parameters, rather than relying on traditional membrane use and associated expenses. Several engineering approaches have been developed to maximize stacked MFC production capabilities through the serial or parallel connection of individual cells [37]. Stacked systems present both scalability possibilities and operational obstacles, including voltage reversal, uneven flow distribution, and system complexity problems that require proper design optimization techniques and flow distribution methods to control these challenges effectively. Scientists have employed microfabrication approaches to build micro-scale microbial fuel cells (MFCs), also known as microfluidic MFCs, which provide precise control over electrode distance, flow rate, and chemical gradients [38]. The small size of these systems delivers effective self-empowered sensor technologies for environmental testing, pathogen detection, and biochemical investigations but presents barriers to wastewater treatment of significant volumes. Scientists developed sediment microbial fuel cells (SMFCs) and benthic microbial fuel cells (BMFCs) by placing the anode in organic-rich sediments beneath the water and using the cathode above them in oxygenated water [39]. The natural geochemical differences within sediments power such devices, leading them to generate electrical energy that supports remote sensors and data loggers while offering sustainable power solutions for distant marine and riverine sites. Research has developed hybrid systems of MFC technology alongside wastewater treatment technologies including constructed wetlands alongside anaerobic digestion and microbial electrolysis cells for enhancing both treatment performance and energy production [40]. Researchers are exploring the potential of combining microbial fuel cells (MFCs) with algae-based photosynthetic organisms to develop biophotovoltaic systems that integrate cathode oxygen production and biomass production [41]. The design configuration for MFC systems determines treatment capabilities and operating constraints in relation to application requirements, wastewater qualities, operational scenarios, and economic feasibility criteria [42]. The optimization of fundamental design variables, such as electrode component selection and chamber measurements alongside flow pattern optimization, along with substrate addition speed control and electrode distance, provides maximum power output in addition to effective wastewater treatment capabilities. Long-term system stability and performance heavily depend on the active or passive use of cathode aeration methods, external resistances, and modular or integrated architectural styles (Table 1). Controlling MFC systems

Table 1 Overview of different wastewater and their key findings

Year	Authors (first author)	Title	Wastewater type	Key findings	References
2024	Attia et al.	Enhancing bioelectricity generation from wastewater in microbial fuel cells using carbon nanomaterials	Domestic wastewater	The incorporation of carbon nanomaterials (CNTs, g-C_3N_4, r-GO) into the electrolyte significantly increased the voltage, power density, and coulombic efficiency.	[43]
2023	Barakat et al.	Synergistic advancements in sewage-driven microbial fuel cells: Novel carbon nanotube cathodes and biomass-derived anodes for efficient renewable energy generation and wastewater treatment	Sewage wastewater	The use of biomass-derived anodes (corncob and mango seeds) and carbon nanotube cathodes achieved power densities of up to 2171 mW/m^2 and COD removal rates of up to 85%	[44]
2024	Arliyani et al.	Constructed wetlands combined with microbial fuel cells (CW-MFCs) as a sustainable technology for leachate treatment and power generation	Leachate	CW-MFC systems offer eco-friendly leachate treatment with energy recovery, utilizing natural materials and plant-microbe interactions.	[45]

(continued)

Table 1 (continued)

Year	Authors (first author)	Title	Wastewater type	Key findings	References
2024	Li	Energy recovery applications of microbial fuel cells in wastewater treatment	Industrial wastewater	Review of MFC principles, mechanisms, and applications in wastewater treatment, highlighting energy recovery potential and comparisons with traditional methods	[46]
2024	Tiwari et al.	A critical review on the advancement of the development of low-cost membranes to be utilised in microbial fuel cells	Various wastewater types	Evaluation of earthen membranes (e.g., clayware, ceramics) in MFCs, noting ceramic membranes' superior power density and COD removal performance	[47]
2023	Esfandyari et al.	Microbial fuel cells for energy production in wastewater treatment plants—a review	Various wastewater types	Examination of MFC principles and applications in wastewater treatment, discussing technological and economic challenges for scaling up	[48]
2022	Paucar and Sato	An overview of microbial fuel cells within constructed wetland for simultaneous nutrient removal and power generation	Nutrient-rich wastewater	Discuss combining constructed wetlands with MFCs (CW-MFC) for nutrient removal and energy recovery, emphasizing low-cost and sustainable treatment	[49]

through optimal decisions between affordability and longevity with reliable operation and maximum power output represents an intense engineering task [50]. Upcoming MFC advancement endeavors will concentrate on producing adaptable modular systems with scalability capabilities which respond effectively to different wastewater content and volume variations and power needs as well as reduce operational complexity and maintenance requirements [51]. The continuous advancement of MFC designs through interdisciplinary research between microbiology and materials science and fluid dynamics and electrochemical engineering sets the stage for maximizing the sustainable functionality of microbial fuel cells as environmental cleanup units and renewable energy generators [22].

6 Performance Parameters and Optimization Strategies

The performance evaluation of microbial fuel cells (MFCs) depends on various critical parameters. These metrics determine their efficiency level, operational stability, and practical suitability in wastewater processes and bioenergy procedures [39]. Effective optimization of these vital parameters enables MFC technology to achieve its highest potential. Power density, along with current density, coulombic efficiency, chemical oxygen demand (COD) removal efficiency, internal resistance, and energy recovery efficiency, constitute essential performance criteria for microbial fuel cells (MFCs) [40, 52]. MFC power generation capacity measurement depends on power density, which represents the power output per unit of electrode surface area, expressed as milliwatts per square meter or milliwatts per cubic meter of reactor volume. The electrical current output per unit area, which determines current density, depends on microbial activity, the properties of the electrode material, and the external circuit resistance. Coulombic efficiency (CE) determines the electron recovery rate through electrical currents compared to theoretical maximum values from substrate degradation, while evaluating electron transfer efficiency and microbial activities that affect other fermentation processes. The efficiency of COD removal in wastewater treatment systems depends on their ability to reduce organic pollutants yet does not guarantee either high energy recovery or maximum substrate utilization because alternative microbial transformations may predominate. The three components of internal resistance (ohmic, activation, and concentration) influence the output voltage and system efficiency; therefore, a superior reactor design, electrode selection, and operational strategies reduce resistance and enhance output power (Fig. 4). Evaluating energy recovery efficiency requires determining how much energy is generated by the system against the total energy expended on operating system components, as this assessment determines the practical implementation potential of MFCs for real-life applications. Improving MFC performance through electrode material innovations is the initial step in optimization strategies.

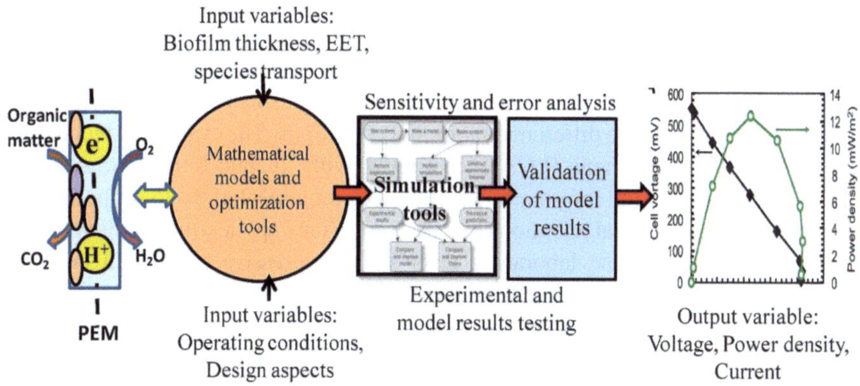

Fig. 4 Performance optimization through machine learning tools [53]

The desired attributes for anodes and cathodes include a high surface area and biocompatibility while also requiring electrical conductivity and chemical stability characteristics to achieve efficient microbial electron transfer and adhesion. Carbon-based materials, such as graphite felt, carbon cloth, carbon nanotubes, and graphene composites, have received extensive research attention for electrode modifications through the addition of catalysts, conductive polymers, or nanostructures. Cathode performance is enhanced through the use of catalysts for oxygen reduction reactions (ORR). Yet, platinum yields the most effective results, but at a high cost that drives scientists toward alternative catalysts, including manganese oxides, iron-based compounds, and bio-cathodes made from photosynthetic organisms or bacterial consortia. Reducing electrode spacing, utilizing flow-through designs, and employing single-chamber air-cathode constructions enable researchers to minimize electrical resistance and enhance material transport within the system. The optimal performance of microbial activity and electrochemical conditions depends on the proper calibration of operational parameters, such as external resistance, temperature, pH value, and substrate concentration level. This can be achieved by conducting microbial operations at near-neutral pH values, combined with mesophilic temperatures in the range of 30–40 °C [54]. The effectiveness of substrate breakdown in electro-bioreactor systems depends on maintaining a proper balance between hydraulic retention time (HRT) and organic loading rate (OLR) to allow an adequate degradation period while preventing system damage due to substrate excess [54]. These critical management strategies in microbial community optimization include pre-exposing microbes to specific wastewater before use while selectively growing electroactive species and culturing genetically modified strains with enhanced electron transfer efficiency [55]. Biofilm management approaches involving systematic electrode motion and periodic removal, or specific shear force control, regulate the biofilm thickness to promote optimal electron transfer rates while preventing mass transfer restrictions [56]. Research has begun using external magnetic fields, ultrasound, and light irradiation to enhance microbial activity and electron transfer rates. MFC systems achieve advanced wastewater treatment performance when they operate in conjunction with complementary systems, such as anaerobic digestion, forward osmosis, and microbial electrolysis cells, as the combination enhances total energy recovery efficiency [57]. Process modeling incorporating computational fluid dynamics (CFD) with multi-scale bioelectrochemical models enables MFC system development and scaling up because it predicts system responses to different operational conditions. The evaluation techniques of Life Cycle Assessment (LCA), combined with Techno-Economic Analysis (TEA), provide essential information about MFC implementations to support scientific growth and industrial adoption (Fig. 4). Real-world application requires further substantial work because laboratory tests show impressive results. In contrast, despite recent advancements, the technology gap between laboratory and real-world performance is a critical hurdle. Systematic optimization of materials, microbiology, engineering design, and operational strategy transformation will enable MFCs to become a competitive and sustainable technology for wastewater treatment and renewable energy generation [58].

7 Challenges in Scaling Up and Commercialization

The commercial adoption of microbial fuel cells (MFCs) for sustainable wastewater treatment and bioenergy generation needs solutions to multiple substantial implementation barriers before earning acceptance as viable alternatives to established treatment methods [59]. The primary obstacle is transforming laboratory-scale MFC systems into large-scale, profitable, and real-world implementations [60]. Laboratory tests yield outstanding MFC results; however, these achievements prove challenging to reproduce at larger scales within industrial treatment systems, which require effective mass transfer management alongside stable electrode function and microbial communities [61]. During upscaling, the power density and coulombic efficiency diminish. At the same time, internal resistance increases due to poor microbial biofilm contact, inadequate electrode surface interaction, and the inability to sustain environments conducive to microbial proliferation and electron transport. When reactors expand in size, the proton and electron flow pathways, as well as substrate transport, lose their uniformity, resulting in a decrease in overall effectiveness. The rise in system dimensions creates increased difficulties in controlling electrode fouling and biofilm management, resulting in reduced system stability and increased maintenance complexity [61]. Implementing MFCs at larger scales presents challenges because wastewater exhibits variable compositions in terms of temperature and pH, contaminant concentrations, and temperature variation, making it difficult to sustain stable microbial operations. Laboratory research utilizes defined synthetic wastewater for testing purposes, although real-world industrial wastewater exhibits a more complex and varied composition [62]. The negative effects of toxic compounds, including heavy metals, pharmaceuticals, and detergents, impede microbial growth and decrease power output, thus necessitating either powerful microbial consortia or pre-treatment solutions to minimize these destructive effects. University-based research has explored genetically modified microbial communities as enhancements to microbial fuel cell (MFC) technology; however, their application in field settings raises questions about stability and ecological concerns [63]. Production expenses for MFCs restrict their commercial adoption because they exceed the current standards of activated sludge or anaerobic digesters in traditional wastewater solutions [64]. Low-cost, high-performance materials, such as carbon-based composites, metal-organic frameworks, and biological cathodes, have reduced the cost of cathode electrode materials despite platinum being a significant cost factor in MFC production. MFC power generation faces challenges because its output capacity remains moderate when treating actual wastewater solutions [65]. An individual MFC device produces insufficient power for industrial purposes because it generates less electricity than conventional fuel cells and photovoltaic devices and, therefore, cannot fulfill large-scale energy requirements. For MFCs to maintain long-term sustainability, the electrodes and proton exchange, membranes require durable components, as fouling, degradation, and performance loss can occur during extended long-term operation [66]. University-based research on MFC technology remains attractive; however, commercial markets tend to avoid it due to the maintenance needs and inevitable decline in system efficiency throughout their operational lifespan. The integration process of MFCs into power distribution networks proves difficult because it requires additions to systems which already have well-established waste management regulations in place. Upgrading treatment plants with MFC systems or placing standalone MFCs in operation requires substantial financial capital, including new hardware, workforce education, and system operation costs, which may prove too expensive without generous economic support. MFC-based technology implementation faces regulatory hurdles because current standards to evaluate their performance as well as their efficiency and environmental footprint remain incomplete. Subscription to undefined standards for MFC deployment and application hampers their commercialization and implementation within industrial facilities. Despite technical hurdles, organizations devise various strategies to make MFC commercialization possible. An efficient approach for MFC commercialization requires system integration between MFCs and other wastewater treatment components, such as anaerobic digestion or constructed wetlands, to achieve improved performance rates and enhanced power generation capabilities [67]. The combined operation enables MFCs to produce more energetic output and achieve higher efficiency by incorporating additional treatment methods [68]. Research interests are focused on the applications of MFCs in remote sensing, decentralized wastewater treatment systems, and off-grid power generation for areas lacking conventional power grids in developing or rural regions [68]. MFC systems can be specifically designed for localized requirements, such as wastewater treatment in remote areas and autonomous sensor systems for environmental assessments, creating new adoption possibilities. Future advancements in research on optimizing microbial consortia and electrodes, combined with reactor optimization, will lead to improved MFC efficiency and scalability, making them more viable for commercial adoption. MFC commercialization will benefit from government incentives and green energy policies alongside carbon credits and sustainable energy solution recognition which will decrease initial costs and promote a favorable market environment [69]. The broader adoption of MFC technologies in both energy and environmental sectors can be supported by emerging circular economy principles which enable the reuse of waste streams for power generation. To successfully scale-up MFC operations microbiologists will need to unite forces with engineers and economists together with policymakers in facing technical and economic and regulatory barriers. Continuous research innovations combined with well-planned investments may make microbial fuel cells an advanced technology which simultaneously addresses worldwide environmental problems and advances renewable energies and sustainable water management [70, 71].

8 Future Directions and Emerging Trends

The development of microbial fuel cells (MFCs) is progressing rapidly due to advancements in all sectors, including materials science, microbial ecology, system integration, and electrochemical engineering [72]. Research developments in MFC technology will enhance their functional capability and improve their practical usefulness for wastewater cleanup and bioenergy manufacturing operations [73]. The discovery and application of new electrode material choices and device configurations hold great potential to enhance MFC operational efficiency and product longevity while reducing the costs associated with MFC production. The increasing interest in carbon-based substances, including graphene, carbon nanotubes, and biochar, stems from their ability to provide a large surface area, electric conductivity, and biocompatible features that enhance electron delivery between microorganisms and electrodes [74]. The development of 3D-printed and flexible electrodes represents a significant advancement in MFC performance, as they enable better microbial connection while improving critical mass transport functions required for large-scale applications. The power output of MFCs is expected to increase substantially through the implementation of metal-organic frameworks (MOFs) with conducting polymers and composite materials used as electrodes, as these materials simultaneously reduce costs and enhance system stability over time. Scientists work to enhance microbial communities while maintaining their focus on this objective [75]. The development of synthetic biology and microbiome engineering techniques enables scientists to create specialized microbial consortia that exhibit greater effectiveness in substrate breakdown and electron transfer. Scientists examine the genetic modifications of electroactive bacteria, including Geobacter sulfurreducens and Shewanella oneidensis, because these alterations enhance their ability to transfer electrons and survive in the presence of toxic pollutants in wastewater [76]. Investigating mixed microbial cultures over monocultures aims to develop stronger and more diversified biofilms that efficiently degrade complex wastewater pollutants with stable power output levels [77]. Planning MFC deployment with wastewater treatment systems, such as anaerobic digestion, constructed wetlands, and membrane bioreactors, presents an exciting method to enhance treatment results and resource recovery. The combination of MFCs and anaerobic digestion offers significant benefits through the COD removal strength of anaerobic processing, enabling the creation of power from previously disposed organic waste material [78]. By implementing this combined method, energy recovery increases and integrates improved sustainability features into water treatment systems. Current research integrates microbial electrochemical cells (MECs) with microbial fuel cells (MFCs) to produce hydrogen as a renewable fuel. MECs use an additional external voltage to split water through their MFC systems, thus producing hydrogen gas at the cathode. Linking MFC technology with hydrogen production capability demonstrates significant potential to create electrochemical systems that produce both hydroelectricity and hydrogen simultaneously, thereby enhancing MFC applications in bioenergy systems. Research into MFCs now focuses on building automated bioelectronic sensors along with monitoring systems that operate without power supply. MFCs show perfect suitability for off-grid applications because they generate power by transforming organic waste into energy especially in resource-deprived and remote locations. Self-powered sensors have gained popularity because they monitor environmental parameters by using MFC technology to measure water quality and nutrient concentrations as well as the detection of particular pollutants [79]. These biosensors are deployed in unserved energy areas to conduct real-time environmental assessments that help monitor pollution control efforts alongside public health surveillance programs. MFC technology combines IoT resources to achieve smart system functionalities that track and enhance wastewater process performance throughout real-time operations [80]. The improved MFC technology for environmental monitoring will constitute a major tool for realizing sustainable water management and pollution control in the coming years. The ability of MFCs to operate as distributed power generation systems is currently being studied with great interest [81]. Widespread research focuses on MFC scaling for powering various electronic devices and monitoring systems, as well as utilizing MFCs to meet the electricity needs of buildings located in areas without access to regular power utilities [81]. Rural and underdeveloped areas that lack reliable energy access would experience a revolutionary change through decentralized organic waste energy harvesting capabilities, such as those derived from wastewater or other waste streams [82]. MFC system design utilizing modular and flexible technologies will facilitate the widespread adoption of MFC technology as an off-grid, sustainable energy solution, as these systems can easily scale to meet specific energy needs. The current focus on circular economy frameworks benefits from advancements in MFC technology, as these systems help remove wastewater and extract usable resources, including biofuels and hydrogen, as well as nutrient-rich agricultural waste materials. MFCs are poised to establish their position within circular economy models since the planet demands enhanced sustainable energy and resource recovery solutions involving waste streams that serve as resources for energy production and both resource extraction and environmental cleanup systems. MFC commercialization encounters multiple barriers which primarily involve minimizing system costs and achieving extended lifetime and seamless hookup to current utility networks. Advanced research in electrode materials together with knowledge of microbial ecosystems and system modifications and hybrid technology development creates auspicious conditions for MFCs to change how we approach wastewater management and sustainable energy and environmental practices. Next-generation MFCs will emerge due to interdisciplinary research on microbiology and engineering and material science and environmental science will lead to advanced MFC solutions for global wastewater treatment and energy generation and pollution management.

9 Conclusion

The sustainable technology harnesses organic pollutants in wastewater to generate electricity, resulting in substantial reductions in pollutant levels. The use of MFCs for waste conversion supports the growing need for sustainable technologies that create circular economy systems. Significant improvements in microbial ecology, electrode materials, and system optimization have not eliminated the obstacles to large-scale MFC implementation in real-world applications. MFC commercial deployment requires solutions for addressable problems, including electrode fouling, fluctuating wastewater chemistry, low power output density, and high operational costs. Significant progress can be made in overcoming MFC limitations by merging MFCs with wastewater treatment technologies, utilizing improved hybrid systems, and developing self-powered biosensors. MFCs show great promise as decentralized off-grid power systems, especially for areas with limited resources, making them a target for future research applications. Research success depends on continued collaboration between microbiologists, engineers, and policymakers, who will solve current barriers and enhance MFC performance. MFCs will assume an increasingly important role in sustainable wastewater management and renewable bioenergy generation as global attention intensifies on environmental sustainability and resource recovery.

References

1. Guo Y, Wang J, Shinde S, Wang X, Li Y, Dai Y et al (2020) Simultaneous wastewater treatment and energy harvesting in microbial fuel cells: an update on the biocatalysts. RSC Adv 10:25874–25887
2. Gul H, Raza W, Lee J, Azam M, Ashraf M, Kim K-H (2021) Progress in microbial fuel cell technology for wastewater treatment and energy harvesting. Chemosphere 281:130828
3. Jalili P, Ala A, Nazari P, Jalili B, Ganji DD (2024) A comprehensive review of microbial fuel cells considering materials, methods, structures, and microorganisms. Heliyon 10:e25439
4. Nawaz A, Haq IU, Qaisar K, Gunes B, Raja SI, Mohyuddin K et al (2022) Microbial fuel cells: insight into simultaneous wastewater treatment and bioelectricity generation. Process Saf Environ Prot 161:357–373
5. Elhenawy S, Khraisheh M, AlMomani F, Al-Ghouti M, Hassan MK (2022) From waste to watts: updates on key applications of microbial fuel cells in wastewater treatment and energy production. Sustainability 14:955
6. Zhang Y, Liu M, Zhou M, Yang H, Liang L, Gu T (2019) Microbial fuel cell hybrid systems for wastewater treatment and bioenergy production: synergistic effects, mechanisms and challenges. Renew Sust Energ Rev 103:13–29
7. Kamali M, Guo Y, Aminabhavi TM, Abbassi R, Dewil R, Appels L (2023) Pathway towards the commercialization of sustainable microbial fuel cell-based wastewater treatment technologies. Renew Sust Energ Rev 173:113095
8. Hamedani EA, Abasalt A, Talebi S (2024) Application of microbial fuel cells in wastewater treatment and green energy production: a comprehensive review of technology fundamentals and challenges. Fuel 370:131855
9. Yaqoob AA, Ibrahim MNM, Umar K, Parveen T, Ahmad A, Lokhat D et al (2021) A glimpse into the microbial fuel cells for wastewater treatment with energy generation. Desalin Water Treat 214:379–389
10. Naha A, Debroy R, Sharma D, Shah MP, Nath S (2023) Microbial fuel cell: a state-of-the-art and revolutionizing technology for efficient energy recovery. Clean Circ Bioecon 5:100050
11. Roy H, Rahman TU, Tasnim N, Arju J, Rafid MM, Islam MR et al (2023) Microbial fuel cell construction features and application for sustainable wastewater treatment. Membranes 13:490
12. Khandaker S, Das S, Hossain MT, Islam A, Miah MR, Awual MR (2021) Sustainable approach for wastewater treatment using microbial fuel cells and green energy generation–a comprehensive review. J Mol Liq 344:117795
13. Apollon W, Rusyn I, Kuleshova T, Luna-Maldonado AI, Pierre JF, Gwenzi W et al (2024) An overview of agro-industrial wastewater treatment using microbial fuel cells: recent advancements. J Water Process Eng 58:104783

14. Zhang Q, Hu J, Lee D-J (2016) Microbial fuel cells as pollutant treatment units: research updates. Bioresour Technol 217:121–128
15. Paucar NE, Sato C (2021) Microbial fuel cell for energy production, nutrient removal and recovery from wastewater: a review. PRO 9:1318
16. Ahmed SF, Mofijur M, Islam N, Parisa TA, Rafa N, Bokhari A et al (2022) Insights into the development of microbial fuel cells for generating biohydrogen, bioelectricity, and treating wastewater. Energy 254:124163
17. Selvasembian R, Mal J, Rani R, Sinha R, Agrahari R, Joshua I et al (2022) Recent progress in microbial fuel cells for industrial effluent treatment and energy generation: fundamentals to scale-up application and challenges. Bioresour Technol 346:126462
18. Stams AJ, De Bok FA, Plugge CM, Van Eekert MH, Dolfing J, Schraa G (2006) Exocellular electron transfer in anaerobic microbial communities. Environ Microbiol 8:371–382
19. Uria N, Ferrera I, Mas J (2017) Electrochemical performance and microbial community profiles in microbial fuel cells in relation to electron transfer mechanisms. BMC Microbiol 17:1–12
20. Nagarajan H, Embree M, Rotaru A-E, Shrestha PM, Feist AM, Palsson BØ et al (2013) Characterization and modelling of interspecies electron transfer mechanisms and microbial community dynamics of a syntrophic association. Nat Commun 4:2809
21. Babauta JT, Atci E, Ha PT, Lindemann SR, Ewing T, Call DR et al (2014) Localized electron transfer rates and microelectrode-based enrichment of microbial communities within a phototrophic microbial mat. Front Microbiol 5:11
22. Rahmani AR, Navidjouy N, Rahimnejad M, Alizadeh S, Samarghandi MR, Nematollahi D (2022) Effect of different concentrations of substrate in microbial fuel cells toward bioenergy recovery and simultaneous wastewater treatment. Environ Technol 43:1–9
23. Rebequi T, Pio Y, Penteado CFA, da Silva Martins LH, Diniz AAR, Komesu A et al (2023) Waste treatment and sustainable bioelectricity generation using microbial fuel cell. J Sci Sustain Eng 1:15460
24. Sani AD, Savla N, Pandit S, Mathuriya AS, Gupta PK, Khanna N et al (2021) Recent advances in bioelectricity generation through the simultaneous valorization of lignocellulosic biomass and wastewater treatment in microbial fuel cell. Sustain Energy Technol Assess 48:101572
25. Jatoi AS, Mallah NB, Ahmed J, Shah SA, Sabzoi N, Ibrahim SM (2025) Enhancing the performance of microbial fuel cell via use of new type of electrode for the bioenergy generation along with the treatment of distillery effluent. Biomass Convers Biorefinery 15:3151–3164
26. Zhou Q, Li R, Li T, Zhou R, Hou Z, Zhang X (2023) Interactions among microorganisms functionally active for electron transfer and pollutant degradation in natural environments. Eco-Environ Health 2:3–15
27. Khamis A, Zulfakar NAM (2024) Bioelectricity production from food waste leachate using double chamber microbial fuel cell: effect of electrolyte pH level, electrodes sizing, and positioning. Int J Environ Waste Manag 33:465–478
28. Cahyono YA, Madurani T, Azzahra WF, Lestari RA (2019) Bioconversion on wastewater of soybeans using microbial fuel cell. Adv Sustain Sci Eng Technol 1:0190106
29. Moqsud MA, Omine K, Yasufuku N, Hyodo M, Nakata Y (2013) Microbial fuel cell (MFC) for bioelectricity generation from organic wastes. Waste Manag 33:2465–2469
30. Singh V, Saxena A, Gupta A, Singh S, Kaul V, Kumar N (2017) Microbial fuel cell-a source of renewable energy: a review. Int J Renew Energy Technol 8:104–118
31. Hisham NSN, Zain SM, Jusoh S, Anuar N, Suja F, Ismail A et al (2013) Microbial fuel cells using different types of wastewater for electricity generation and simultaneously removed pollutant. J Eng Sci Technol 8:316–325
32. Choudhury P, Ray RN, Bandyopadhyay TK, Tiwari ON, Bhunia B (2021) Kinetics and performance evaluation of microbial fuel cell supplied with dairy wastewater with simultaneous power generation. Int J Hydrog Energy 46:16815–16822
33. Chonde Sonal G, Mishra A, Raut P (2013) Bioelectricity production from wastewater using microbial fuel cell (MFC). Int J Adv Res Eng Technol 4:62–69

34. Gude VG (2016) Wastewater treatment in microbial fuel cells–an overview. J Clean Prod 122:287–307
35. Aghababaie M, Farhadian M, Jeihanipour A, Biria D (2015) Effective factors on the performance of microbial fuel cells in wastewater treatment–a review. Environ Technol Rev 4:71–89
36. Sonawane JM, Mahadevan R, Pandey A, Greener J (2022) Recent progress in microbial fuel cells using substrates from diverse sources. Heliyon 8:e12353
37. Pant D, Van Bogaert G, Diels L, Vanbroekhoven K (2010) A review of the substrates used in microbial fuel cells (MFCs) for sustainable energy production. Bioresour Technol 101:1533–1543
38. Min B, Kim J, Oh S, Regan JM, Logan BE (2005) Electricity generation from swine wastewater using microbial fuel cells. Water Res 39:4961–4968
39. Mohyudin S, Farooq R, Jubeen F, Rasheed T, Fatima M, Sher F (2022) Microbial fuel cells a state-of-the-art technology for wastewater treatment and bioelectricity generation. Environ Res 204:112387
40. Gezginci M, Uysal Y (2016) The effect of different substrate sources used in microbial fuel cells on microbial community. JSM Environ Sci Ecol 4:1035
41. Min B, Logan BE (2004) Continuous electricity generation from domestic wastewater and organic substrates in a flat plate microbial fuel cell. Environ Sci Technol 38:5809–5814
42. Meylani V, Surahman E, Fudholi A, Almalki WH, Ilyas N, Sayyed R (2023) Biodiversity in microbial fuel cells: review of a promising technology for wastewater treatment. J Environ Chem Eng 11:109503
43. Attia YA, Samer M, Mohamed MS, Salah M, Moustafa E, Hameed RA et al (2024) Enhancing bioelectricity generation from wastewater in microbial fuel cells using carbon nanomaterials. J Chem Technol Biotechnol 99:1172–1180
44. Barakat NA, Gamal S, Kim HY, Abd El-Salam NM, Fouad H, Fadali OA et al (2023) Synergistic advancements in sewage-driven microbial fuel cells: novel carbon nanotube cathodes and biomass-derived anodes for efficient renewable energy generation and wastewater treatment. Front Chem 11:1286572
45. Arliyani I, Noori MT, Ammarullah MI, Tangahu BV, Mangkoedihardjo S, Min B (2024) Constructed wetlands combined with microbial fuel cells (CW-MFCs) as a sustainable technology for leachate treatment and power generation. RSC Adv 14:32073–32100
46. Li M (2024) Energy recovery applications of microbial fuel cells in wastewater treatment. J Energy Biosci 15:186–196
47. Tiwari A, Yadav N, Jadhav DA, Saxena D, Anghan K, Sandhwar VK et al (2024) A critical review on the advancement of the development of low-cost membranes to be utilized in microbial fuel cells. Water 16:1597
48. Esfandyari M, Jafari D, Azami H (2024) Microbial fuel cells for energy production in wastewater treatment plants-a review. Biofuels 15:743–753
49. Paucar NE, Sato C (2022) An overview of microbial fuel cells within constructed wetland for simultaneous nutrient removal and power generation. Energies 15:6841
50. Singh HM, Pathak AK, Chopra K, Tyagi V, Anand S, Kothari R (2019) Microbial fuel cells: a sustainable solution for bioelectricity generation and wastewater treatment. Biofuels 10:11–31
51. Pandey P, Shinde VN, Deopurkar RL, Kale SP, Patil SA, Pant D (2016) Recent advances in the use of different substrates in microbial fuel cells toward wastewater treatment and simultaneous energy recovery. Appl Energy 168:706–723
52. Chaudhary TN, Ahmed S, Usman M, Maka AO, Rasool S, Ghaleeh M et al (2025) Utilizing different food cultures for wastewater treatment and enhanced power generation in microbial fuel cells. Biomass Bioenergy 195:107706
53. Jadhav DA, Carmona-Martínez AA, Chendake AD, Pandit S, Pant D (2021) Modeling and optimization strategies towards performance enhancement of microbial fuel cells. Bioresour Technol 320:124256
54. Lovley DR (2017) Happy together: microbial communities that hook up to swap electrons. ISME J 11:327–336

55. Akram J, Song C, El Mashad HM, Chen C, Zhang R, Liu G (2024) Advances in microbial community, mechanisms and stimulation effects of direct interspecies electron transfer in anaerobic digestion. Biotechnol Adv 76:108398
56. Zhi W, Ge Z, He Z, Zhang H (2014) Methods for understanding microbial community structures and functions in microbial fuel cells: a review. Bioresour Technol 171:461–468
57. Kang H-J, Lee S-H, Lim T-G, Park H-D (2019) Effect of inoculum concentration on methanogenesis by direct interspecies electron transfer: performance and microbial community composition. Bioresour Technol 291:121881
58. Ma W-J, Zhang H-M, Tian Y (2024) Rapid start-up sulfur-driven autotrophic denitrification granular process: extracellular electron transfer pathways and microbial community evolution. Bioresour Technol 395:130331
59. Koroglu EO, Yoruklu HC, Demir A, Ozkaya B (2019) Scale-up and commercialization issues of the MFCs: challenges and implications. In: Microbial electrochemical technology. Elsevier, Amsterdam, pp 565–583
60. Aldieri L, Vinci CP (2021) Scalability and commercialization in support of sustainable development goals. Ind Innov Infrastruct 21:979–988
61. Sagaram US, Chugh N, Gautam K, Dixit R, Menon D, Rajvanshi M et al (2021) Challenges in scale-up and commercialization of microalgae products. Elsevier, Amsterdam
62. Dodson BP, Levine AD (2015) Challenges in the translation and commercialization of cell therapies. BMC Biotechnol 15:1–15
63. Breslauer DN (2025) Current progress on scale-up and commercialization of microbially-produced silk. Adv Funct Mater 35:2408386
64. Jeevahan JJ, Chandrasekaran M, Venkatesan S, Sriram V, Joseph GB, Mageshwaran G et al (2020) Scaling up difficulties and commercial aspects of edible films for food packaging: a review. Trends Food Sci Technol 100:210–222
65. Gujjala LKS (2024) Techno-economic feasibility, scaling up, commercialization and hurdles of waste management strategies. NIT, Rourkela
66. Govindaraju C (2010) R&D commercialization challenges for developing countries: the case of Malaysia
67. Anuradha M, Balasubramanya S, Subbalakshmi G, Shilpa P (2025) Commercialization of in vitro secondary metabolite production: challenges and opportunities. In: In vitro production of plant secondary metabolites: theory and practice. IntechOpen, London, pp 321–346
68. da Silva EF (2021) Science and technology innovation commercialization: success factors and challenges for entrepreneurship university. Universidade do Porto, Porto
69. Richard TL (2010) Challenges in scaling up biofuels infrastructure. Science 329:793–796
70. Yan J, Savenije TJ, Mazzarella L, Isabella O (2022) Progress and challenges on scaling up of perovskite solar cell technology. Sustain Energy Fuels 6:243–266
71. Bhati M, Rai R (2018) Commercialization of large-scale perovskite solar energy technology and scaling-up issues. In: Perovskite photovoltaics. Elsevier, Amsterdam, pp 387–445
72. Gill SS, Xu M, Ottaviani C, Patros P, Bahsoon R, Shaghaghi A et al (2022) AI for next generation computing: emerging trends and future directions. Internet Things 19:100514
73. Koc E, Pantazis E, Soibelman L, Gerber DJ (2020) Emerging trends and research directions. Construction 4:460–476
74. Vansteenkiste M, Niemiec CP, Soenens B (2010) The development of the five mini-theories of self-determination theory: an historical overview, emerging trends, and future directions. Decade Ahead 16:105–165
75. Alivisatos P, Barbara PF, Castleman AW, Chang J, Dixon DA, Klein ML et al (1998) From molecules to materials: current trends and future directions. Adv Mater 10:1297–1336
76. Coates H, McCormick AC (2014) Emerging trends and perspectives. In: Engaging university students: international insights from system-wide studies. Springer, Cham, pp 151–158
77. Mahto KU, Das S (2025) Electroactive biofilm communities in microbial fuel cells for the synergistic treatment of wastewater and bioelectricity generation. Crit Rev Biotechnol 45:434–453

78. Sharma M, Salama E-S, Zhang P, Zhang L, Xing X, Yue J et al (2022) Microalgae-assisted microbial fuel cells for electricity generation coupled with wastewater treatment: biotechnological perspective. J Water Process Eng 49:102966
79. Tong KTX, Tan IS, Foo HCY, Show PL, Lam MK, Wong MK (2023) Sustainable circular biorefinery approach for novel building blocks and bioenergy production from algae using microbial fuel cell. Bioengineered 14:246–289
80. Jatoi AS, Akhter F, Mazari SA, Sabzoi N, Aziz S, Soomro SA et al (2021) Advanced microbial fuel cell for waste water treatment—a review. Environ Sci Pollut Res 28:5005–5019
81. Jatoi AS, Unar IN, Aziz S, Soomro SA, Abro M, Khoso JA et al (2018) Microbial fuel cell for bio-electricity generation via bio-electrochemical conversion of fermentation industry sludge. Abasyn J Life Sci 2:42–53
82. Jatoi AS, Baloch A, Jadhav A, Nizamuddin S, Aziz S, Soomro SA et al (2020) Improving fermentation industry sludge treatment as well as energy production with constructed dual chamber microbial fuel cell. SN Appl Sci 2:1–8

Conversion of CO_2 into Methanol Using BES

Nibedita Nath, Subhendu Chakroborty, and Sangeeta Jangid

Abstract Among the most promising strategies for reducing carbon dioxide (CO_2) buildup in the environment and addressing the need for sustainable energy is converting CO_2 into fuels and value-added products. To reduce CO_2 emissions, renewable carbon could serve as a substitute for fossil fuels like coal and petroleum. The newest technology and processes must be implemented to develop innovative carbon recycling techniques for CO_2. One possible solution for CO_2 sequestration is the electrochemical transformation of the stable molecule into useful fuel or products. In addition to homogeneous and heterogeneous catalysts, photocatalysis, electrochemical, thermochemical, and bioelectrochemical reduction methods can be used to create value-added products. Among these methods, bioelectrochemical systems (BESs) play a crucial role in addressing the energy issue and mitigating global warming. Potential applications for BESs include waste biorefineries, pollution treatment, CO_2 capture, and the electrochemical synthesis of eco-friendly biofuels or byproducts. This chapter covers the use of BESs in producing methanol from CO_2.

Keywords Carbon dioxide · Biofuel · Electrochemical

N. Nath
Department of Chemistry, D.S Degree College, Laida, Sambalpur, Odisha, India

S. Chakroborty (✉)
School of Basic Sciences, Department of Chemistry, Chandigarh University Uttar Pradesh, Unnao, Uttar Pradesh, India

S. Jangid
Department of Basic Science, IES College of Technology, Bhopal, Madhya Pradesh, India

A. Sattar Jatoi, N. M. Mubarak (eds.), *Bio-Electrochemical Conversion of Waste to Fuel*, Green Energy and Technology,
https://doi.org/10.1007/978-981-95-4310-6_10

1 Introduction

The finite availability of fossil fuels and their pollution have increased demand for alternative energy sources. Global energy consumption currently exceeds 13 terawatt-hours (TW) per year; by 2050, it is projected to reach approximately 23 TW [1]. In the meantime, the extensive use of resources in industrial, agricultural, and municipal operations will continue to deteriorate the environment, posing a threat to global warming and its subsequent effects. Due to this, developing renewable energy sources has become increasingly crucial for maintaining a sustainable ecosystem, particularly through expanding energy production sources and reducing reliance on energy imports [2].

Rapid economic development and industrialization have increased global energy demand, resulting in a rise in the use of traditional fossil fuels, primarily natural gas, oil, and coal. As a result, more CO_2 has been escaping into the atmosphere. Anthropogenic emissions amount to 53.8 Gt CO_2 equivalent per year; CO_2 is one of the primary contributors to global greenhouse gases (GHGs) and a major trigger of present-day climate change issues [3, 4]. Various strategies are being used to reduce the amount of CO_2 in the atmosphere [5]. The development of innovative CO_2 collection and utilization technology, along with the substitution of sustainable and renewable energy sources for fossil fuels, are two important coordinated strategies that can lower the CO_2 load in the atmosphere. Due to its abundance, low toxicity, and nonflammability, CO_2 is considered a C-1 building block for the second fuel synthesis approach and other industrially important value-added compounds [6]. Even while studies relating CO_2 to fuels and value-added chemicals began a decade ago, none of these "zero-carbon" options have reached the commercial stage. This is partly because activating CO_2 and further reducing it requires a lot of energy [7]. Significant efforts are being made to develop sustainable technologies that can capture and sequester CO_2 using less energy-intensive chemical, photochemical, electrochemical, or microbiological processes, which utilize renewable energy sources to balance energy needs [8–10].

The BESs, a hybrid system comprising electrochemistry and biotechnology, have emerged recently in this transdisciplinary field of study. Among the many uses for the new BESs are the following: removing salinity (microbial desalination cell, MDC) [11], treating wastewater in conjunction with energy generation microbial fuel cell (MFC) [12–14], microbial electrolysis cell, or MEC, which produces hydrogen with minimal external energy input [15–17], bioelectrochemical treatment, or BET, which treats harmful and resistant pollutants [18–20], and more. A rapidly developing area of biocatalysis and electrochemistry (BES), microbial electrosynthesis (MES) utilizes minimal electrical energy in a microbial catalytic system to convert CO_2 into fuels and value-added chemicals [21, 22].

In addition, using microbes, BESs offer novel possibilities for the effective and environmental-friendly production of valuable compounds and fuels. Over the past decade, the use of BESs, an emerging sustainable technology, has increased significantly, potentially producing electricity while treating wastewater simultaneously [14, 23]. Because the BES operates in a fully multidisciplinary manner, a significant amount of research has been conducted globally in environmental science, materials science, biotechnology, electrochemistry, bioelectrochemistry, microbiology, and other related fields. In these electrochemical systems, a potential difference created by the redox potential of an oxidation reaction at an anode and a reduction at the cathode drives electrons to spontaneously transition from a low to a high potential [24]. Electric current is a measure of the movement of electrons via an external circuit. The microbial electrochemical system (MXC), also known as a bioelectrochemical system (BES), is a system in which microbes or enzymes participate in either the reduction or oxidation process or both [25]. Microbes are used as bioanodes when they catalyze anodic processes and as biocathodes when they catalyze cathodic reactions. One type of MXC that utilizes the breakdown of organic material in an anode chamber to generate energy is known as an MFC. Bioelectricity is typically generated when there is a low redox potential for oxidation at the anode for organic matter and a high redox potential for oxygen reduction at the cathode. In microbial electrolysis cells (MECs), the cathode potentials that drive the generation of valuable compounds are subjected to an external voltage [26]. MES is a type of BES where organic molecules or CO_2 are cathodically converted to value-added organic compounds [22]. Additionally, BESs have been utilized in various applications, such as microbial desalination cells (MDCs), which are employed to desalinate water by combining microorganisms with plant roots [27–29]. Additionally, BESs are effective bioreactors for treating toxic wastewater and stubborn pollutants; this process is known as microbial electroremediation or bioelectrochemical treatment (BET) [30].

2 CO_2 and Its Utility

Carbon dioxide is extremely stable, non-flammable, and harmless. Since CO_2 is produced by numerous power plants, oil refineries, fermenters, and various other industrial processes, all of which contribute to the problems associated with greenhouse gases (GHGs), it should not be viewed as a liability but rather as an excellent raw material to produce commodities materials, chemicals and fuels through the use of creative, economical catalytic processes [31].

Massive volumes of greenhouse gases, particularly CO_2, the primary contributor to global warming and climate change, are released when fossil fuels are burned [32]. CO_2 alone accounts for 20% of greenhouse gas emissions, which have increased by over 15% in the past decade to approximately 30 billion tons and are expected to double by 2050 [33, 34]. Therefore, to bring the Earth's temperature down to 6 °C by the end of this century, CO_2 emissions must be reduced by 20 gigatons/year over the next 30 years to almost zero [35]. Today, this poses a serious risk to humankind. As a result, carbon dioxide has drawn the attention of all specialists and researchers [36].

Since it allows for the valorisation of CO_2 through thermochemical, biochemical, and electrochemical processes, the use of CO2 to create a range of end products has

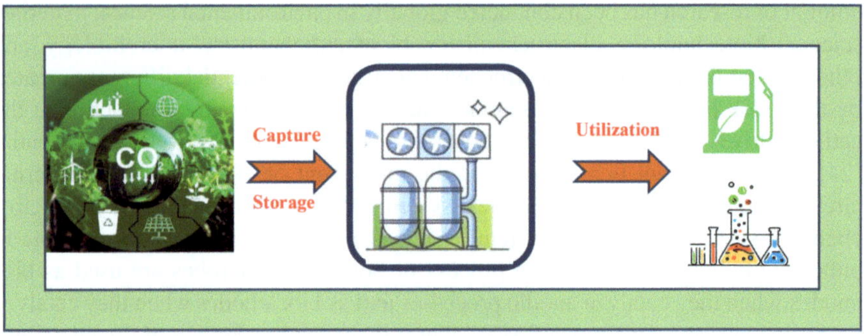

Fig. 1 Capturing, storing, and utilizing CO_2 to produce valuable products

the potential to be transformative (Fig. 1). The amount of CO2 used and the economic value of the final product will differ depending on the usage technologies.

One readily accessible feedstock for synthesizing numerous fuels and chemicals with added value is CO_2, which offers a way to address the problems associated with energy supply and CO_2 emissions [37]. One promising strategy to lower CO_2 emissions is CO_2 capture, utilization, and sequestration (CCUS). One technology that reduces CO_2 and fixes it into a long-lasting chemical is called carbon capture and storage (CCS) [38]. However, technological and financial obstacles must be addressed before carbon capture and storage (CCS) can be widely adopted. The primary financial barrier is the high cost of CCS, primarily due to capture and compression, which account for roughly 75% of the total cost [39]. Since carbon capture and utilization (CCU) can convert CO2 into valuable products rather than permanently sequestering it, it has garnered significant attention globally. The benefits of CCU over CCS include that using CO_2 is typically profitable because the products have a wide range of uses [39]. To lower CO_2 emissions, CCS and CCU technologies are utilized to capture and store emissions from sources such as power plants and various industrial operations [37]. In CCU, CO_2 is captured and transformed into various beneficial products [37, 39], whereas in CCS, CO_2 is captured and stored in a suitable location for an extended period [40–45]. Figure 1 shows the capture, storage, and utilization of CO_2.

Under normal circumstances, CO_2 is a non-polar and chemically unreactive molecule. Due to its low reactivity, a catalyst lowers the reaction's activation energy, transforming it into other valuable products.

3 Bioelectrochemical Systems (BESs)

By generating energy and useful chemicals while breaking down organic and inorganic substrates, BESs offer a potential hybrid strategy to address multiple aspects of human-environment compatibility [46] concurrently. BESs utilize the metabolic activities of collective microorganisms to transform the chemical energy contained in materials into electrical power, in addition to employing the heaviest and hazardous substances as oxygen receptors within the cathode compartment, which lessens their toxicity and generates biofuels as well as valuable chemicals [47, 48]. Electrically active microorganisms (EAM) generate value-added chemicals (such as ethanol, butanol, biopolymers, and volatile fatty acids (VFA) and fuels (such as electricity, H_2, CH_4, and ethanol) using systems such as MFCs, microbial electrolysis cells (MEC), and biological electrosynthesis cells (BESC), among others. To generate bioelectricity, MFCs, a type of BES, use EAM to oxidize various biomass and organic waste as a potential substrate [49, 50]. BES is a novel technique based on electrochemical altering techniques that generate electricity by combining the biological energy stored in recyclable organic materials with the catalytic activities of microorganisms (Fig. 2) [51].

Technologies for producing renewable energy hold considerable promise thanks to BESs. BES can utilize electric current to create valuable resources in MECs or employ MFCs to produce power. BESs, or bioelectrochemical systems, are comparatively recent technological advancements. Using MFC or MEC, these machines convert substrates into either power or valuable chemical compounds [52, 53]. The

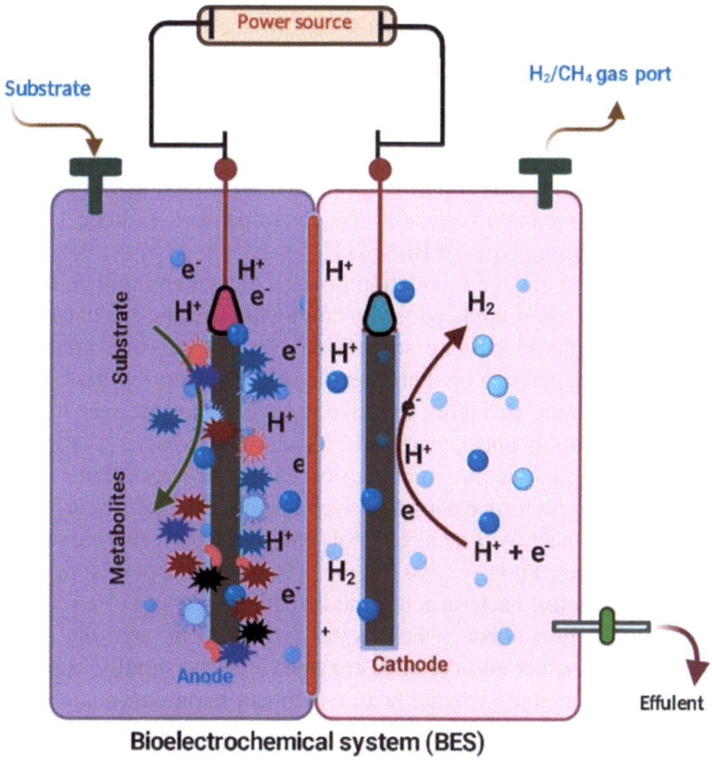

Fig. 2 The BES schematically [51]

anode and cathode are the two electrodes of the traditional BES reactor, which are connected by either a resistance-matched fuel cell (MFC) or an externally powered electric cell (MEC). Separating the anodic and cathodic spaces is necessary, and this can be done with the right membrane or membranes [54]. The reactor's body design concept reflects this. Research in the MEC reactors has been rising recently. By either direct or indirect contact with the electrode, microorganisms are employed by at least one electrode to catalyze the process of redox. The anode, where oxidation occurs, and the cathode, where reduction occurs, make up BES. The electrode and surrounding bacteria, usually arranged in biofilm, are termed bioelectrode. Although not necessary for BES, a membrane can separate the anode from the cathode. Commonly used interchangeable terms include microbial electrochemical systems (MES) and microbial electrochemical technology (MET) [55–59].

An external power source is used to augment the energy generated by the bioanode via biomass fermentation in MECs, a special type of BES design. At the cathode, valuable commodities are produced by overcoming the thermodynamically unfavourable reduction reactions. MECs can also work using abiotically evolved H_2 in the cathodic chamber. As an alternative, electroactive microorganisms and soluble electron acceptors capture electrons from the cathode to create H_2, CH_4, or other chemicals [60–62]. Electroactive microorganisms utilize CO_2 reduction in microbial electrochemical systems (MES), a cathode-related process, to transform electricity into chemicals. A MES reactor can produce methane, several types of alcohol, terpenoids of volatile fatty acids, bioplastics, and other useful commodities [63–65]. In some literature reports, MES is referred to as "Electro fermentation" (EF) [66]. MFC is a type of BES in which exoelectrogenic microbes, located near the anode act as a terminal electron acceptor and break down organic matter. Electric current is produced in a circuit by the spontaneous passage of electrons from the negatively charged bioanodes towards the electropositive cathode [12, 67, 68]. Figure 3 displays various types of BESs [11].

EFCs facilitate the catalytic oxidation process of fuel and initiate specific processes required for various applications by employing specific enzymes on the electrode surface or in the electrolyte suspension. Various enzymes are used on the anode and cathode depending on their specific redox function. If the enzymes used at the anode and cathode are highly selective, the membrane between the anode and cathode compartments is unnecessary. Electrically active bacteria utilize the solar energy captured by photoautotrophic microorganisms or larger plants in microbial solar cells (MSCs) to perform electrode-driven operations. Electric current and molecules such as methane, hydrogen, ethanol, and hydrogen peroxide are produced during these processes. Photosynthesis is the primary source of organic compounds oxidized by electroactive bacteria at the anode, which generates electrons. The electrons then flow to the cathode, where water is formed through the reduction of oxygen [69]. It is typically observed that enzyme catalysts partially oxidize the fuel, causing side reactions that generate heat, which can harm enzymatic activity [70, 71]. Plant microbial fuel cells (PMFCs) incorporate the root system of the living plant into the MFC anode, allowing it to utilize organic compounds known as rhizodeposits that the roots release into the soil, which serve as the substrate for

Fig. 3 Several kinds of BESs

electroactive microorganisms to generate power [72]. Thus, in a broader sense, PMFCs absorb solar energy and convert it into efficient and clean green energy. Most *rhizodeposits* from the roots contain exudates, secretions, lysates, and gases [72]. This chapter covered the use of BESs to produce methanol from CO_2.

4 Conversion of CO_2 to Methanol Using BESs

Compared to single-stage systems, the consecutive operation of MFCs to produce current has demonstrated advantages due to improved operational performance. Methanotrophic effluents containing methanol can only be converted to power with MFCs. Consequently, the effluents of methanotrophic reactors have been utilized in an air-cathode MFC for the first time to produce energy. A maximum methanol concentration of 6.45 mM was achieved in the methanotrophic reactors with Methylosinus sporium, simulating biogas (4:1 (v/v) CH4:CO2) with a 50% CH4 content. Methanotrophic reactor effluents employing pure CH_4 (MFC-1) and simulated biogas (4:1 (v/v) CH_4: CO_2) (MFC-2) achieved maximum power densities of 235 and 270 mW/m2. Figure 4 displays two cubic single-chamber air cathode microbial fuel cells (MFCs) [73].

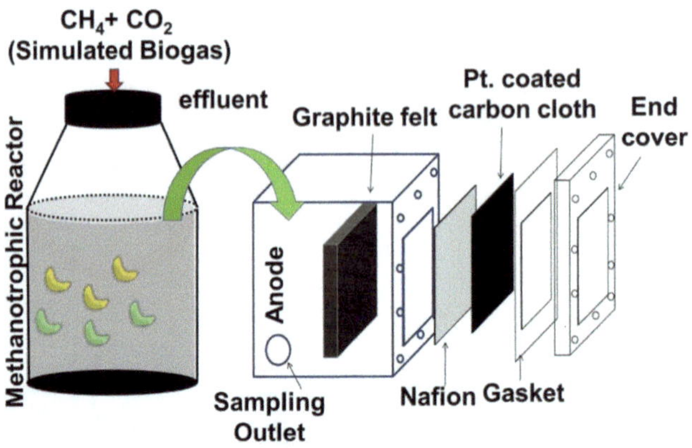

Fig. 4 The methanotrophic reactor using *M. sporium,* which simulates biogas, features a single-chamber air cathode MFC [73]

The hybrid electrocatalytic/bioelectrocatalytic reaction cascade for the selective multi-step reduction of CO_2 to methanol was studied by the Wang research group. For the effective and selective electrochemical reduction of CO_2 to formate with approximately 95% Faradaic efficiency at neutral pH, the system incorporates a gas-diffusion electrode modified with the inorganic catalyst Ag-Bi_2O_3. This ensures both the integrity of the enzymes and compatibility with the coupled enzymatic reactions. Through the successive action of the two NADH-dependent enzymes, FaDH and ADH, formate was converted to formaldehyde, which was then converted to methanol using an enzymatic cascade.

A NADH-regenerating photoelectrochemical (PEC) cell with a hybrid photosynthetic platform, combining inorganic and bioorganic components, and a multienzyme cascade was described. Formaldehyde dehydrogenase (FaldDH), formate dehydrogenase (FDH), and alcohol dehydrogenase (ADH) are three different enzymes that may progressively convert CO_2 to formaldehyde, methanol, and formate. By transferring electrons generated by photosynthesis through the oxidation of water upon being exposed to visible light into the enzyme cascades used in the biocatalytic conversion of CO_2 to methanol, a tandem PEC cell using an interconnected enzyme cascade (TPIEC) technology replicates the natural photosynthetic Z-scheme. To construct a multienzymatic cascade system that produces methanol, we used FDH from *Clostridium carboxidivorans* P7T (CcFDH), as well as PcFaldDH and the commercial enzyme YADH. An estimated 27 mg of CcFDH protein was produced for every liter of culture. For the reduction of CO_2, CcFDH outperformed the commercial enzyme CbFDH in terms of catalytic activity and binding constants. CcFDH worked with PcFalDH to convert CO_2 into formate and formaldehyde sequentially. The CcFDHPcFaldDH-YADH cascading system demonstrated higher activity for the transformation of CO2 to methanol in comparison to other multienzyme cascades [74].

The electro-enzymatic process for converting CO_2 to methanol can be significantly enhanced in efficiency by Zhang et al., who recommend using deep eutectic

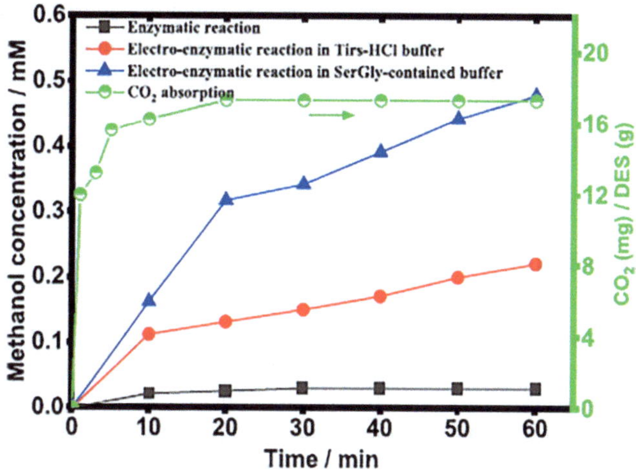

Fig. 5 Electro-enzymatic conversion of CO_2 to methanol using various concentrations of CO_2 in SerGly solution [75]

solvents (DESs) as the electrolyte in the reaction. Four natural deep eutectic solvents (NADESs) were produced for the first time: histidine glycerol (HisGly), glutamate glycerol (GluGly), serine glycerol (SerGly), and arginine glycerol (ArgGly). To improve the process, a sustainable method of manufacturing methanol was developed using the enzymatic reaction in conjunction with NADH electroregeneration in the SerGly and Tris-HCl buffers. The control and enzymatic reactions devoid of NADH regeneration were also compared. Consequently, Fig. 5 shows the design and assessment of three systems. The enzymatic response without NADH electroregeneration quickly reached equilibrium at a low methanol concentration (0.03 mM), as this type of multi-enzymatic reaction is reversible.

Additionally, the accumulation of the byproduct NAD+ over time precluded CO_2 conversion. Following the electro-reduction of NAD^+ into NADH, the methanol concentration was 0.22 mM, seven times higher than that of the control reaction. It demonstrated that NAD^+ was efficiently changed into the coenzyme NADH, which increased the concentration of NADH and eliminated NAD^+. To further boost CO_2 electro-enzymatic conversion.

The effectiveness of CO_2 enzyme-mediated conversion to methanol is limited by its low solubility in water (33 mM), and commercial CO2 enzymatic conversion is not feasible due to the expensive nature of the cofactor (NADH). To tackle these two issues, a bioelectrocatalytic system was developed. To increase the CO_2 concentration and pre-concentrate NADH, enzymes were encapsulated within the metal-organic framework ZIF-8. A Rh-complex-grafted electrode was developed to sustainably regenerate NADH. Three hours after the enzymes were encapsulated in ZIF-8, the methanol concentration increased from 0.061 to 0.320 mM. Industrial CO_2 enzymatic conversion may not be practical because of the high cost of the cofactor (NADH), and the low solubility of CO_2 enzymes in water (33 mM) limits their efficacy when converted to methanol. The development of a bio-electrocatalytic system addressed these two problems. Using the metal-organic framework ZIF-8, enzymes have been in situ encapsulated to raise the CO_2 content and pre-concentrate NADH. An electrode grafted with the Rh complex was created to regenerate NADH in a sustainable manner. When these enzymes were encapsulated in ZIF-8, the methanol concentration rose from 0.061 to 0.320 mM in 3 h [76].

5 Conclusion and Future Perspective

BESs are multipurpose devices that utilize bacteria as catalysts to convert chemical energy into electrical energy and vice versa. MES is the most recent application of BESs in chemical synthesis and the production of electrically powered fuels. Nevertheless, knowledge of the multidisciplinary elements of MES processes is still in its early stages. More knowledge from various fields, such as electrochemistry, bioelectrochemistry, and materials science, is needed for MES than for traditional industrial microbiology. Generally speaking, a biological system's performance is influenced by several factors, including redox conditions, substrate transformation, biological catalyst activity, and the specificity of the enzymes utilized in the conversion process. Electrochemical processes are influenced by the cathode and anode potentials, mass and charge transfer, mediators, electrical conductivity of the electrolytes, and overpotentials (voltage losses). These two processes combine to form MES, which makes its overall limitations particularly intricate and challenging to pinpoint. The rapidly rising amount of CO_2 emitted into the environment is the main driver of global warming. Capturing and utilizing CO_2 to produce value-added products can help reduce net CO_2 emissions and global warming.

To maximize conversion efficiency and use energy efficiently, several reduction procedures require further development. A vast amount of energy is available to help reduce CO_2, and the best non-carbon and plentiful alternative is renewable energy sources, such as solar energy. Due to CO_2's high stability and poor diffusing capacity, reduction processes are not feasible, which motivates researchers and engineers to find a catalyst that can effectively address these issues. Effective separation techniques must be developed and enhanced, as solvent and other chemical regeneration during separation procedures pose significant financial challenges. The creation of synthetic fuel and the reduction of carbon dioxide will significantly benefit from overcoming the present obstacles, and these products will have the most significant market potential.

References

1. Chae KJ, Choi MJ, Kim KY, Ajayi FF, Chang IS, Kim IS (2009) A solar-powered microbial electrolysis cell with a platinum catalyst-free cathode to produce hydrogen. Environ Sci Technol 43(24):9525–9530

2. Ellabban O, Abu-Rub H, Blaabjerg F (2014) Renewable energy resources: current status, future prospects and their enabling technology. Renew Sust Energ Rev 39:748–764
3. Pachauri RK, Allen MR, Barros VR, Broome J, Cramer W, Christ R, Church JA, Clarke L, Dahe Q, Dasgupta P, Dubash NK (2014) Climate change 2014: synthesis report. In: Contribution of working groups I, II and III to the fifth assessment report of the intergovernmental panel on climate change. IPCC, Geneva
4. IEA (2022) Greenhouse gas emissions from energy. www.iea.org
5. Solomon S, Plattner GK, Knutti R, Friedlingstein P (2009) Irreversible climate change due to carbon dioxide emissions. Proc Natl Acad Sci 106(6):1704–1709
6. Yang L, Wang H (2014) Recent advances in carbon dioxide capture, fixation, and activation by using N-heterocyclic carbenes. ChemSusChem 7(4):962–998
7. Lim X (2015) How to make the most of carbon dioxide. Nature 526(7575):628–631
8. Aresta M, Dibenedetto A (2007) Utilisation of CO2 as a chemical feedstock: opportunities and challenges. Dalton Trans 28:2975–2992
9. Ma S, Kenis PJ (2013) Electrochemical conversion of CO2 to useful chemicals: current status, remaining challenges, and future opportunities. Curr Opin Chem Eng 2(2):191–199
10. Centi G, Perathoner S (2009) Opportunities and prospects in the chemical recycling of carbon dioxide to fuels. Catal Today 148(3-4):191–205
11. ElMekawy A, Hegab HM, Pant D (2014) The near-future integration of microbial desalination cells with reverse osmosis technology. Energy Environ Sci 7(12):3921–3933
12. Bajracharya S, Sharma M, Mohanakrishna G, Benneton XD, Strik DP, Sarma PM, Pant D (2016) An overview on emerging bioelectrochemical systems (BESs): technology for sustainable electricity, waste remediation, resource recovery, chemical production and beyond. Renew Energy 98:153–170
13. Mohan SV, Velvizhi G, Modestra JA, Srikanth S (2014) Microbial fuel cell: critical factors regulating bio-catalyzed electrochemical process and recent advancements. Renew Sust Energ Rev 40:779–797
14. ElMekawy A, Srikanth S, Bajracharya S, Hegab HM, Nigam PS, Singh A, Mohan SV, Pant D (2015) Food and agricultural wastes as substrates for bioelectrochemical system (BES): the synchronized recovery of sustainable energy and waste treatment. Food Res Int 73:213–225
15. Pasupuleti SB, Srikanth S, Mohan SV, Pant D (2015) Development of exoelectrogenic bioanode and study on feasibility of hydrogen production using abiotic VITO-CoRE™ and VITO-CASE™ electrodes in a single chamber microbial electrolysis cell (MEC) at low current densities. Bioresour Technol 195:131–138
16. Babu ML, Subhash GV, Sarma PN, Mohan SV (2013) Bio-electrolytic conversion of acidogenic effluents to biohydrogen: an integration strategy for higher substrate conversion and product recovery. Bioresour Technol 133:322–331
17. Logan BE, Call D, Cheng S, Hamelers HV, Sleutels TH, Jeremiasse AW, Rozendal RA (2008) Microbial electrolysis cells for high yield hydrogen gas production from organic matter. Environ Sci Technol 42(23):8630–8640
18. Mohan SV, Velvizhi G, Krishna KV, Babu ML (2014) Microbial catalyzed electrochemical systems: a bio-factory with multi-facet applications. Bioresour Technol 165:355–364
19. Mohan SV, Babu PS, Srikanth S (2013) Azo dye remediation in periodic discontinuous batch mode operation: evaluation of metabolic shifts of the biocatalyst under aerobic, anaerobic and anoxic conditions. Sep Purif Technol 118:196–208
20. Velvizhi G, Mohan SV (2011) Biocatalyst behavior under self-induced electrogenic microenvironment in comparison with anaerobic treatment: evaluation with pharmaceutical wastewater for multi-pollutant removal. Bioresour Technol 102(23):10784–10793
21. Nevin KP, Woodard TL, Franks AE, Summers ZM, Lovley DR (2010) Microbial electrosynthesis: feeding microbes electricity to convert carbon dioxide and water to multicarbon extracellular organic compounds. MBio 1(2):10–128
22. Rabaey K, Rozendal RA (2010) Microbial electrosynthesis—revisiting the electrical route for microbial production. Nat Rev Microbiol 8(10):706–716

23. Mohan SV, Mohanakrishna G, Velvizhi G, Babu VL, Sarma PN (2010) Bio-catalyzed electrochemical treatment of real field dairy wastewater with simultaneous power generation. Biochem Eng J 51(1-2):32–39
24. Patil SA, Gildemyn S, Pant D, Zengler K, Logan BE, Rabaey K (2015) A logical data representation framework for electricity-driven bioproduction processes. Biotechnol Adv 33(6):736–744
25. Harnisch F, Schröder U (2010) From MFC to MXC: chemical and biological cathodes and their potential for microbial bioelectrochemical systems. Chem Soc Rev 39(11):4433–4448
26. Singh A, Sevda S, Abu Reesh IM, Vanbroekhoven K, Rathore D, Pant D (2015) Biohydrogen production from lignocellulosic biomass: technology and sustainability. Energies 8(11):13062–13080
27. Cao X, Huang X, Liang P, Xiao K, Zhou Y, Zhang X, Logan BE (2009) A new method for water desalination using microbial desalination cells. Environ Sci Technol 43(18):7148–7152
28. Chiranjeevi P, Mohanakrishna G, Mohan SV (2012) Rhizosphere mediated electrogenesis with the function of anode placement for harnessing bioenergy through CO2 sequestration. Bioresour Technol 124:364–370
29. ElMekawy A, Hegab HM, Vanbroekhoven K, Pant D (2014) Techno-productive potential of photosynthetic microbial fuel cells through different configurations. Renew Sust Energ Rev 39:617–627
30. Mohanakrishna G, Mohan SV, Sarma PN (2010) Bio-electrochemical treatment of distillery wastewater in microbial fuel cell facilitating decolorization and desalination along with power generation. J Hazard Mater 177(1-3):487–494
31. Yadav GD (2023) In pursuit of the net zero goal and sustainability: hydrogen economy, carbon dioxide refineries, and valorization of biomass & waste plastic. AsiaChem Magaz 3(1):110–123
32. Jones MW, Peters GP, Gasser T, Andrew RM, Schwingshackl C, Gütschow J, Houghton RA, Friedlingstein P, Pongratz J, Le Quéré C (2023) National contributions to climate change due to historical emissions of carbon dioxide, methane, and nitrous oxide since 1850. Scientific Data 10(1):155
33. Zhu DD, Liu JL, Qiao SZ (2016) Recent advances in inorganic heterogeneous electrocatalysts for reduction of carbon dioxide. Adv Mater 28(18):3423–3452
34. Jacobson TA, Kler JS, Hernke MT, Braun RK, Meyer KC, Funk WE (2019) Direct human health risks of increased atmospheric carbon dioxide. Nat Sustain 2(8):691–701
35. Mac Dowell N, Fennell PS, Shah N, Maitland GC (2017) The role of CO_2 capture and utilization in mitigating climate change. Nat Clim Chang 7(4):243–249
36. Chakroborty S, Nath N, Soren S, Barik A, Kaur K (2024) Plasmonic-based TiO_2 and TiO_2 nanoparticles for photocatalytic CO_2 to methanol conversion in energy applications: current status and future prospects. Top Catal 67(1):232–245
37. Markewitz P, Kuckshinrichs W, Leitner W, Linssen J, Zapp P, Bongartz R, Schreiber A, Müller TE (2012) Worldwide innovations in the development of carbon capture technologies and the utilization of CO2. Energy Environ Sci 5(6):7281–7305
38. Ma J, Li L, Wang H, Du Y, Ma J, Zhang X, Wang Z (2022) Carbon capture and storage: history and the road ahead. Engineering 14:33–43
39. Styring P, Jansen D, De Coninck H, Reith H, Armstrong K (2011) Carbon capture and utilisation in the green economy. Centre for Low Carbon Futures. http://co2chem.co.uk/wp-content/uploads/2012/06/CCU%20in%20the%20green%20economy%20report.pdf
40. Metz B, Davidson O, De Coninck H, Loos M, Meyer L (2005) IPCC special report on carbon dioxide capture and storage. Intergovernmental Panel on Climate Change, Geneva
41. Hertwich EG, Aaberg M, Singh B, Strømman AH (2008) Life-cycle assessment of carbon dioxide capture for enhanced oil recovery. Chin J Chem Eng 16(3):343–353
42. Cooper C (2009) A technical basis for carbon dioxide storage. Energy Procedia 1(1):1727–1733
43. Nagashima S, Miyagawa T, Matsumoto M, Suzuki S, Komaki H, Takagi M, Murai S (2011) Life cycle assessment performed on a CCS model case in Japan and evaluation of improvement facilitated by heat integration. Energy Procedia 4:2457–2464

44. Zapp P, Schreiber A, Marx J, Haines M, Hake JF, Gale J (2012) Overall environmental impacts of CCS technologies—a life cycle approach. Int J Greenhouse Gas Contr 8:12–21
45. Wang W, Chang JS, Lee DJ (2022) Integrating anaerobic digestion with bioelectrochemical system for performance enhancement: a mini review. Bioresour Technol 345:126519
46. Chandrasekhar K, Kumar G, Mohan SV, Pandey A, Jeon BH, Jang M, Kim SH (2020) Microbial electro-remediation (MER) of hazardous waste in aid of sustainable energy generation and resource recovery. Environ Technol Innov 19:100997
47. Jung S, Lee J, Park YK, Kwon EE (2020) Bioelectrochemical systems for a circular bioeconomy. Bioresour Technol 300:122748
48. Wilberforce T, Sayed ET, Abdelkareem MA, Elsaid K, Olabi AG (2021) Value added products from wastewater using bioelectrochemical systems: current trends and perspectives. J Water Process Eng 39:101737
49. Wu R, Yu YY, Wang Y, Wang YZ, Song H, Ma C, Qu G, You C, Sun Z, Zhang W, Li A (2021) Wastewater-powered high-value chemical synthesis in a hybrid bioelectrochemical system. Iscience 24(12):103401
50. Saratale GD, Banu JR, Nastro RA, Kadier A, Ashokkumar V, Lay CH, Jung JH, Shin HS, Saratale RG, Chandrasekhar K (2022) Bioelectrochemical systems in aid of sustainable biorefineries for the production of value-added products and resource recovery from wastewater: a critical review and future perspectives. Bioresour Technol 359:127435
51. Wang J, Ren K, Zhu Y, Huang J, Liu S (2022) A review of recent advances in microbial fuel cells: preparation, operation, and application. Biotech 11(4):44
52. Qin X, Lu X, Cai T, Niu C, Han Y, Zhang Z, Zhu X, Zhen G (2021) Magnetite-enhanced bioelectrochemical stimulation for biodegradation and biomethane production of waste activated sludge. Sci Total Environ 789:147859
53. Roy M, Aryal N, Zhang Y, Patil SA, Pant D (2022) Technological progress and readiness level of microbial electrosynthesis and electrofermentation for carbon dioxide and organic wastes valorization. Curr Opin Green Sustain Chem 35:100605
54. Villano M, Aulenta F, Ciucci C, Ferri T, Giuliano A, Majone M (2010) Bioelectrochemical reduction of CO2 to CH4 via direct and indirect extracellular electron transfer by a hydrogenophilic methanogenic culture. Bioresour Technol 101(9):3085–3090
55. Ceballos-Escalera A, Molognoni D, Bosch-Jimenez P, Shahparasti M, Bouchakour S, Luna A, Guisasola A, Borràs E, Della PM (2020) Bioelectrochemical systems for energy storage: a scaled-up power-to-gas approach. Appl Energy 260:114138
56. Yu J, Kim S, Kwon OS (2019) Effect of applied voltage and temperature on methane production and microbial community in microbial electrochemical anaerobic digestion systems treating swine manure. J Ind Microbiol Biotechnol 46(7):911–923
57. Saratale RG, Saratale GD, Pugazhendhi A, Zhen G, Kumar G, Kadier A, Sivagurunathan P (2017) Microbiome involved in microbial electrochemical systems (MESs): a review. Chemosphere 177:176–188
58. Lee M, Reddy CN, Min B (2019) In situ integration of microbial electrochemical systems into anaerobic digestion to improve methane fermentation at different substrate concentrations. Int J Hydrog Energy 44(4):2380–2389
59. Ye Y, Wang L, Chen Y, Zhu S, Shen S (2010) High yield hydrogen production in a single-chamber membrane-less microbial electrolysis cell. Water Sci Technol 61(3):721–727
60. Park SG, Rhee C, Shin SG, Shin J, Mohamed HO, Choi YJ, Chae KJ (2019) Methanogenesis stimulation and inhibition for the production of different target electrobiofuels in microbial electrolysis cells through an on-demand control strategy using the coenzyme M and 2-bromoethanesulfonate. Environ Int 131:105006
61. Zhang Z, Song Y, Zheng S, Zhen G, Lu X, Kobayashi T, Xu K, Bakonyi P (2019) Electro-conversion of carbon dioxide (CO2) to low-carbon methane by bioelectromethanogenesis process in microbial electrolysis cells: the current status and future perspective. Bioresour Technol 279:339–349

62. Marshall CW, Ross DE, Fichot EB, Norman RS, May HD (2012) Electrosynthesis of commodity chemicals by an autotrophic microbial community. Appl Environ Microbiol 78(23):8412–8420
63. Enzmann F, Holtmann D (2019) Rational scale-up of a methane producing bioelectrochemical reactor to 50 L pilot scale. Chem Eng Sci 207:1148–1158
64. Jourdin L, Freguia S, Flexer V, Keller J (2016) Bringing high-rate, CO2-based microbial electrosynthesis closer to practical implementation through improved electrode design and operating conditions. Environ Sci Technol 50(4):1982–1989
65. Li X, Liu G, He Z (2020) Flexible control of biohythane composition and production by dual cathodes in a bioelectrochemical system. Bioresour Technol 295:122270
66. Call D, Logan BE (2008) Hydrogen production in a single chamber microbial electrolysis cell lacking a membrane. Environ Sci Technol 42(9):3401–3406
67. Hassanein A, Witarsa F, Lansing S, Qiu L, Liang Y (2020) Bio-electrochemical enhancement of hydrogen and methane production in a combined anaerobic digester (AD) and microbial electrolysis cell (MEC) from dairy manure. Sustainability 12(20):8491
68. Pawar AA, Karthic A, Lee S, Pandit S, Jung SP (2022) Microbial electrolysis cells for electromethanogenesis: materials, configurations and operations. Environ Eng Res 27(1):200484
69. Minteer SD, Liaw BY, Cooney MJ (2007) Enzyme-based biofuel cells. Curr Opin Biotechnol 18(3):228–234
70. Palmore GT, Whitesides GM (1994) Microbial and enzymatic biofuel cells. Enzym Convers 566:271–290
71. Strik DP, Hamelers HV, Snel JF, Buisman CJ (2008) Green electricity production with living plants and bacteria in a fuel cell. Int J Energy Res 32(9):870–876
72. Kondaveeti S, Patel SK, Pagolu R, Li J, Kalia VC, Choi MS, Lee JK (2019) Conversion of simulated biogas to electricity: sequential operation of methanotrophic reactor effluents in microbial fuel cell. Energy 189:116309
73. Wang P, Wang X, Chandra S, Lielpetere A, Quast T, Conzuelo F, Schuhmann W (2025) Hybrid enzyme-Electrocatalyst Cascade modified gas-diffusion electrodes for methanol formation from carbon dioxide. Angew Chem Int Ed 64:e202422882
74. Kuk SK, Singh RK, Nam DH, Singh R, Lee JK, Park CB (2017) Photoelectrochemical reduction of carbon dioxide to methanol through a highly efficient enzyme cascade. Angew Chem Int Ed 56(14):3827–3832
75. Zhang Z, Wang H, Nie Y, Zhang X, Ji X (2022) Natural deep eutectic solvents enhanced electro-enzymatic conversion of CO_2 to methanol. Front Chem 10:894106
76. Zhang Z, Li J, Ji M, Liu Y, Wang N, Zhang X, Zhang S, Ji X (2021) Encapsulation of multiple enzymes in a metal–organic framework with enhanced electro-enzymatic reduction of CO_2 to methanol. Green Chemistry 23(6):2362–71

Biohydrogen Production Using Bioelectrochemical System

Alok Tiwari, Vishal Kumar Sandhwar, Diksha Saxena, Shivendu Saxena, and Raj Patel

Abstract The bioelectrochemical system provides a biological pathway to harness energy through electroactive microorganisms, and BES biohydrogen production appears very promising. This review article aims to provide a comprehensive analysis of the current scientific literature on biohydrogen production using BES systems, focusing on mechanisms, types, and related issues. The principles underlying the design and construction of BES, including their components and electron transfer processes mediated by electroactive microbes, are reviewed. Microbial fuel cells, microbial electrolysis cells, and their associated bioelectrochemical systems are examined in terms of their effectiveness and the range of operational parameters. The interaction and coordination between the oligotrophic heterotrophic bacteria and hydrogen producers from complex substrates in biohydrogen production are highlighted, especially exoelectrogenic bacteria and their hydrogen production potential. Various biochemical pathways of hydrogen production and their influencing environment are studied, too. This chapter also emphasises the design and optimisation of various BESs, focusing on how those could enhance the cell's performance. Scalability, economics, and sustainability are discussed alongside the challenges, innovations, and future possibilities that the scope presents. By collating data from existing literature and knowledge, the article provides insights into future research directions that can help propel the development of biohydrogen production technology applications.

Keywords Biohydrogen production · Bioelectrochemical cells · Electroactive microorganisms · Microbial fuel cells · Microbial electrolysis cells · Exoelectrogenic bacteria

---

A. Tiwari · V. K. Sandhwar (✉) · D. Saxena · S. Saxena · R. Patel
Department of Chemical Engineering, Parul Institute of Technology, Parul University, Vadodara, Gujarat, India
e-mail: vishal.sandhwar8850@paruluniversity.ac.in

A. Sattar Jatoi, N. M. Mubarak (eds.), *Bio-Electrochemical Conversion of Waste to Fuel*, Green Energy and Technology,
https://doi.org/10.1007/978-981-95-4310-6_11

1 Introduction

Biohydrogen, a renewable energy source, has garnered significant attention due to its potential as a clean alternative to fossil fuels. Producing hydrogen through biological processes, particularly from organic waste, presents an environmental-friendly solution to the growing energy crisis. Biohydrogen production is characterised by its sustainability, as it utilises renewable biomass resources, thus contributing to the reduction of greenhouse gas emissions and reliance on fossil fuels [1]. The biologically generating hydrogen can be achieved through various methods, including dark fermentation, photofermentation, and bioelectrochemical systems (BESs) [2, 3]. Among these methods, bioelectrochemical systems have emerged as an up-and-coming technology for sustainable hydrogen production.

Bioelectrochemical systems (BES) play a crucial role in sustainable hydrogen production by harnessing the metabolic activities of microorganisms. These systems utilise electroactive microbes that can transfer electrons to electrodes, facilitating the conversion of organic substrates into hydrogen gas [4]. Integrating BES into energy recovery processes enhances hydrogen production and allows for simultaneous wastewater treatment, thereby addressing two critical environmental challenges: energy generation and waste management [5]. The versatility of BES makes them suitable for various applications, including the treatment of agricultural waste, food waste, and industrial effluents, which are rich in organic matter and can serve as substrates for hydrogen production [3].

The history and evolution of bioelectrochemical systems can be traced back to the early twentieth century when the first microbial fuel cells (MFCs) were developed. These systems initially focused on electricity generation from organic matter through microbial metabolism [6]. Over time, research has expanded to include microbial electrolysis cells (MECs), specifically targeting hydrogen production by applying an external voltage to drive the electrochemical reactions [7]. Advancements in electrode materials, reactor designs, and microbial consortia have significantly improved the efficiency and scalability of these systems, making them more viable for practical applications [2, 7]. The transition from microbial fuel cells (MFCs) to microbial electrochemical cells (MECs) represents a pivotal development in bioelectrochemical technologies, highlighting the potential for optimising hydrogen production processes.

This review article aims to provide a comprehensive overview of biohydrogen production through bioelectrochemical systems, focusing on the mechanisms, advancements, and challenges associated with this technology. By synthesising recent research findings, this article aims to elucidate the role of BES in sustainable hydrogen production and its potential contributions to the circular bioeconomy. Furthermore, the review will explore the integration of BES with other renewable energy technologies and their implications for future energy systems, ultimately highlighting the importance of continued research and innovation in this field.

In summary, biohydrogen represents a promising renewable energy source that can be efficiently produced by bioelectrochemical systems. The evolution of these technologies has paved the way for innovative approaches to energy recovery and waste treatment, underscoring their significance in addressing global energy and environmental challenges. This review aims to contribute to the understanding of biohydrogen production through BES and inspire further research and development in this vital area of renewable energy.

2 Fundamentals of Bioelectrochemical Cells

BESs offer a novel approach for harvesting biohydrogen by employing electroactive microorganisms for the electrochemical conversion of organic substrates. Such systems typically feature an anode site for oxidation, a cathode site for reduction, and a proton exchange membrane separator positioned between the two [8]. The electron transfer mechanisms in BESs are fundamental to hydrogen synthesis since they are responsible for either delivering electrons from the microorganisms to the electrodes or acting as intermediaries for electron transfer [9]. Biohydrogen production can utilise electricity generated through these processes, where electroactive organisms, such as Shewanella oneidensis and Geobacter sulfurreducens, provide electrons to the anode as part of their metabolic activity, allowing circuit flows to be generated as they produce hydrogen [10]. Many factors, including microbial composition, the type of electrodes employed, and the operation mode, have been shown to affect the productivity of BESs, thus necessitating additional work to optimise these systems for the sustainable production of biohydrogen [11].

3 Types of Bioelectrochemical Systems

Biohydrogen can be produced by two systems: microbial fuel cells (MFCs) and microbial electrolysis cells (MECs). MFCs are electrochemical devices that directly convert organic substrates into electricity, whereas microbial electrolysis cells employ bioengineering methods for producing hydrogen gas as a biofuel. Both systems depend on microbial activity to convert electrons using mediated electron transfer (MET) and direct electron transfer (DET) mechanisms within a proton exchange membrane. MFCs generate electricity, while MECs require electricity to produce hydrogen.

Electricity can be generated through electroactive microorganisms that aid in the transfer of electrons. MFCs accomplish this by converting organic constituents directly into electricity using microbes that treat wastewater by transferring electrons to an anode [12].

However, organic properties are congruent with greater kinetic hydrogen production from the electrode substrate, and MECs can generate hydrogen gas by applying cathode voltage [13]. The MECs are more effective than MFCs for hydrogen

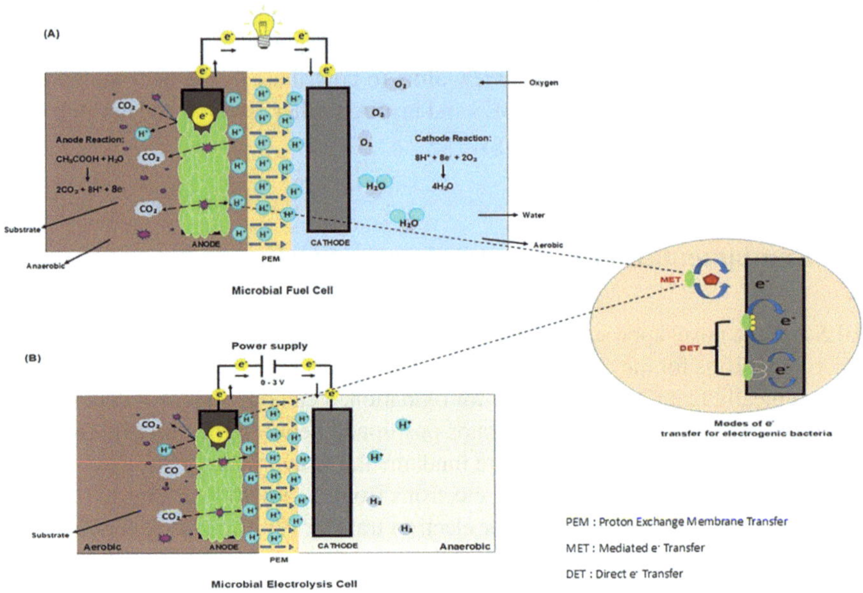

Fig. 1 (**a**) Microbial fuel cell, (**b**) Microbial electrolysis cell

production, and Most contemporary researchers thought that water electrolysis was the best option for hydrogen gas production [14] (Fig. 1).

The type of substrate, the structure of the microbial community, and the materials used for the electrodes significantly impact the efficiency of both systems. When considering energy recovery and the treatability of wastewater, microbial fuel cells (MFCs) have great potential. However, in the case of biohydrogen, integrating MECs with renewable energy sources shows promising forecasts for the technology [15].

4 Microbial Communities in Biohydrogen Production

Biohydrogen production, a promising avenue for renewable energy, relies heavily on the intricate interactions within microbial communities. The diversity of microorganisms involved in this process is vast, encompassing various bacterial and archaeal species contributing to hydrogen generation through different metabolic pathways. This note explores the diversity of microorganisms involved in biohydrogen production, the role of exoelectrogenic bacteria in enhancing hydrogen yield, and the influence of microbial consortia on biohydrogen production efficiency.

4.1 Diversity of Microorganisms Involved in Biohydrogen Production

The microbial diversity in biohydrogen production systems is critical for optimising hydrogen yields. Various species of bacteria, particularly those from the genera Clostridium, Klebsiella, and Bacillus, are known for their ability to produce hydrogen. For instance, *Klebsiella oxytoca* ATCC 13182 has been identified as a robust hydrogen producer due to its facultative anaerobic nature, allowing it to thrive in environments with varying oxygen levels [16]. This adaptability is crucial for biohydrogen production, especially in wastewater treatment applications where oxygen levels fluctuate.

Moreover, the metabolic pathways employed by these microorganisms vary significantly. Some bacteria, such as *Clostridium diollis*, utilise acetate and lactate to produce hydrogen, carbon dioxide, and butyrate, demonstrating the complexity of metabolic interactions within microbial communities [17]. The presence of hydrogenotrophic methanogens, which utilise hydrogen to produce methane, also plays a significant role in the overall dynamics of hydrogen production, as they can compete with hydrogen-producing bacteria for available substrates [18].

The identification of specific microbial strains capable of performing all steps of biohydrogen production remains a challenge. Many pure bacterial strains exhibit incomplete hydrogen production capabilities, necessitating the use of mixed cultures to achieve higher yields [19]. The synergistic interactions among diverse microbial populations can enhance the overall efficiency of biohydrogen production, as different species can complement each other's metabolic functions.

4.2 Role of Exoelectrogenic Bacteria in Enhancing Hydrogen Yield

Exoelectrogenic bacteria, which can transfer electrons to external electron acceptors, play a pivotal role in enhancing hydrogen production. These bacteria can facilitate electron transfer processes crucial for the metabolic pathways leading to hydrogen generation. For example, integrating exoelectrogenic bacteria in biohydrogen production systems can improve substrate utilisation efficiency, thereby increasing hydrogen yields [20].

The mechanisms through which exoelectrogenic bacteria enhance hydrogen production involve complex electron transfer pathways. These bacteria can utilise organic substrates and convert them into hydrogen while transferring electrons to an external electron acceptor, such as a metal oxide or an electrode in bioelectrochemical systems. This dual functionality not only enhances hydrogen production but also facilitates the degradation of organic pollutants, making it a sustainable approach for wastewater treatment [20].

Furthermore, exoelectrogenic bacteria can influence the overall microbial community structure of biohydrogen production systems. Their ability to interact with other microbial species can lead to the establishment of stable consortia that optimise hydrogen production processes. For instance, the co-cultivation of exoelectrogenic bacteria with hydrogen-producing bacteria can create a synergistic environment that maximises hydrogen yields while minimising the production of inhibitory by-products [20].

5 Mechanisms of Hydrogen Production in BESs

The primary sources of hydrogen are the organic substrates that microbial organisms ferment to produce hydrogen as a byproduct or endpoint of metabolism. During dark fermentation, anaerobic bacteria, such as Clostridium and Enterobacter, consume carbohydrate substrates and produce hydrogen gas (H2), organic acids, and alcohols. This reaction can be summarised in a simplified overall response:

$$C_nH_{2n+1}O_n \rightarrow nCO_2 + \frac{n}{2}H_2 + \text{other products} \quad (1)$$

Equation (1) presents the reaction of carbohydrates with their products, which are both the end products of the reaction, emphasising their stoichiometric proportion concerning the substrate [21]. In this process, several factors influence hydrogen production during dark fermentation, including the type of substrate, the composition of the microbial community, and specific environmental parameters such as pH and temperature [22]. In Fig. 2, the metabolic processes leading to the production of biohydrogen, such as fermentation (photo- and dark fermentation), biophotolysis (direct and indirect via photosynthesis), and electrochemical methods (photoelectrochemical and electrolytic), are illustrated. Additionally, it encompasses water splitting, acidogenic fermentation, and microbial electrolysis cells (MECs), types of hydrogen-generating methods that are largely anaerobic or light-assisted.

Environmental conditions, such as temperature, pH, and substrate concentration, can significantly impact microbial metabolism and hydrogen yield, as these parameters alter the microbial community and enzymatic activities [22, 23]. For example, specific organic acids can stimulate hydrogen generation by serving as substrates for fermentative bacteria [24]. Additionally, mixing cultures in which fermentative and respiring bacteria coexist may enhance hydrogen production, as these communities can efficiently utilise fermentation metabolites [1]. These mechanisms must be well understood when optimising BESs to produce sustainable hydrogen.

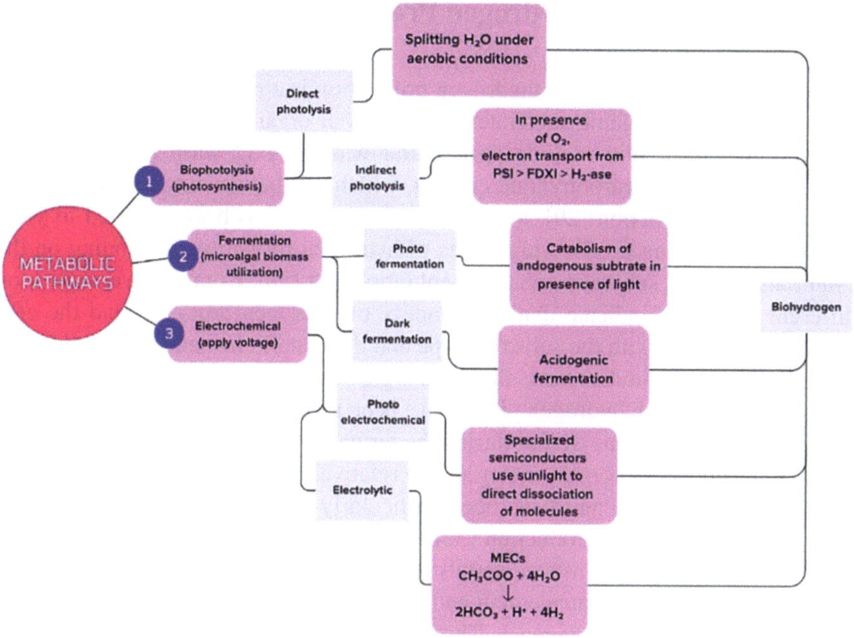

Fig. 2 Metabolic pathways of biohydrogen production

6 Electron Transfer Mechanism

Electron transfer mechanisms, like direct and mediated, play significant roles in BESs, influencing Biohydrogen production technology. Nanowires enable the direct, natural transfer of electrons to the electrodes by utilising cytochromes located in the bacterial outer membrane for electroactive bacteria, such as Geobacter and Shewanella species [25]. On the other hand, mediated turnover, predicated on redox mediators such as flavins, operates by engaging electrons that can enhance the turnover rate, transferring electrons between bacteria and the electrode surface [26]. Researchers have noted that factors such as biofilm thickness and composition can significantly contribute to the electron transfer rate during electroactive biofilm interactions [27]. Moreover, engineering strategies and improvements in conductive materials have enhanced BES, as observed in biofilm-optimising density, which has led to a considerable enhancement in biohydrogen production [28]. Ultimately, it has been demonstrated that improving the processes of these mechanisms accelerates the development of more effective biohydrogen production technologies.

7 Substrates for Biohydrogen Production

The application of BESs in biohydrogen production is a relatively new technology involving different organic substrates. Examples of organic substrates suitable for biohydrogen generation include agricultural wastes, food waste, and industrial effluents with a high carbohydrate content, which can be converted into hydrogen through microbial action [29]. Meanwhile, food waste has been reported to yield good results and can produce high rates of hydrogen formation, depending on the extent and type of substrate enhancement employed [30]. Comparative studies on different substrates indicate that the quantity of hydrogen produced and the efficiency tend to vary with the substrate type and the conditions under which they are used. For instance, studies have shown that glucose fermentation has been widely investigated, and high hydrogen yields have been recorded for specific pH and temperature combinations, as well as during the fermentation process [31]. On the other hand, lignocellulosic biomass materials are relatively more difficult to convert due to their inherent nature and have been particularly noted for their applications in sustainable biohydrogen production [32].

The need to employ waste materials as substrates must not be underestimated because it aids in waste management and, at the same time, fosters sustainability by minimising fossil fuel dependency. The principle of a circular economy is realised through the conversion of organic waste into biohydrogen, transforming waste into a valuable form of energy. At the same time, environmental effects are lessened [33]. This approach addresses energy needs and enhances ecological benefits by reducing greenhouse gas emissions associated with waste disposal [34].

8 Issues Affecting the Production of Biohydrogen

BESs are the backbone of biohydrogen production, but they pose some risks to their economic and operational undertakings. To use biohydrogen more effectively, an investment needs to be made in ameliorating microbes incorporated into BECs. Now and then, the pH and temperature rapidly fluctuate, making it difficult to maintain efficient substrate management [35]. The distribution of BESs also raises some concerns as movements from experimental phases to trade floors seem to lack the ability to persist in optimal conditions.

So far, BESs and biohydrogen production are in tough competition regarding cost estimations. Biofuel costs range from 7.2 euros to 2.1 euros and are not within the correct parameters compared to hydrogen production strategies. This usage barrier appears to evolve when considering the cheaper designs of reactors currently available on the market and those that are yet to be developed [35, 36]. Using BESs with organic waste can efficiently tackle such costs while producing renewable energy in the process when they are combined with already existing waste management systems to address deployment costs. Environmental considerations are also crucial to the biohydrogen generation process. BESs should also consider the environmental impact of the feedstock so that the impact of biohydrogen as an environmentally friendly fuel can be optimised as much as possible.

9 Future Perspectives and Innovations

New technologies in this area aim to improve the biohydrogen production rate through enhanced design approaches for microbial fuel cells (MFCs) and microbial electrolysis cells (MECs). Such systems promote the electron transfer processes aimed at increasing hydrogen production from organic products through the cooperative effects of microorganisms [37]. More recently, nanotechnology has provided additional boosts for biohydrogen production, with the development of nanoparticles facilitating electron transfer [38].

Biohydrogen can be utilised in energy systems and as an industrial fuel. Khan et al. [39], for instance, note that biohydrogen can be used in hydrogen-powered vehicles and as a substitute for fossil fuels in energy systems. Additionally, producing biohydrogen from organic waste materials may help address energy issues while facilitating waste management, aligning with the principles of the circular economy. Future research directions should emphasise a multidisciplinary approach by integrating concepts and ideas from disciplines such as microbiology, engineering, and environmental science to overcome the barriers to biohydrogen production. This also includes improving the ratio of the bioreactor, optimising microbial consortia, and industrialising cost-effective processes. If these barriers are removed, multidisciplinary collaboration can improve biohydrogen production technologies, thus significantly contributing to sustainable energy systems.

10 Conclusion

Biohydrogen production through BECs represents a promising avenue for sustainable energy solutions. We noted that this technology is emerging for environmentally sustainable and biofuel production. The research that is currently available demonstrates the effectiveness of BESs in converting organic waste to hydrogen through microbial electrolysis and fermentation. It has also been noted that BESs can have hydrogen production rates exceeding 90%, significantly higher than those achieved by traditional production methods. The reason for this efficiency has been noted as organic substrates being directly electrochemically converted to hydrogen, which significantly reduces, if not eliminates, energy loss associated with thermal conversion.

Including BESs in waste management systems meets energy requirements and helps manage organic waste, contributing to environmental cleanup. Additionally, the evolution of MFC technology demonstrates that BESs can operate under various conditions, thereby broadening their usability for other types of waste.

References

1. Lewis AJ, Campa MF, Hazen TC, Borole AP (2018) Unravelling biocomplexity of electroactive biofilms for producing hydrogen from biomass. Microb Biotechnol 11(1):84–97
2. Karthikeyan R, Cheng KY, Selvam A, Bose A, Wong JW (2017) Bioelectrohydrogenesis and inhibition of methanogenic activity in microbial electrolysis cells-a review. Biotechnol Adv 35(6):758–771
3. Yasin NH, Mumtaz T, Hassan MA (2013) Food waste and food processing waste for biohydrogen production: a review. J Environ Manag 130:375–385
4. Zhou H, Mei X, Liu B, Xie G, Xing D (2019) Magnet anode enhances extracellular electron transfer and enrichment of exoelectrogenic bacteria in bioelectrochemical systems. Biotechnol Biofuels 12:1–9
5. Borole AP (2015) Sustainable and efficient pathways for bioenergy recovery from low-value process streams via bioelectrochemical systems in biorefineries. Sustainability 7(9):11713–11726
6. Logan B, Rossi R, Ragab A, Saikaly P (2019) Electroactive microorganisms in bioelectrochemical systems. Nat Rev Microbiol 17(5):307–319. https://doi.org/10.1038/s41579-019-0173-x
7. Kadier A, Simayi Y, Abdeshahian P, Azman NF, Chandrasekhar K, Kalil MS (2016) A comprehensive review of microbial electrolysis cells (MEC) reactor designs and configurations for sustainable hydrogen gas production. Alex Eng J 55(1):427–443
8. Alfadaly RA, Elsayed A, Hassan RY, Noureldeen A, Darwish H, Gebreil AS (2021) Microbial sensing and removal of heavy metals: bioelectrochemical detection and removal of chromium (VI) and cadmium (II). Molecules 26(9):2549
9. Jiao AY, Li YF, Liu B, Yang CP, Guo BB (2020) Fermentative biohydrogen production by microbes: process design and engineering control. In: 2010 international conference on mechanic automation and control engineering. IEEE, New York, pp 1833–1837
10. Pirbadian S, Chavez MS, El-Naggar MY (2020) Single-cell electrophysiological measurements reveal bacterial membrane potential dynamics during extracellular electron transfer bioRxiv
11. Gimkiewicz C, Hunger S, Harnisch F (2016) Evaluating the feasibility of microbial electrosynthesis based on Gluconobacter oxydans. ChemElectroChem 3(9):1337–1346
12. Cheng WL, Erbay C, Sadr R, Han A (2018) Dynamic flow characteristics and design principles of laminar flow microbial fuel cells. Micromachines 9(10):479
13. Dange P, Pandit S, Jadhav D, Shanmugam P, Gupta PK, Kumar S, Kumar M, Yang YH, Bhatia SK (2021) Recent developments in microbial electrolysis cell-based biohydrogen production utilizing wastewater as a feedstock. Sustainability 13(16):8796
14. Rousseau R, Etcheverry L, Roubaud E, Basséguy R, Délia ML, Bergel A (2020) Microbial electrolysis cell (MEC): strengths, weaknesses and research needs from electrochemical engineering standpoint. Appl Energy 257:113938
15. Foley JM, Rozendal RA, Hertle CK, Lant PA, Rabaey K (2010) Life cycle assessment of high-rate anaerobic treatment, microbial fuel cells, and microbial electrolysis cells. Environ Sci Technol 44(9):3629–3637
16. Thakur V, Tiwari KL, Jadhav SK (2015) Production of biohydrogen from wastewater by Klebsiella oxytoca ATCC 13182. Water Environ Res 87(8):683–686
17. Abreu AA, Alves JI, Pereira MA, Sousa DZ, Alves MM (2011) Strategies to suppress hydrogen-consuming microorganisms affect macro and micro scale structure and microbiology of granular sludge. Biotechnol Bioeng 108(8):1766–1775

18. Spagni A, Casu S, Farina R (2010) Effect of the organic loading rate on biogas composition in continuous fermentative hydrogen production. J Environ Sci Health A 45(12):1475–1481
19. Li D, Wang JQ, Bu DP (2012) Ruminal microbe of biohydrogenation of trans-vaccenic acid to stearic acid in vitro. BMC Res Notes 5:1–8
20. Aldaby ES, Mahmoud AH, El-Bery HM, Ali MM, Shoreit AA, Mawad AM (2023) Microalgal upgrading of the fermentative biohydrogen produced from Bacillus coagulans via non-pretreated plant biomass. Microb Cell Factories 22(1):190
21. Teng Y, Xu Y, Wang X, Christie P (2019) Function of biohydrogen metabolism and related microbial communities in environmental bioremediation. Front Microbiol 10:106
22. Qiu C, Yuan P, Sun L, Wang S, Lo S, Zhang D (2017) Effect of fermentation temperature on hydrogen production from xylose and the succession of hydrogen-producing microflora. J Chem Technol Biotechnol 92(8):1990–1997
23. Qiu C, Zheng Y, Zheng J, Liu Y, Xie C, Sun L (2016) Mesophilic and thermophilic biohydrogen production from xylose at various initial pH and substrate concentrations with microflora community analysis. Energy Fuel 30(2):1013–1019
24. Chen P, Wang Y, Yan L, Wang Y, Li S, Yan X, Wang N, Liang N, Li H (2015) Feasibility of biohydrogen production from industrial wastes using defined microbial co-culture. Biol Res 48:1–8
25. Mayer F, Stöckl M, Krieg T, Mangold KM, Holtmann D (2018) Adsorption of Shewanella oneidensis MR-1 to the electrode material activated carbon fabric. J Chem Technol Biotechnol 93(10):3000–3010
26. Hou L, Cortez R, Hagerman M, Hu Z, Majumder EL (2024) Co-occurrence of direct and indirect extracellular electron transfer mechanisms during electroactive respiration in a dissimilatory sulfate reducing bacterium. Microbiol Spectr 5:e01226
27. Zhao J, Li F, Kong S, Chen T, Song H, Wang Z (2023) Elongated riboflavin-producing Shewanella oneidensis in a hybrid biofilm boosts extracellular electron transfer. Adv Sci 10(9):2206622
28. Wang H, Jing X, Feng K, Tremblay PL, Wang J, Liu X, Zhou S (2024) In situ tuning the structure of Geobacter biofilm for bioelectricity enhancement. Environ Sci Technol Lett 11(2):106–113
29. Jensen LS, Kaul C, Juncker NB, Thomsen MH, Chaturvedi T (2022) Biohydrogen production in microbial electrolysis cells utilizing organic residue feedstock: a review. Energies 15(22):8396
30. Jais FK, Hassan SH, Zamri MF, Zulkornain MF, Shamsuddin AH (2023) Biohydrogen production from kitchen organic waste via effective pre-treatment process of dark fermentation. IOP Conf Ser 1205(1):012021
31. Amin MM, Ghasemian M, Bina B, Taheri E, Fatehizadeh A (2019) Biohydrogen production as a clean fuel by acid and alkaline pretreated mixed culture during glucose fermentation. Health Scope 8(2):e12903
32. Bhatia L, Sarangi PK, Singh AK, Prakash A, Shadangi KP (2022) Lignocellulosic waste biomass for biohydrogen production: future challenges and bio-economic perspectives. Biofuels Bioprod Biorefin 16(3):838–858
33. Materazzi M, Chari S, Bajwa S, Sebastiani A (2023) Waste-to-hydrogen: challenges and opportunities in the UK scenario. Detritus 23:65–75
34. Rambabu K, Bharath G, Thanigaivelan A, Das DB, Show PL, Banat F (2021) Augmented biohydrogen production from rice mill wastewater through nano-metal oxides assisted dark fermentation. Bioresour Technol 319:124243
35. Ahmed SF, Rafa N, Mofijur M, Badruddin IA, Inayat A, Ali MS, Farrok O, Yunus Khan TM (2021) Biohydrogen production from biomass sources: metabolic pathways and economic analysis. Front Energy Res 9:753878
36. Abba GI, Uchenna UK (2023) Review of techniques for biohydrogen production from palm oil mill effluent. World J Adv Eng Technol Sci 10(1):98–102

37. Raturi S, Singh P, Jha SK, Karn SK (2022) Biohydrogen: opportunities and challenges as an alternative energy resource. J Appl Nat Sci 14(2):692–701
38. Biswal T, Shadangi KP, Sarangi PK (2023) Application of nanotechnology in the production of biohydrogen: a review. Chem Eng Technol 46(2):218–233
39. Khan MA, Ngo HH, Guo W, Liu Y, Zhang X, Guo J, Chang SW, Nguyen DD, Wang J (2018) Biohydrogen production from anaerobic digestion and its potential as renewable energy. Renew Energy 129:754–768

Bioelectricity Generation from Different Organic Waste Using BES

Satesh Kumar Devrajani and Sania Kanwal

Abstract Bioelectrochemical systems (BES) represent an innovative approach that merges renewable energy generation with efficient waste treatment, enabling the transformation of organic waste into electrical energy. This chapter explores the fundamental scientific principles underlying BES, highlighting their functional mechanisms and electrochemical operations. It examines various BES designs and system configurations, highlighting recent advancements that have contributed to improved efficiency. A thorough examination of different BES models showcases their operational versatility. This chapter also examines the utilization of diverse organic waste sources, including agricultural by-products, food industry residues, and municipal wastewater, as feedstocks for electricity generation, with a focus on how factors such as biodegradability influence energy output. Comparative analyses across different substrates provide valuable insights into system effectiveness. Additionally, this chapter outlines the practical applications of BES, including decentralized wastewater treatment, energy recovery, and sustainable waste utilization. A particular focus is placed on electrogenic microbes that facilitate the biodegradation of organic matter and the transfer of electrons. Ultimately, this chapter highlights the integrated advantages of using BES for both power generation and wastewater purification, positioning the technology as a promising tool for promoting circular economic practices and sustainable environmental solutions.

Keywords Bioelectricity · Organic waste · BES · Cathode · Anode · Fuel Cell · Microbial

---

S. K. Devrajani (✉)
Department of Civil, Environmental, Architectural Engineering and Mathematics, University of Brescia, Brescia, Italy

S. Kanwal
DICEA, Department of Civil, Architectural and Environmental Engineering, University of Padova, Padova, Italy

A. Sattar Jatoi, N. M. Mubarak (eds.), *Bio-Electrochemical Conversion of Waste to Fuel*, Green Energy and Technology,
https://doi.org/10.1007/978-981-95-4310-6_12

1 Introduction

In an era when environmental issues are becoming increasingly urgent and global energy demands are rising, the pursuit of sustainable energy alternatives has intensified research efforts across various scientific and engineering disciplines. Traditional energy generation methods, heavily reliant on fossil fuels, have been identified as major contributors to climate change and environmental degradation [1, 2]. Consequently, the development and implementation of renewable and environmentally benign energy technologies have become paramount. Among the promising emerging technologies in this domain are bioelectrochemical systems (BESs), which offer a unique and sustainable approach to energy production while simultaneously addressing the growing challenge of waste management [3, 4].

Innovative approaches for resource recovery from organic waste and sustainable energy production are provided by bioelectrochemical systems (BESs), a developing field at the nexus of biotechnology and electrochemistry [5, 6]. Fundamentally, BESs utilize the catalytic activity of enzymes or the metabolic capacities of microorganisms to promote electrochemical reactions at electrode interfaces. A primary application of BESs lies in bioelectricity generation, where microorganisms, acting as biocatalysts, oxidize organic substrates present in waste materials [5]. An electric current is produced when this oxidation process emits electrons, which are subsequently moved to an anode, pass through an external circuit, and finally decrease an electron acceptor at the cathode [1]. This direct conversion of chemical energy into electrical energy bypasses the limitations of traditional combustion processes, offering a potentially more efficient and environmentally benign route to power production.

The adaptability of BESs is underscored by their capacity to use a diverse range of organic waste streams as fuel for bioelectricity generation. These include agro-food waste and wastewater, such as chicken and dairy industry effluents, organic-rich chocolateries' wastewater, composite vegetable waste, acidogenic food waste leachate, and brewery wastewater [1, 7–9]. Furthermore, BESs can valorise municipal solid waste (MSW), which has a significant organic fraction comprising kitchen waste, fruits, and vegetables, as well as industrial wastewater from sectors such as the petroleum industry (including petroleum sludge and petroleum refinery effluent), the paper industry, and distillery wastewater. Even agricultural residues and lignocellulosic biomass hold promise as substrates for generating bioelectricity in bioelectrochemical systems (BESs). The importance of organic waste in building energy systems (BESs) is multifaceted. Firstly, it serves as the elementary source of chemical energy for microorganisms to drive bioelectrochemical reactions. Secondly, utilizing waste not only provides a renewable energy source but also offers an environmentally benign alternative to energy-intensive chemical degradation approaches used in wastewater treatment, contributing to resource recovery [8]. By converting waste into electricity, BESs contribute to a "waste-to-energy conversion" approach, aligning with the principles of a circular economy [10, 11]. Biomass-based BESs need to break down biomass and produce biofuel simultaneously through microbial metabolism. The primary components of the lignocellulosic material are lignin (5%–30%), cellulose (35%–50%), and hemicellulose (20%–35%), whereas the main biochemicals in algae are proteins (6%–71%), lipids (7%–23%), and carbohydrates (4%–57%). Vegetable cooking oil, abattoir waste, fats, oils, and grease (commonly referred to as FOG) are types of biowaste rich in lipids, with lipid content ranging from 38% to 84%. They also have high levels of total solids (26%–99%) and volatile solids (29%–99%). The foundation for utilizing these feedstocks as substrates for biofuel production is provided by their biochemical composition[12].

Bioelectrochemical systems (BESs) encompass a broad spectrum of technologies that leverage the interaction between biological and electrochemical processes for diverse applications, including the generation of bioelectricity [13]. Although the production of bioelectricity relies heavily on microbial fuel cells (MFCs), other important configurations include microbial electrolysis cells (MECs) for the production of biohydrogen, microbial desalination cells (MDCs) for the desalination of water, and microbial electrosynthesis (MES) for the production of valuable biochemicals [1, 5, 14, 15]. These systems typically consist of at least two electrodes, a cathode and an anode, connected by an external circuit, often with a separator, such as a membrane, to maintain distinct anodic and cathodic environments [4]. The flexibility of BESs enables the treatment of various organic pollutants in effluent while simultaneously recovering energy or valuable products. The development and optimization of BESs require a multidisciplinary approach, integrating principles from environmental engineering, microbiology, materials science, and electrochemistry.

The historical background and evolution of BES technology date back to the early twentieth century, with initial experiments demonstrating electricity production using microorganisms like *E. coli* and fungi [16]. A significant early contribution was made by Cohen and his group in 1931, who developed stacked bacterial fuel cells. Further exploration in the mid-twentieth century included NASA's work in 1963 on utilizing human waste for electricity generation in space missions [17, 18]. However, it is in decades that the field has witnessed rapid advancement and diversification, driven by growing concerns over energy security, climate change, and sustainable waste management. The focus has expanded from fundamental understanding and performance improvement to the development of various BES configurations for specific applications, including enhanced biogas production, biohydrogen generation, and the synthesis of value-added chemicals from waste materials. Recent field tests and industrial applications of technologies such as bioelectric toilets, pilot-scale microbial electrolysis cells (MECs), and microbial desalination cells (MDCs) indicate an increasing readiness of biogas energy systems (BES) for real-world implementation, marking a crucial step toward achieving sustainable development goals related to clean energy and environmental management.

2 Fundamentals of Bio Electrochemical System (BES)

Using a process called BES, organic waste can be converted into environmentally acceptable electricity. Bioelectrochemical systems (BES) operate through a process where microorganisms or other biobased catalysts drive oxidation and/or reduction reactions at the cathode and/or anode. BES utilizes the ability of microbes to produce and accept electrons [19, 20]. Ex-electrogenic bacteria typically carry out this task. Through the use of microbes as catalysts, BESs can transform chemical energy found in organic matter bonds into electrical energy or vice versa (Fig. 1). BES offers a combinatorial benefit by reducing the load of organic contaminants in waste and dirty water in addition to producing electrical energy [21].

In a Bioelectrochemical System (BES), the anode and cathode are typically positioned in separate chambers, one cathodic and the other anodic, and are linked by an external electrical circuit. At the anode, substances such as organic waste or sulfides act as electron donors and undergo oxidation. Meanwhile, at the cathode, electron acceptors, such as oxygen or nitrates, are reduced. The external circuit facilitates the flow of electrons between the electrodes, allowing energy to be generated or utilized without the need for combustion [22]. BES is separated into microbial fuel cells (MFCs) and microbial electrolysis cells (MECs) based on whether electricity is being produced or supplied from the circuit. MFCs show significant promise as a sustainable approach for extracting energy from waste. The process of producing hydrogen or methane from organic waste is reversed in a microbial electrolysis cell by supplying an electric current. Current can then be produced by using the resultant hydrogen or methane [23]. Anaerobic bacteria and waste organics are introduced to the anodic chamber of a microbial fuel cell. The cathodic chamber is filled with PBS or acidic water. When microorganisms eat organic materials, they release electrons and protons. Protons travel through the proton exchange membrane, while electrons move through a resistive load in the external circuit before reaching the cathode, where energy is produced. In the cathodic chamber, the combination of

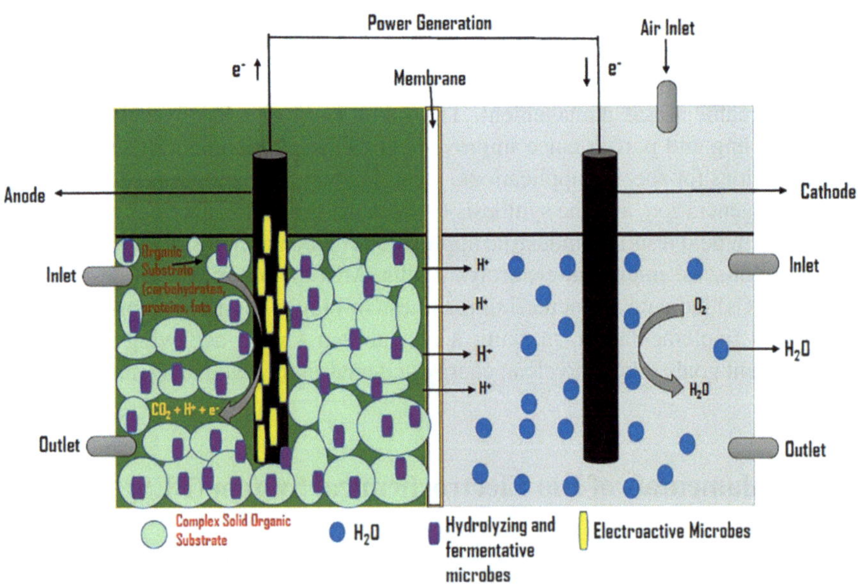

Fig. 1 Working principle of BES (adapted form [4])

protons, electrons, and oxygen results in the formation of water. The electricity generated in this process can be captured through the external circuit [24].

Through migration and ammonium conversion at high pH, the cathodic reduction process recovers nutrients (such as ammonia) and converts organic matter into electrical energy. Metals can also be extracted or recovered using BES. MFCs were also shown to be capable of producing wastewater that can be reused or irrigation [25]. Similar to an MFC, a microbial electrolysis cell uses an external voltage to reduce protons, thereby producing hydrogen gas. An MEC can potentially generate methane, depending on the microbes in the cathode [11].

Environmental sensing, effluent treatment and the production of bioenergy are just a few of the many uses for BES systems. The utilization of microorganisms as electrocatalysts, such as bacteria or archaea, is central to BES. Specialized outer membrane proteins found in certain electroactive bacteria enable direct extracellular electron transport to electrodes. Microorganisms often form biofilms on electrode surfaces in bioelectrochemical systems (BES). These biofilms improve the stability and effectiveness of metabolic and electron transport activities [24]. The ionic conductivity, which permits free movement of ions while blocking direct electron transport, is maintained between the anode and cathode by an electrolyte solution. An organic substrate, which provides microbes with energy, is necessary for BES. Sewage, different organic substances, and organic wastes are examples of common substrates. Utilizing microbial activity, BES can treat wastewater rich in organic materials while producing energy or other valuable products. BES is a clean and renewable energy source that may convert organic feedstocks into biofuels (such as biohydrogen or bioelectricity). BES can be used to clean up polluted areas by encouraging microbial activity that breaks down contaminants. Biosensors that monitor environmental conditions or identify specific analytes in real time can be developed using BES [24]. Challenges for BES research include enhancing electron transfer efficiency, optimizing microbial consortia, scaling up systems for real-world applications, and addressing issues such asouling and biofilm control [24]. Because BES can treat waste and produce renewable energy simultaneously, they are regarded as sustainable technologies that support resource recovery and environmental preservation [23]. In the domains of sustainable energy production, environmental science, and biotechnology, BES is an intriguing area of research and development because it has the potential to address environmental issues and energy concerns simultaneously [23].

2.1 BES Configuration

A BES consists of two chambers, each containing an electrode, with a membrane separating the chambers. The power is collected or invested by an external circuit that connects the two electrodes, which are separated by a membrane [26]. Factors influencing the performance of the bioelectrochemical system include substrate type, concentration, particle size, pH, electrode material, structure, availability of electrons and protons, electron and proton flow, distance between electrodes, substrate oxidation, presence of microorganisms, and type of electron acceptor [27]. Generally, solid waste contains a higher organic content than other types of waste. The Exoelectrogenic bacteria produce protons and electrons as they oxidize the organic materials in solid waste in the anodic chamber under anaerobic conditions. The proton exchange membrane allows protons to pass through. At the same time, electrons are directed to the cathode chamber via the external circuit, where they combine with oxygen (the electron acceptor) to form water [28].

2.1.1 Anode Chamber Process

The anode chamber, which houses the anode electrode, is maintained under anaerobic conditions to enable microbial activity and the breakdown of organic matter. Oxygen must be excluded, as its presence inhibits bacterial energy production [11]. Solid organic waste is pre-mixed with water before being introduced into the chamber to preserve anolyte conductivity and ensure sufficient moisture. Without water, the transport of electrons and protons is hindered. Complex organic matter in solid waste must first undergo hydrolysis to convert macromolecules into simpler compounds that are suitable for microbial utilization [29]. This breakdown process consists of four key stages. Initially, during hydrolysis, extracellular enzymes solubilise recalcitrant organic materials, such as polysaccharides, lipids, and proteins, to sugars, fatty acids, and amino acids [30]. These hydrolysed products are then processed through acidogenesis, where fermentative microbes convert them into intermediate compounds, including ethanol, volatile fatty acids (VFAs), lactate, and gases such as CO_2, H_2, H_2S, and NH_3. Acetogenesis follows, during which acetogenic bacteria convert VFAs and alcohols into hydrogen and acetic acid. The final step, methanogenesis, involves the conversion of acetogenesis by-products into methane (CH_4) and carbon dioxide by methanogenic archaea [31]. The carbon-rich organic matter acts as an electron donor. Through microbial oxidation at the anode, electrons and protons are released, along with secondary metabolites like CO_2, CH_4, and VFAs. Prominent electroactive bacteria involved include *Geobacter sulfurreducens*, *Geobacter metallireducens*, and *Shewanella putrefaciens* [32–35].

Several factors influence the performance of the anode chamber: substrate concentration, pH, particle size, electrode configuration and material, substrate oxidation rate, microbial activity, and adequate hydration to support ionic flow. Additionally, the inoculum source plays a crucial role in determining the microbial consortia on the anode, thereby influencing the formation of electroactive biofilms and energy output. In MFCs inoculated with anaerobic sludge, the *predominant bacteria* are *Bacteroidetes*, *Firmicutes*, and *Proteobacteria*. Sewage sludge yields species such as *Methanolinea*, *Clostridium*, *Methanosphaera*, *Lactobacillus*, *Methanosarcina*, *Flavobacterium*, and *Methanospirillum*, while dairy manure leads to a prevalence of *Desulfobulbus*, *Ochrobactrum*, *Comamonas*, *Clostridium*, and *Pseudomonas* [36].

2.1.2 Cathode Chamber Process

A cathode and aerobic conditions are maintained in the cathodic chamber. Cathode chamber solutions commonly used are phosphate-buffered saline (PBS) solutions and acidic water. It is essential to maintain this chamber in an aerobic environment, allowing protons and electrons to combine and generate water. The cathode and anode chambers continuously generate electrical energy in the bioelectrochemical system [37, 38]. The performance of the cathode chamber is influenced by the electrode's form and composition, the type and amount of the electron receiver, and the availability of protons and oxygen. Due to its high availability, robust oxidation potential, and non-toxic by-products, oxygen serves as a beneficial electron acceptor in the cathodic chamber [11].

3 Types of BES

3.1 Microbial Fuel Cell (MFC)

MFCs are the type of BES that generate electricity by converting the chemical energy of biodegradable organic matter through microbially mediated oxidation–reduction reactions under anaerobic and ambient conditions [39, 40]. In these systems, electrogenic microorganisms oxidize substrates such as carbohydrates, organic acids, hydrogen, and methane at the anode, releasing electrons via extracellular electron transfer (EET). These electrons travel through an external circuit to the cathode, typically where oxygen serves as the terminal electron acceptor. A proton exchange membrane (PEM) separates the anode and cathode chambers, enabling proton migration while restricting oxygen crossover [41, 42].

As protons pass through the membrane and combine with electrons and oxygen at the cathode, water is formed. MFCs are capable of achieving high-energy conversion efficiencies, recovering over 80% of the energy in organic matter and removing up to 90% of COD and BOD from wastewater [41, 43]. Diverse microorganisms, such as Rhodoferax ferrireducens, Shewanella putrefaciens, Clostridium beijerinckii, Clostridium cellulolyticum, Enterobacter cloacae, Clostridium butyricum, and Geobacter sulfurreducens, as well as fungi like *Aspergillus awamori* and *Phanerochaete chrysosporium*, are known to act as effective electron donor [44]. MFC performance is influenced by several factors, including membrane and electrode characteristics—such as durability, porosity, surface area, and stability [45, 46]. Optimal temperature supports microbial biofilm formation, enhancing EET efficiency. The cathodic oxygen reduction reaction (ORR) is also crucial in determining the system's power output and efficiency. Hence, electrode material selection is crucial for minimizing electron loss and enhancing energy recovery [47].

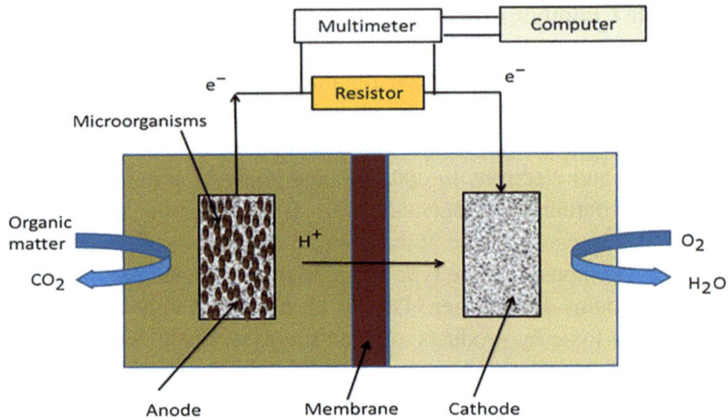

Fig. 2 Diagram of a microbial fuel cell (MFC)

Carbon-based materials, such as graphite rods, carbon fabric, and carbon mesh, are commonly used as electrodes due to their cost-effectiveness, chemical stability, and biocompatibility. These materials are often pretreated—using ammonia, acids, or chemical modifiers to enhance the effectiveness of electron transfer and microbial adhesion. Non-carbon alternatives, including stainless steel, cobalt, titanium, silver, gold, and copper, are also explored for specific applications. Ideal electrodes should offer a combination of high mechanical strength, corrosion resistance, thermal stability, affordability and electrical conductivity [45, 46]. The proton exchange membrane (PEM), typically Nafion, plays a crucial role in maintaining cathode-anode separation while facilitating the transfer of protons. However, Nafion presents challenges, including low conductivity and limited stability, which make it less suitable for large-scale applications. Ongoing research is focused on identifying cost-effective, high-performance alternatives that enhance proton transport and improve overall system efficiency [48].

Despite the potential of MFCs in renewable energy production, several challenges remain. Chief among these is the cathodic oxygen reduction, which often limits performance. Additionally, the instability and high cost of current electrode materials hinder scalability for industrial applications (Fig. 2). To address these issues, future investigations should concentrate on the development of economical, durable, and biocompatible electrode materials [49].

3.2 Microbial Electrolysis Cell (MEC)

MECs are an advanced and modified form of MFCs designed to enhance the anaerobic biodegradation of organic waste by utilizing electrodes as electron acceptors. MECs utilize electroactive microbes that release protons and electrons when they oxidize organic matter at the anode. These electrons are transferred to the cathode

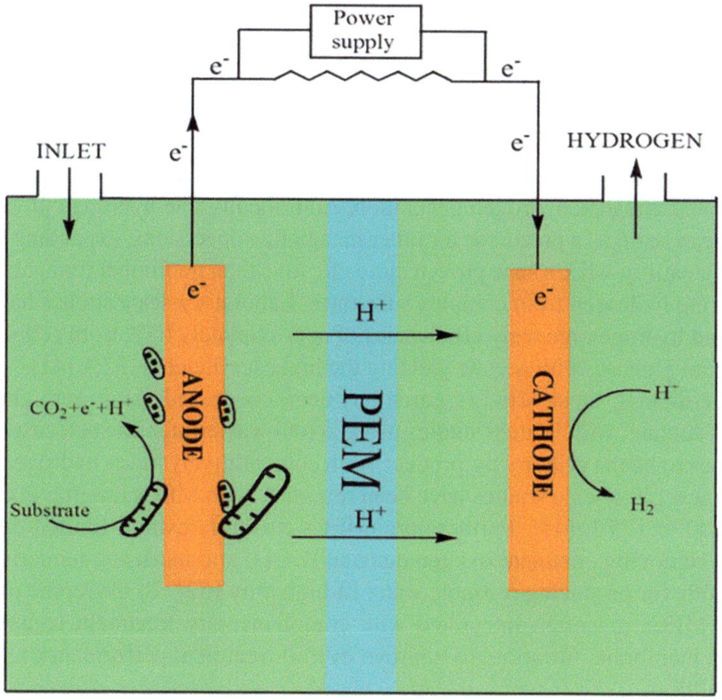

Fig. 3 Diagram of microbial electrolysis cell (MEC) (adapted from [10])

through an external circuit, where they undergo a redox reaction to produce hydrogen or other molecule values [50]. Unlike MFCs, the cathodic chamber in MECs operates under anaerobic conditions, which promotes hydrogen evolution (Fig. 3). In the presence of CO_2 and methanogenic microbes, methane can also be generated under these anoxic conditions [51]. The electroactive biofilm on the anode acts as a biological catalyst, accelerating oxidation reactions. Electrons derived from the microbial breakdown of substrates at the anode are transferred to the cathode through an external circuit, where they facilitate the generation of hydroxide ions through the reduction of water and protons.

Because the cathodic reaction in MECs is thermodynamically unfavorable, the process must be driven by a tiny external voltage. However, this voltage requirement is significantly lesser than that for conventional water electrolysis, as part of the energy demand is met by the microbial breakdown of organics. When other compounds, apart from protons, undergo cathodic reduction, additional bioproducts can also be synthesized [11, 51]. This makes MECs a promising platform for renewable energy recovery, especially as the required voltage for hydrogen evolution in MECs is around −0.41 V, substantially lower than the energy demand of traditional electrolysis. MECs are uniquely capable of overcoming fermentation-related barriers to hydrogen production and may use various organic substrates. However, despite technological advancements, numerous challenges need to be addressed before the commercialization of MECs. One limitation is that MECs cannot generate current spontaneously and require a continuous external voltage supply [50]. The anodic processes in MECs resemble those in MFCs, involving the generation of microbial electricity. Because the cathode operates anaerobically, the system is often referred to as a "bioelectrochemical assisted microbial reactor." The development of microbial communities on carbon-based cathodes can expedite system startup and enhance hydrogen generation. Additionally, the hydrogen produced in MECs can serve as a precursor for other metabolic conversions, expanding the system's versatility. MECs have proven more effective than traditional fermentation for recovering hydrogen from complex substrates. Laboratory-scale studies have demonstrated hydrogen recovery efficiencies of approximately 63% from cellulose and 64% from glucose, with acetate yielding the highest efficiency, 82% [41].

Nevertheless, most MEC research has been conducted using pure chemicals such as acetate, with limited studies using actual wastewater. In such cases, methane tends to be the primary by-product, with comparatively lower hydrogen yields. Only a few pilot-scale studies have been reported, with MECs ranging in capacity from 100 to 175 L [41]. Furthermore, MECs currently exhibit limited effectiveness in removing chemical oxygen demand (COD) and nutrients, rendering them unsuitable for producing reusable water of high ality [52]. To overcome this limitation, MECs are often integrated with complementary treatment technologies, such as membrane filtration, to improve overall treatment performance and effluent quality.

3.3 Microbial Desalination Cell (MDC)

MDCs represent a novel evolution of MFC technology, integrating wastewater treatment, electricity generation, and desalination into a single system [41]. MDCs operate on the same basic principle as MFCs, utilizing the potential difference between the anode and cathode. However, they feature an additional desalination chamber positioned between the two electrodes. This central chamber is separated from the anode and cathode by a cation exchange membrane (CEM) and an anion exchange membrane (AEM), respectively [51]. In MDCs, electroactive microorganisms at the anode oxidize organic substrates, releasing electrons and protons in the process. The electrons travel through an external circuit to the cathode, generating electrical power. Simultaneously, ions from saline water in the desalination chamber migrate across the ion exchange membranes to maintain charge balance: cations (e.g., Na^+, K^+) move toward the cathode through the CEM, and anions (e.g., Cl^-, SO_4^{2-}) move toward the anode through the AEM. This ionic migration results in the effective desalination of the central chamber. At the same time, it enhances the ionic conductivity of both the anolyte and catholyte, thereby improving the system's overall energy output [51]. By coupling multiple environmental and energy functions, MDCs present a multifunctional and sustainable solution for treating saline wastewater, recovering energy, and reducing water salinity in a single integrated process.

The desalination performance of MDCs is significantly affected by the salt concentration of the feedwater. Higher salinity reduces the ohmic resistance, enabling increased current density and accelerating the desalination rate. However, extreme salinity can impair the metabolic activity of the electroactive microbial communities at both electrodes. To mitigate this, pretreatment methods, including reverse osmosis (RO), can be employed to lower the salt content of the influent, which in turn helps reduce energy consumption and microbial stress. MDCs have shown promising results in both laboratory and pilot-scale studies. Reports indicate salt removal efficiencies ranging between 90% and 99% [41]. Moreover, 58% of the electricity required to power downstream reverse osmosis (RO) systems is generated by municipal distribution companies (MDCs). For instance, Jacobson et al. [53] found that while reverse osmosis (RO) requires approximately 2.2 kWh to desalinate 1 m^3 of seawater, membrane distillation combined with desalination (MDCs) can desalinate the same volume with a net energy gain of 4 kWh and produce around 1.8 kWh of electrical energy (Fig. 4). This results in an energy recovery that is 180%–231% greater than the energy input required [51].

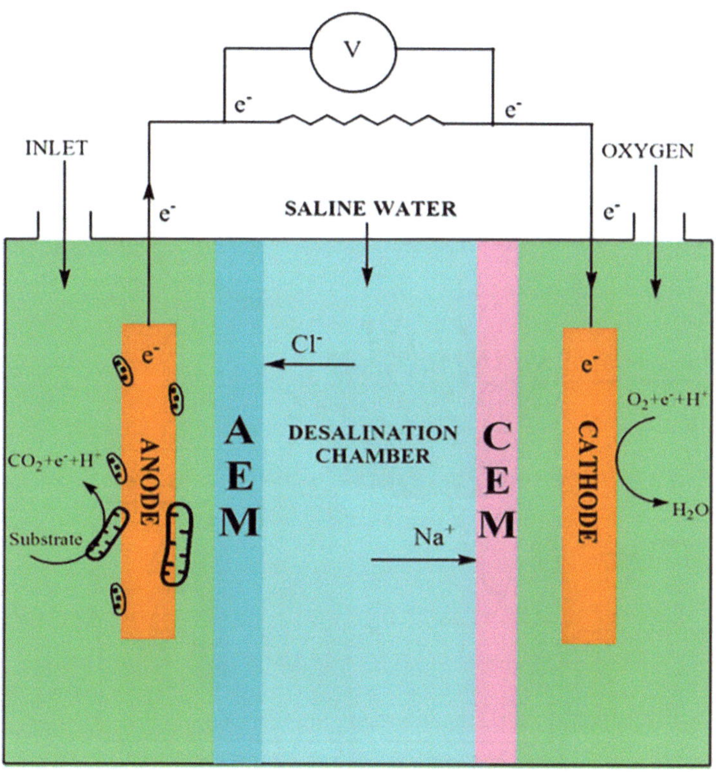

Fig. 4 Diagram of microbial desalination cell (MDC) (adopted from [10])

3.4 Microbial Solar Cell (MSC)

MSCs also referred to as bio-photovoltaic systems are innovative biotechnological platforms that integrate photosynthetically active organisms with electroactive microbes to generate biofuels and bioenergy products such as hydrogen, methane, ethanol, and hydrogen peroxide [51]. These systems harness solar energy through photoautotrophic organisms—such as green plants or cyanobacteria—and convert it into chemical energy, which is then utilized by electroactive bacteria for electricity generation. In MSCs, photosynthesis drives the production of organic compounds, which are subsequently transported to the anode compartment. Here, electroactive microorganisms oxidize these organic substrates, releasing electrons in the process. These electrons then travel through an external circuit to the cathode, where they participate in reduction reactions—typically reducing oxygen to water (Fig. 5). Cyanobacteria and green plants play a dual role: they use solar energy to split water into protons, electrons, and oxygen, and they also fix carbon dioxide using the released electrons and protons to synthesize multi-carbon compounds. These compounds form the foundation for microbial metabolism and serve as substrates for further energy conversion within the microbial symbiotic community (MSC) system [55]. Overall, MSCs offer a promising route for sustainable bioenergy generation by coupling solar energy capture with microbial electrochemical conversion, effectively linking photosynthetic carbon fixation with electrical energy production in an integrated system.

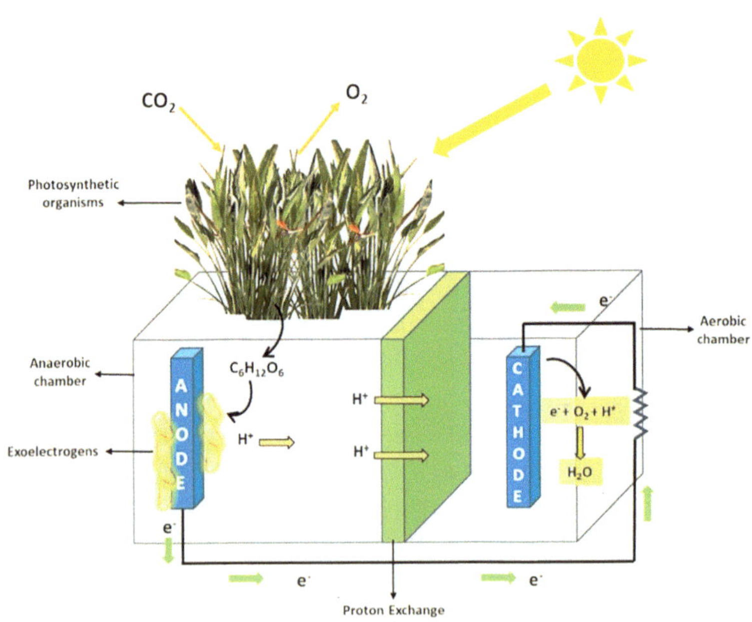

Fig. 5 Diagram of microbial solar cell (MSC) (adapted from [54])

3.5 *Microbial Electrosynthesis (MES)*

MES is an advanced approach that merges electrochemical and biological systems to convert CO_2 and other low-value compounds into valuable chemicals using microbial catalysts and externally supplied electrical energy [41]. The process utilizes electrogenic bacteria that absorb electrons directly from a cathode to reduce carbon dioxide and synthesize organic products (Fig. 6). In MES, oxidation occurs at the anode, producing electrons and protons, which then travel to the cathode. Here, specific electroactive microorganisms behave as biocatalysts, utilizing the incoming electrons for the reductive transformation of CO_2 into value-added products, including butyrate, methane, ethanol, and acetate [56]. For example, *Methanococcus maripaludis* can utilize self-secreted mediators for CO_2 reduction. At the same time, Sporomusa ovata has demonstrated the ability to enhance acetate production through a specialized cathode design that facilitates direct CO_2 delivery.

The technology enables the conversion of organic matter or CO_2 into fuels, providing a sustainable pathway for resource recovery and utilization. However, common industrial microbes like *E. coli* are not naturally electro-trophic. Instead, chemolithoautotrophy, such as that of Ralstonia, is better suited due to its metabolic adaptability, even though it often lacks efficiency in extracellular electron transfer (EET). Genetic engineering is being increasingly explored to enhance microbial electro-trophic capabilities and expand the metabolic pathways for synthesisg

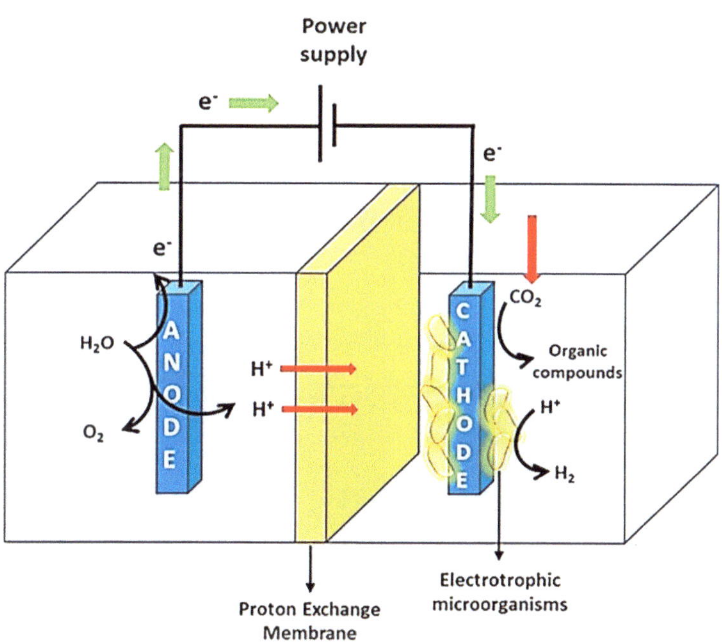

Fig. 6 Diagram of Microbial electrosynthesis (MES) (adopted from [54])

commercially important compounds [57]. MES enables the initiation of targeted biochemical conversions—such as transforming CO_2 into acetic acid or glycerol into ethanol and 1,3-propanediol—by controlling the addition or removal of electrons. However, *homoacetogens*, which are frequently employed for CO_2 conversion, often exhibit limited selectivity, typically favoring acetate as the dominant product regardless of the desired outcome [58]. This limitation affects the overall electron utilization efficiency and cost-effectiveness of the process. Although MES holds great potential for sustainable chemical production from waste or carbon dioxide, current production rates and system efficiencies remain inadequate for large-scale, economically viable deployment. Ongoing research is focused on enhancing microbial specificity, electron transfer, and process scalability to make microbial electrosynthesis (MES) a practical solution for industrial applications [41, 58].

3.6 Enzymatic Fuel Cell (EFC)

EFCs, a variant of bioelectrochemical systems (BES), utilize enzymes as biocatalysts to facilitate redox reactions at the electrodes, serving as electron donors or acceptors [56]. These enzymes can either be immobilized on the electrode surfaces or suspended in the electrolyte, allowing the oxidation of organic substrates and the generation of electrical current (Fig. 7). Compared to MFCs, EFCs can generate higher power densities, but they suffer from enzyme instability, limiting their operational lifespan to approximately 7–10 days. These devices function optimally in mild conditions, with a neutral pH and room temperature. The enzymatic oxidation

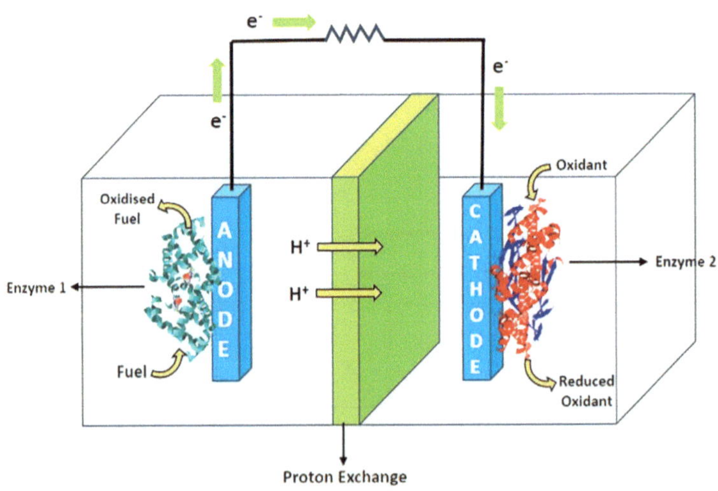

Fig. 7 Diagram of enzymatic fuel cell (EFC) (adapted from [54])

is typically partial, leading to heat generation that can inhibit enzyme activity. To improve enzyme durability and performance, techniques like encapsulation in micelle polymers are employed. This provides pH control, protection against denaturation, and a hydrophobic environment. Immobilizing microbial communities on nanostructured electrodes or within conductive polymers also enhances long-term stability and mass transfer efficiency [51].

Bioanodes employing flavin adenine dinucleotide (FAD)-dependent glucose dehydrogenases, in conjunction with electrodes modified with osmium polymers, demonstrate high glucose sensitivity and robust current output in glucose-fueled EFCs. Similarly, pyrroloquinoline quinone (PQQ)-dependent alcohol dehydrogenases naturally found in acetic acid bacteria demonstrate excellent electron transfer on carbon-based electrodes, supporting both biosensor and power generation applications [41]. EFCs are especially suitable for portable and wireless energy systems because of their high-energy density and low power output. Stability and reaction selectivity has been demonstrated by using a copper-based laccase from fungi as a cathodic catalyst and FAD-dependent fructose dehydrogenase from *Gluconobacter* as an anodic catalyst. These enzymes, especially multicopper oxidases, exhibit a high affinity for oxygen reduction, thereby further enhancing the cell's efficiency. Amperometric enzyme electrodes developed using oxidoreductase enzymes offer a wide range of biosensing applications. When combined with systems offering higher energy density, EFCs can serve self-regulating devices such as video surveillance sensors. Multi-enzyme EFC configurations have been explored to enhance substrate conversion and performance, although they are costlier due to the need for purified enzymes [59].

4 Organic Wastes Used in BES for Bioelectricity Generation

BESs represent a promising pathway for converting the chemical energy inherent in organic matter, particularly from waste streams, into usable bioelectricity and other valuable products. [5, 11, 60]. This technology suggests substantial advantages for the sustainable generation of energy and environmental management by simultaneously treating waste and recovering resources. Bioelectricity production in BESs can be facilitated by a variety of organic wastes and effluents originating from agricultural, industrial, and municipal sources [60, 61]. The selection and effectiveness of a particular waste stream as a substrate are heavily influenced by its characteristics, biodegradability, and the specific BES configuration employed [1, 15, 61].

BESs are cutting-edge technologies that use microbes to transform organic materials into useful chemicals or power electrochemically. Biodegradable compounds can be found in organic waste, which includes sewage sludge, food scraps, and agricultural residues. This organic matter is directly or indirectly transformed into electricity utilizing BES, MFCs, or MECs. As microorganisms break down the organic chemical electrons are produced that can be transformed into electrical energy [11, 21, 62–64].

BES is an effective and sustainable method for handling organic waste. These systems' microorganisms facilitate the breakdown of organic waste, thereby reducing its volume and mitigating its negative environmental impacts. Additionally, BES can assist in cleaning organic waste of contaminants and pathogens. Apart from producing energy and hydrogen, BES can also turn organic waste into valuable chemicals and bioproducts. BES, for instance, might be made to make alcohol, organic acids, or other beneficial substances with commercial uses. BES can employ some solid waste products, including organic sludges, as substrates to create hydrogen gas (H_2). Among its many uses, hydrogen serves as a clean energy source that can be utilized in industrial operations and fuel cells [62–64]. By recovering resources from solid waste streams, the principles of the circular economy, which turn trash into valuable resources, can be supported by integrating BES into waste management systems. By turning trash into valuable goods, BES can help increase the value of garbage, lessen the demand for incineration and landfilling, and possibly increase sales of chemicals or electricity [11, 21, 62–64]. Utilizing trash in BES is a viable approach to managing waste sustainably, producing renewable energy, and recovering resources, which leads to more cost-effective and environmentally beneficial waste treatment methods. Several kinds of organic waste are frequently employed in bioelectrochemical systems [60, 62, 64].

4.1 Municipal Solid Waste (MSW)

MSW, generated from industrial, commercial, and residential activities in urban areas, is a valuable feedstock for BES because of its large amount of biodegradable organic material. Globally, organic waste accounts for approximately 47% of total municipal solid waste (MSW), with the proportion even higher in countries like India due to the significant amount of kitchen waste generated [65]. This organic fraction, known as OFMSW, includes food scraps, fruits, and vegetables from households, canteens, hotels, and yard waste. Chemically, the OFMSW is composed of complex biodegradable substances, including lipids, proteins, fats, inorganic salts, carbohydrates, and trace elements [4]. Food waste alone accounts for approximately 18% of the global waste stream and is characterized by a high concentration of macromolecules, including lipids, carbohydrates, and proteins. Due to its rich organic content, OFMSW is regarded as an ideal substrate for BES applications [66]. However, to utilize the solid organic fraction efficiently, phase conversion processes like acidogenic fermentation are typically employed, often using leach bed reactors (LBRs) to generate volatile fatty acid (VFA)-rich leachate [65, 67, 68]. For instance, treating diluted OFMSW leachate in an air-cathode MFC produced a maximum power density of 5.9 W/m^3 [65]. Similarly, domestic waste processed in a BES configuration (BES-C) achieved an ultimate power density of approximately 24 ± 0.5 mW/cm^2 [1]. Food waste has shown particular promise for bioelectricity and biohydrogen generation in integrated bioreactor systems (BES) that combine microbial fuel cells (MFCs) and microbial electrolysis cells (MECs).

In one study, MFCs fed with food waste as a substrate attained a peak power density of 2.4 ± 0.1 W/m^2 at an open-circuit voltage (OCV) of 690 mV. At the same time, subsequent MEC treatment of the digestate yielded a biohydrogen production rate of 55.1 mL/g COD removed—higher than results obtained from sewage sludge or compost leachate [3, 65]. Additionally, compost leachate has also been successfully used as a substrate in such integrated systems. Field demonstrations, including bio-electric toilets and other municipal-scale biogas energy systems (BES) projects, further validate the practical feasibility of using municipal solid waste (MSW) for bioenergy recovery.

4.2 *Agricultural Waste*

It comprises crop waste, animal excrement, and other leftovers from agricultural activities [69, 70]. Because they contain a high amount of organic components and nutrients, these wastes are suitable for BES. Agricultural practices generate substantial amounts of organic waste and by-products, including animal manure and crop residues, which are rich in nutrients and organic matter. These are considered abundant and cheap renewable energy resources [61]. Crop residues, including maize stover, rice straw, wheat straw, bamboo, grass, leaves, and wood, primarily consist of lignocellulosic materials [6]. While they offer potential as feedstocks, the presence of lignin makes them recalcitrant to easy microbial degradation, often requiring pretreatment, such as hydrolysis, to release fermentable sugars. Research findings indicate promising prospects for utilizing lignocellulosic biomass in microbial fuel cells (MFCs). For instance, research involving single-chamber MFCs demonstrated power densities of 343 mW/m^2 with steam-exploded maize stover and 296 mW/m^2 with raw maize stover [6]. MESs can also utilize agricultural and forestry waste to produce various valuable compounds and fuels, including syngas, biohydrogen, biomethane, ethanol, and diesel fuel.

4.3 *Food Processing Waste (FPW)*

Food production waste (FPW) and food waste generated from food industries, households, and restaurants account for a significant portion of the global waste stream, approximately 18%, and over 1.3 billion tons produced annually [66, 71]. These wastes are typically landfilled or incinerated despite their rich organic content and high biodegradability. Their composition varies significantly with geography, climate, cultural habits, and economic status, but they commonly contain large quantities of lipids, proteins, carbohydrates, and other macromolecules. Due to this high organic load, food waste is highly suitable for anaerobic biological processes. During carcinogenesis, the breakdown of these carbon-rich materials yields soluble compounds that can serve both as carbon and energy sources for microbes. These products are also useful in the biosynthesis of biohydrogen and biodegradable polymers such as polyhydroxyalkanoates. Moreover, fatty acids derived from the anaerobic digestion of food waste are considered highly effective electron donors for biological nitrogen and phosphorus removal processes. Food processing wastes, in particular, are ideal substrates for BESs, especially MFCs, owing to their high levels of readily degradable organic compounds and consistent availability [2, 61]. Various industrial waste streams have been explored, including effluents from breweries, wineries, dairies, canteens, starch and whey industries, and meat, vegetable, and fish processing plants. Other examples include rice mill effluent, chocolate factory wastewater, wine lees, yogurt waste, fermented apple juice, and wastewater from cereal processing [72, 73]. The efficiency of energy generation and waste treatment in BESs depends mainly on the composition and biodegradability of the substrate. Generally, the more biodegradable the waste, the higher the potential for conversion and power output. Food waste leachate—particularly the acidic fraction—has also shown promise as an effective substrate for microbial fuel cells (MFCs), further demonstrating the potential of food-related wastes in energy recovery applications.

4.4 Industrial Waste

Industrial wastewater from sectors like petroleum, distillery, molasses processing, paper recycling, textile, and pharmaceuticals has been effectively treated in Bioelectrochemical Systems (BESs), often with concurrent electricity generation [5, 11, 60, 74]. These waste streams often contain complex or recalcitrant compounds that are difficult to degrade using conventional biological methods. Wastewater from the petroleum industry, such as refinery effluent, formation water, and petroleum sludge, contains high concentrations of hydrocarbons and other persistent pollutants. Nonetheless, hybrid BESs have demonstrated substantial removal efficiencies for such contaminants—up to 96.5% for COD, 99% for phenanthrene, 94% for pyrene, and 80% for total dissolved solids (TDS) [60]. Similarly, BESs have demonstrated success in treating distillery effluents by achieving simultaneous decolourization, desalination, and power generation. Molasses-based wastewater, though rich in organics, is more complex and often yields lower energy outputs when processed through BESs. Despite these challenges, BESs can be applied to treat industrial wastewater containing refractory pollutants, such as phenols. They have also enabled the dichlorination of chlorinated phenols and decolourization of azo dyes via electroactive microbial communities functioning at the anode.

Wastewater treatment plants (WWTPs) typically handle large volumes of organic-laden water and generate significant sludge as a by-product [18, 75, 76]. Traditional methods, including anaerobic digestion (AD) and activated sludge process, are inefficient and energy-intensive for low-strength waste streams [51]. BESs offer an energy-efficient alternative with an additional advantage of producing electricity, and they can potentially be integrated into existing WWTP infrastructure. Sewage sludge is a nutrient-rich material that holds promise as a feedstock in biogas energy systems (BESs) for energy and resource recovery. BESs also produce less sludge than conventional systems, such as anaerobic digestion (AD), which may reduce sludge management costs. However, treating sludge poses challenges, including odor control, dewatering, and pathogen removal. Sludge may contain hazardous components, including heavy metals, microplastics, and pathogenic microorganisms, as well as recoverable resources such as phosphorus [54].

Electrochemical techniques such as electro-dewatering are being explored for sludge processing due to their compact design and reduced need for chemical inputs. The high salinity (1–3 mS/cm) typical of some sludges makes them suitable for electrochemical treatment [77–79]. However, phosphorus recovery is complicated by the presence of iron, which can cause it to precipitate. Alkaline treatment has been used to enhance phosphorus solubility, offering a gentler alternative to acidic extraction. In integrated BES applications, sewage sludge used in MFCs has shown promising results, achieving a power density of 1.8 ± 0.1 W/m^2 and an open-circuit voltage of 595 mV. Further treatment in microbial electrolysis cells (MECs) led to effective COD removal and high biohydrogen production [3, 54].

4.5 Other Organic Waste

In addition to the primary waste categories, various other organic matter and by-products have been explored as prospective substrates for bioelectrochemical systems (BESs). Lipid-rich wastes, including FOG, which is commonly found in kitchen wastewater, represent a significant but underexplored resource for BES applications [12, 80]. Although traditionally targeted for biodiesel production via anaerobic digestion, FOG has also demonstrated promise in BESs. A constructed sediment MFC treating FOG-rich effluent attained the highest power density of 114.44 mW/m^2 [80].

Organic matter is increasingly regarded as a valuable feedstock not only for electricity generation but also for producing biofuels and other high-value products through biorefineries and integrated systems. MECs have demonstrated strong potential for biohydrogen production from various biomass sources and liquid waste. MESs further extend this capability by converting agricultural and forestry residues into products such as biohydrogen, biomethane, ethanol, and diesel. Anaerobic digestion remains a well-established process for producing biogas (a mixture of methane and carbon dioxide) from a wide range of organic wastes. Looking ahead, the biodegradable organic fraction of MSW could also be harnessed to generate biofuels, ethanol and hydrogen. The exploitation of lipid- and protein-rich biowastes presents a promising new frontier for BESs, particularly for enhancing bioelectricity and biohydrogen generation (Table 1).

Table 1 The various kinds of organic materials that can be employed for BES, together with possible applications of each type

Organic waste material	Potential applications in BES
General organic waste	Bioelectricity generation, biohydrogen production, biogas production (in combination with AD), biochemical value-added output (MES), wastewater treatment, nutrient recovery, metal recovery, desalination, bioremediation, biosensors, biochar production
MSW/OFMSW	Bioelectricity generation, biogas production, biohydrogen production, organic acid production
Food waste	Bioelectricity generation, biohydrogen production, ethanol production
Agricultural waste	Bioelectricity generation, biohydrogen production, syngas production, biomethane production
Animal dung	Biogas production, bioelectricity generation
Crop waste (e.g., Rice straw)	Bioelectricity generation
Lignocellulosic waste	Bioelectricity generation, biohydrogen production, biofuel production
Industrial wastewater	Bioelectricity generation, wastewater treatment
Dairy wastewater	Bioelectricity generation (co-substrate)
Petroleum refinery wastes	Bioelectricity generation, wastewater treatment
Swine waste	Bioelectricity generation
Slaughterhouse waste	Bioelectricity generation, biogas production
Pineapple waste	Bioelectricity generation
Sewage sludge	Bioelectricity generation, biohydrogen production, bioplastic production
Landfill leachate	Electricity generation, leachate treatment

4.6 *Influence of Waste Characteristics Biodegradability*

To promote their breakdown, solid waste that is difficult to biodegrade may be combined with various substrates. We refer to this process as co-digestion. It's crucial to remember that several variables, including the type and composition of organic matter, the BES's design, and the surrounding environment, affect the effectiveness of the BES. BES technology and waste characteristics must be thoroughly understood for adoption to be successful.

The efficiency of BESs in producing energy and waste treatment is highly reliant on the biodegradability and characteristics of the input waste. Co-digestion, which involves blending poorly degradable solid waste with more biodegradable substrates, can enhance the degradation process. The effectiveness of this process is entirely dependent on numerous factors, including the type of waste, its organic composition, the design of the BES, and environmental conditions, emphasizing the need for a comprehensive understanding of waste profiles and system configurations. Substrates rich in simply biodegradable organic compounds, such as dairy, vegetable, and food processing wastes, tend to yield higher power outputs and improved treatment efficiency. The organic fraction of municipal solid waste (OFMSW), particularly kitchen waste, exhibits high biodegradability and high moisture content, enabling rapid microbial degradation and a high biochemical methane potential. Leachates from the anaerobic digestion of OFMSW contain significant levels of VFAs like acetic, propionic, butyric, and valeric acids, which serve as effective energy and carbon sources for electroactive microbes in BESs. A higher ratio between total volatile fatty acid (TVFA) and total chemical oxygen demand (TCOD) is a strong indicator of substrate biodegradability.

Biodegradable solid wastes, such as yard waste, agricultural residues, and food waste, are valuable for BESs due to their nutrient-rich organic content. However, these wastes may also contain hazardous components, including pathogens, heavy metals, or toxins, which require proper treatment. Anaerobic digestion and composting provide sustainable alternatives to landfilling and incineration, enabling the recovery of methane and the recycling of nutrients. Solid waste, which encompasses domestic, commercial, industrial, and animal-derived residues, exhibits significant variation in composition based on source and location [27, 76, 81–83]. Before use in BESs, non-biodegradable components (e.g., plastics, metals) must be separated. Pretreatment of the organic fraction can enhance the degradation process. Moisture content—an essential factor for substrate degradation and electron flow—varies widely by waste type and origin [27]. Optimal water levels are essential for supporting microbial activity and electrochemical performance in BESs.

The biodegradability of waste is defined by its capacity to undergo microbial decomposition, as seen with paper or food scraps [84–86]. Conversely, materials like plastic and metal resist degradation. Chemically, solid waste may contain contaminants such as pesticides and heavy metals, which, if improperly handled, can pose significant health and environmental risks. Understanding the detailed composition of waste is crucial for designing sustainable waste management systems, enhancing recycling, and optimizing resource recovery. Odor issues caused by the degradation of organic material, as well as potential health risks from infectious agents in solid waste, highlight the importance of proper handling and disposal [87]. Waste-to-energy technologies should be selected based on local regulations, infrastructure availability, waste composition, and the desired end-products. Energy recovery from organic matter offers environmental benefits, including greenhouse gas (GHG) mitigation and reduced landfill use, provided these processes are well-managed.

Organic waste types suitable for BES applications encompass a range of materials with varying characteristics. MSW, composed mainly of food, fruit, and vegetable scraps, is rich in complex organic compounds, including carbohydrates, proteins, and lipids, and possesses a high BOD/COD ratio; however, its low nitrogen content may limit anaerobic digestion. Food waste (FW), generated from food industries, restaurants, and households, is highly biodegradable due to its rich organic matrix and releases oligosaccharides and monosaccharides during acidogenesis, making it suitable for microbial processes; examples include yogurt residues, vegetable waste, and wine lees. Agricultural waste, such as rice straw, corn stover, and animal dung, contains cellulose and lignocellulose, which require pretreatment due to microbial resistance but offer the potential for conversion into syngas, hydrogen, and methane. Industrial wastewater varies in complexity and biodegradability, ranging from low-energy dairy and sugar mill effluents to high-energy wastes from wineries, olive mills, and cheese production. Petroleum refinery wastewater (PRW) is notable for its high total dissolved solids (TDS) and organic load. Animal-based wastewaters, particularly from slaughterhouses and swine operations, are rich in fats, proteins, and biodegradable carbon, with swine sludge serving as an economical nutrient source. Other specific wastes include pineapple waste, which contains fiber, amino acids, and sugars; cassava starch wastewater with high COD; fermentation effluents rich in VFAs and alcohols, especially dark fermentation effluent (DFE); and molasses-based distillery wastewater, which presents both high-energy potential and treatment challenges due to its intense COD and colouration (Table 2).

Table 2 Key characteristics of different organic wastes: due to the wide variability within each waste type and the different parameters reported across sources, constructing a table with precise quantitative data for all types is challenging. However, the following table summarizes the key characteristics discussed

Waste type	Key organic components	Biodegradability	Potential for BES	Pretreatment needs	Examples
OFMSW	Carbohydrates, fats, proteins, VFAs	High to medium	High	Size reduction, sorting	Kitchen waste (food, fruits, vegetables)
Food waste	Lipids, carbohydrates, proteins, sugars, organic acids	High	High	Size reduction, sorting, and sometimes fermentation	Restaurant/ household waste, canteen waste, vegetable waste, yogurt waste, wine lees, food processing wastewater
Agricultural waste	Cellulose, hemicellulose, lignin, sugars (after hydrolysis)	Variable (often lower)	Moderate to high	Hydrolysis (acidic, enzymatic), grinding	Animal dung (swine, cattle), crop residues (rice straw, corn Stover)
Industrial wastewater	Sugars, starches, proteins, fats, oils, phenols, etc.	Variable	Moderate to high	Depends on the pollutant	Brewery, food processing, paper recycling, petroleum refinery, winery, olive mills, cheese wastewater

Table 2 (continued)

Waste type	Key organic components	Biodegradability	Potential for BES	Pretreatment needs	Examples
Animal-based wastewater	Biodegradable organic carbon, fats, proteins	High	High	Solids separation	Slaughterhouse wastewater, swine wastewater, animal carcass waste, wastewater sludge
Pineapple waste	Monosaccharides, polyalcohols, amino acids, organic acids, fiber	High	High	Size reduction	Pineapple processing waste
Cassava starch wastewater	Starch, sugars	High	High	Solids separation	Cassava starch processing effluent
Fermentation effluent (DFE)	VFAs, sugars, alcohols	High	High	pH adjustment	Effluent from dark fermentation for hydrogen production
Molasses distillery wastewater	High organic load, color	High	High	Dilution, pretreatment for color	Distillery effluent

This table highlights the diverse potential of BES technology in utilizing various organic waste materials for energy production, resource recovery, and environmental remediation. The specific application and efficiency are based on the type and characteristics of organic waste, as well as the design and operative parameters of BES.

4.7 *Comparative Performance and Efficiency*

Studies comparing various waste types as substrates for BESs reveal notable differences in performance. For example, an integrated BES system (MFC followed by MEC) utilizing different organic substrates demonstrated that Food Waste (FW) generated the highest OCV of 690 mV and ultimate power density of 2.4 ± 0.1 W/m^2, surpassing Sewage Sludge (SS) (OCV of 595 mV and power density of 1.8 ± 0.1 W/m^2) and Compost Leachate (CL) (OCV of 505 mV and power density of 1.3 ± 0.1 W/m^2) [3], indicating FW as a potent substrate for bioelectricity generation. An MFC treating diluted leachate from the acidogenic fermentation of OFMSW attained a peak power density of 5.9 W/m^3 [65], while domestic wastewater in MFCs has been shown to generate up to 12 W/m^3, though still relatively low compared to the energy potential in the organic waste itself [88]. Performance also varies across different biomass substrates: *E. prolifera* algal biomass attained a high-power density of 3810 mW/m^2, whereas bamboo hydrolysate (lignocellulosic) reached 578 mW/m^2 [12]. In contrast, the power density of 73.78 mW/m^2 was significantly lower for unhydrolyzed sago hampas biomass [89], and kitchen wastewater (FOG effluent) in a sediment MFC yielded a power density of 114.44 mW/m^2. Regarding COD removal, the integrated BES treating FW, SS, and CL achieved overall COD removals of 83.25%, 88.23%, and 73.41%, respectively [3]. COD removal has also been observed in BESs treating industrial wastewater, including those from dairy, textile, and petroleum industries, highlighting the influence of waste complexity on both power generation and treatment efficiency.

5 Factors Affecting Bioelectricity Generation from Organic Waste

BESs are promising technologies that simultaneously treat organic matter and generate renewable energy and bioelectricity. The efficient operation and performance of BESs are influenced by multiple interconnected elements, including the nature of the organic substrate, the microbial communities involved, the system design and configuration, and various environmental conditions. These factors are essential for optimizing BES performance and facilitating their practical application. Recent investigations continue to explore and refine the understanding of these elements to enhance bioelectricity generation and waste treatment efficiency.

5.1 Characteristics of Organic Waste

Organic waste, serving as a substrate or fuel in the anodic chamber of a BES, significantly impacts the system's ability to generate bioelectricity. The composition and degradability of the waste material are key determinants of performance.

5.1.1 Carbon Content

Microorganisms in BESs obtain energy from the organic substrate. Microorganisms, particularly electrochemically active bacteria (EAB), oxidize the substrate in the anaerobic anodic chamber, producing electrons and protons [60]. The concentration and type of organic compounds which contribute to COD directly influence the accessibility of electron donors for these reactions. The type and amount of carbon sources in the waste directly influence microbial metabolic activity and the resulting bioelectricity output. Simple organic molecules, such as VFAs, sugars, and alcohols, are readily utilized by electroactive bacteria. Studies have shown that the substrate and its concentration are crucial factors in determining the outcome. For instance, increasing the operational temperature in an MFC to a certain level can enhance the power output by increasing the substrate degradation rate. Different organic wastes, including agricultural by-products, food waste, dairy effluent, and domestic and industrial wastewaters, have been utilized, and their performance varies based on their composition and concentration.

5.1.2 Nutrient Composition

Beyond just carbon sources, the presence of phosphorous, nitrogen and other trace elements is vital for microbial growth and metabolism. Nutrient-enriched leachate and volatile fatty acids (VFA), obtained from the anaerobic digestion of organic substrates, can serve as a carbon source and help recover valuable nutrients for the microbial community in biogas energy systems (BESs). Proper nutrient availability supports healthy microbial consortia and efficient substrate degradation, which in turn impacts power generation.

5.1.3 Waste Degradability

The complexity and biodegradability of the organic matter directly influence how readily the microbial community can break it down. Complex organic compounds undergo various metabolic processes, including hydrolysis, acidogenesis, acetogenesis, and methanogenesis, to be converted into simpler molecules and gases [3, 29]. Hydrolysis is often a rate-limiting step, particularly for complex substrates like lignocellulosic materials. Microorganisms cannot utilize insoluble macromolecules in organic substrates until they have been hydrolysed into simple molecules. The efficiency of substrate degradation by microbes in the anodic chamber directly affects current production. Using readily consumable compounds, such as volatile fatty acids (VFAs), which are products of hydrolysis and acidogenesis, can be advantageous. Different types of organic waste have varying degradability, which influences the overall performance of the BES.

5.2 Microbial Consortia and Activity

Microorganisms, acting as biocatalysts, are at the heart of BESs, driving the electrochemical reactions that generate energy and treat waste. The composition, activity, and spatial arrangement of these microbes are crucial for system performance.

5.2.1 Microbial Diversity

Diverse microbial communities are often needed for the oxidation of intricate organic materials in wastewater. A mixed culture, enriched with anaerobic microbes, is typically used in BESs, which develop biofilm on the anode electrode surface [90]. The diversity of microbial species can enhance substrate degradation and electron transfer. In the anode, the microbial community can be considerably impacted by the source of inoculum. Different inoculum sources result in dominant species, including Proteobacteria, Firmicutes, Bacteroidetes, Acidobacteria, Clostridium, Shewanella, Geobacter, and others, depending on the waste source. Optimizing microbial consortia is a key challenge in bioremediation and environmental sustainability (BES) research.

5.2.2 Biofilms

On the surfaces of electrodes, microorganisms frequently create biofilms that function as electrocatalytic units. For BESs to remain stable and effective over the long term, these biofilms are essential because they improve the stability and effectiveness of metabolic and electron transfer activities [5]. These biofilms are created by exoelectrogenic bacteria that can transfer electrons from their metabolism to an external solid electron acceptor, such as the anode. One important aspect of the anode is the capacity of microorganisms to facilitate the formation of biofilms, thereby enhancing the likelihood of extracellular electron transfer (EET). Direct electron transport between microorganisms and the surface of the electrode of the anode is enhanced by biofilm that develops on the anode electrode [60, 91]. Biofilm development and efficiency are considered the limiting factors for overall BES performance. Strategies for immobilizing exoelectrogenic microorganisms onto the anode are being explored to enhance anode activity, energy output, robustness, and biocompatibility.

5.2.3 Microbial Electrochemical Activity

The primary function of microbes in BES is to catalyze electrochemical reactions, either oxidizing substrates at the anode or reducing compounds at the cathode. Exoelectrogenic microbes perform electron generation and acceptance [1]. They transfer electrons via direct contact (e.g., surface c-type cytochromes), electron shuttles, or conductive pili (in conductive biofilms) [57, 77]. Achieving high efficiency depends on how effectively electrons are moved from the electrode to the biocatalyst. Optimizing microbial consortia entails selecting or developing microbes that can efficiently break down specific organic compounds and transport electrons to the anode, thereby increasing the efficiency of waste treatment and energy production. Specific species, such as Geobacter sulfurreducens and Shewanella oneidensis, are known for their high current density and efficient electron transfer.

5.3 System Design and Configuration

The physical design and components of a BES have a significant influence on its performance. The structure of the electrodes, the reactor configuration, and the arrangement of chambers all play a role.

5.3.1 Electrode Materials

The selection of cathode and anode materials is crucial for BES performance, as it affects power output, conductivity, electron transfer efficiency, and system cost. Anode electrodes must have strong biocompatibility and amplify the electron transfer rate. BES performance is considered to be limited by the bacterial anode's activity and effectiveness. The electrode material is one of the key factors influencing anode bacterial growth. Carbonaceous materials, including carbon mesh and graphite rods, are commonly used due to their stability, biocompatibility, and cost-effectiveness. Additionally utilized are non-carbon materials, including copper, silver, titanium, stainless steel, and gold. Recent developments focus on anode amendment and immobilization techniques. Cathode materials also affect efficiency and energy output. At the cathode, the oxygen reduction process (ORP) occurs and affects overall performance. Biocathodes, where microorganisms catalyze cathodic reactions, are alternatives to conventional cathodes and can yield good power density. The synergy between microbes and optimized operating conditions, including electrode material, is key to BES effectiveness.

5.3.2 Cell Design

BESs generally consist of at least two electrodes (anode and cathode), often housed in separate chambers. An ion exchange membrane (often a proton exchange membrane, PEM) separates the anaerobic cathodic and anodic chamber in a dual-chamber system, which is a common design. Single-chamber systems also exist. The reactor design, size, and configuration, including the surface area of the electrode and the spacing between electrodes, are important factors. The geometry of MEC reactors is also considered an influencing parameter. While dual-chamber systems with membranes allow proton transport, membrane-less designs are also being explored. Reactor design that ensures less internal resistance is crucial for optimal performance.

5.3.3 Operational Parameters

Various parameters controlled during operation significantly affect BES performance. These include applied voltage (especially in MECs), ionic strength and conductivity of electrolytes, external resistance and loads, organic loading rate (OLR), substrate particle size, and hydraulic retention time (HRT). Controlling these parameters is crucial for optimizing microbial growth, electron transference, and overall system output.

5.4 Environmental Factors

Environmental conditions within the BES reactor have a direct impact on microbial activity and electrochemical processes.

5.4.1 Temperature and pH

Temperature is a crucial condition for the growth of microorganisms and enzymes, the formation of persistent biofilms, and the rate at which substrates degrade. For pertinent microorganisms, the optimum temperature range typically falls between 25 °C and 40 °C, and performance generally improves with increasing temperature within this range. However, excessive heat can damage microbial enzymes and reduce performance [92]. For the development and functioning of microorganisms, an optimal pH range is also essential. A pH outside of the proper range significantly reduces microbial activity. pH variations can disturb biological activity and affect nutrient absorption. Identifying the suitable pH for exoelectrogenic microorganisms is crucial, as pH levels can vary within the BES, resulting in an acid-to-alkaline gradient from the anode to the cathode due to proton accumulation and transport issues [93].

5.4.2 Oxygen Concentration

BESs, particularly the anodic chamber, where anaerobic microbes oxidize organic material, are typically maintained under anaerobic conditions. These microbes are unable to produce electricity when oxygen is present. At the aerobic cathode of MFCs, oxygen is frequently employed as the terminal electron acceptor (TEA), where it combines with protons and electrons to produce water [4]. However, oxygen diffusion from the cathode to the anode can negatively affect anaerobic processes. In MECs, the cathodic chamber is generally maintained without oxygen to favor hydrogen production. While facultative bacteria can utilize extra oxygen, its presence can reduce proton availability for methane production, decreasing the calorific value of the biogas [51]. Oxygen reduction at the cathode, while crucial for cell potential, is often a limiting factor in MFCs.

5.4.3 Contaminants

The presence of inhibitors or toxic substances in the organic waste substrate can negatively impact the microbial community and BES performance. While the sources mention the potential for BESs to treat complex and even toxic wastewater, they do not explicitly detail the impact of specific contaminants, such as heavy metals or antibiotics, on bioelectricity generation [94]. Nevertheless, given the reliance on microbial growth and the sensitivity of microorganisms to their environment, it can be inferred that such inhibitors would likely hinder microbial performance, substrate degradation, and, consequently, bioelectricity output. Further investigation on the impact of numerous contaminants on different microorganisms and their electrochemical activity in BESs is an important area of study.

6 Potential Applications of BES Using Organic Waste Material

Bioelectricity Generation MFCs are widely studied for their ability to convert chemical energy into electrical energy. Various organic wastes, including MSW, particularly OFMSW, food waste, agricultural waste, animal dung, crop waste, lignocellulosic waste, industrial wastewater, dairy wastewater, petroleum refinery wastes, swine waste, slaughterhouse waste, pineapple waste, brewery wastewater, winery wastewater, starch wastewater, whey, vegetable waste, meat processing wastewater, fish processing wastewater, canteen waste, rice mill wastewater, domestic wastewater, pharmaceutical wastewater, and textile industry wastewater have been tested for electricity generation in MFCs. The power production efficiency is dependent on the biodegradability and nature of the organic components present in the wastewater. Even low-strength wastewater can be used.

Biohydrogen Production MECs, another type of BES, can utilize organic waste to produce biohydrogen. This can be achieved from the digestate of MFCs or directly from various biomass and biowastes, such as fruit waste and *S. Japonica*. Combining anaerobic digestion with MECs can also enhance methane yield.

Biogas Production BES can be combined with anaerobic digestion (AD) to enhance biogas production, particularly in terms of methane concentration. For instance, bioelectrochemical treatment (BET) through BES systems can lead to a simultaneous decrease in toxic products and potentially enhance biogas production.

Value-Added Biochemical Production (Microbial Electrosynthesis: MES) MES is a promising application for generating significant biochemicals, such as biodiesel, acetic acid, butyric acid, and ethanol, utilizing organic-rich food and agricultural waste and wastewater. MES can also convert CO_2 (which can be generated from organic waste treatment) into biofuels and commercially valuable compounds, including drug intermediates. Examples include the production of syngas, biohydrogen, biomethane, ethylene, methanol, dimethyl ether, urea, and succinic acid from agricultural and forestry waste. Even short-chain organics in wastewater effluents can be converted into long-chain and other useful products using membrane separation technology (MES).

Wastewater Treatment BES, particularly MFCs, are effective in treating a variety of industrial and wastewater municipal wastewater by degrading organic pollutants. High levels of COD removal up to 92% and total suspended solids (TSS) removal (>99.0%) can be achieved. BES can also be incorporated with currently available wastewater treatment technologies for a more realistic and feasible approach.

Nutrient Recovery BES can be utilized for the recovery of valuable nutrients, including ammonia, phosphorus, and nitrogen, from organic waste and wastewater. For instance, microbial recovery cells (MRCs), an advanced version of MDCs, are used for nutrient recovery.

Metal Recovery Compared to conventional approaches, MECs and MESs have demonstrated potential in recovering metals from polluted areas, wastewater, and waste streams in an environmentally friendly and cost-effective approach.

Desalination Microbial desalination cells (MDCs), another type of BES, can be used for water desalination while treating wastewater containing organic matter.

Bioremediation BES can be applied to bioremediate polluted environments, including soils contaminated with petroleum hydrocarbons and toxic metals, through a process known as bioelectrochemical remediation.

Biosensors BES are self-powered biosensors for monitoring environmental parameters or detecting specific analytes.

Production of Biochar BES can be involved in producing biochar, which has various applications (Table 3).

Table 3 Different BES types used for generating bioelectricity from various organic waste substrates

BES type	Organic waste substrate	Key outcomes	References
MFC	Organic substrate	Harnesses electrical current from the microbial oxidation of organic matter. Directly converts chemical energy from the organic substrate to electrical energy—Electricity generation	[4, 6, 95]
MFC	Domestic wastewater, organic wastewater	Bioelectricity generation and power production, often coupled with wastewater treatment	[6, 95]
MFC	Municipal sewage sludge	Electricity generation, energy output, bioelectricity generation, electricity or hydrogen production	[3, 18]

Table 3 (continued)

BES type	Organic waste substrate	Key outcomes	References
MFC	Food waste (FW), composite vegetable waste, food waste leachate, wine lees, fermented apple juice, cereal processing wastewater, chocolate industry wastewater	Bioelectricity generation, power generation, direct conversion to electricity	[96–100]
MFC	Agricultural waste, agricultural by-products, cellulosic and lignocellulosic materials, agricultural waste hydrolysate, raw Agri-residue, sugarcane bagasse, agricultural residues	Solid waste degradation and electric current generation (when integrated with AD/fermentation/ composting), power generation, and electricity generation	[101–104]
MFC	Animal manure, slaughterhouse wastewater, animal carcass wastewater	Electricity generation can also produce biogas simultaneously	[105–107]
MFC	Dark fermentation effluent (from sucrose, containing VFA, glucose, fructose, acetic acid, propionic acid, butyric acid, isobutyric acid, caproic acid, lactic acid, ethanol, n-butanol)	Power production, evaluated for wastewater treatment	[108, 109]
MFC	Winery wastewater, brewery wastewater, white wine lees wastewater, beer brewery wastewater	Electric power generation, bioelectricity production	[110, 111]
MFC (air-cathode)	Leachate (from OFMSW/ simulated food waste), VFA-laden leachate	Electricity, power generation	[112, 113]
MDC	Organic wastewater, wastewater, municipal solid waste (MSW), organic waste	Can use organic wastewater to generate electricity. It can enable simultaneous water desalination and electric power generation. Part of waste-to-energy systems. Primarily designed for desalination	[61]
Plant microbial fuel cell (PMFC)	Organic substrates. Associated with rhizosphere. (specific plant substrates like rice, Spartina, lupines, and tomato have been mentioned in the context of MSC/bio-photovoltaic systems in our conversation history)	Produces electricity through the decomposition of organic substrates by microorganisms. Related to microbial electricity generation in rice paddy fields	[54, 114, 115]
Enzymatic fuel cell (EFC)	Simple organic molecules, glucose and related compounds	Produces electrochemical energy using enzymes as biocatalysts. Produces amperage for portable power devices	[54, 60]

7 Microbes Engaged in Organic Waste Treatment Utilizing BES

BESs utilize microbes as biocatalysts to promote electrochemical reactions. Numerous BES types have been designed for various applications, and specific microorganisms play crucial roles in their functioning. MFCs are designed to produce bioelectricity by harnessing the metabolic activity of microorganisms. Exoelectrogenic bacteria can transmit electrons extracellularly to the anode surface via a variety of methods, including mediated electron transfer (MET) with redox-active chemicals and direct electron transfer) via outer membrane cytochromes or nanowires (e-pili) [5]. Common examples of exo-electrogene mentioned in the sources include *Geobacter sulfurreducens* [51, 116], known for high current density and efficient electron transfer [117], and *Shewanella oneidensis* [116], have the capability of transporting electrons to electron acceptors. Other electrogenic microbes cited are *Clostridium cellulolyticum*, *Rhodoferax ferrireducens*, *Shewanella putrefaciens*, *Clostridium butyricum*, *Clostridium beijerinckii*, and *Enterobacter cloacae* [54]. Fungi like *Aspergillus awamori* and *Phanerochaete chrysosporium* are also mentioned as electrogenic microbes. The development of electroactive biofilms on the anode is critical for efficiently transferring electrons. Diverse microbial communities, including anodophilic bacteria, are implicated in the oxidation of complex organic material in wastewater. In some cases, while attempts are made to maximize electron transmission to the anode, methanogenic archaea may also be present and produce methane as a by-product [118].

MECs are primarily used for the production of hydrogen. Similar to MFCs, MECs utilize exoelectrogenic microorganisms at the anode for the oxidation of organic waste, emitting protons and electrons. However, MECs require an externally applied voltage to drive the reduction of protons to hydrogen gas at the cathode. The sources highlight that research into the mechanisms underlying microbial extracellular electron transport in microbial electrochemical cells (MECs) is still ongoing. Various microorganisms can be utilized in MECs for hydrogen generation by degrading different substrates, including polysaccharides, monosaccharides, acetate, proteins, and complex waste materials [119]. Examples of microorganisms involved in MECs include anaerobic sludge for ethanol production, hydrogenophilic dechlorinating cultures, *Actinobacillus succinogenes* and *Geobacter metallireducens* for succinate production, *Geobacter lovleyi* for cis-DCE reduction, *Dechloromonas agitata* for Cl^- reduction, *Acinetobacter johsonii* for H_2O formation, and hydrogenotrophic methanogens that reduce CO_2 to methane [5].

MES, in which cathodic biocatalysts (with cathodic biofilms attached) employ the reducing power produced by anodic oxidation to create value-added products at the cathode [60]. This process can be highly specific, depending on the biocatalyst and electron acceptor. Microorganisms on the biocathode contribute to the production of lower-value substances, such as butyrate, ethanol, and acetate. A diverse range of microorganisms have been applied in microbial electrosynthesis systems (MES) to produce useful biochemical compounds. *Sporomusa ovata*, for instance, is capable of converting carbon dioxide into acetate and 2-oxobutyrate through bioelectrochemical processes [11, 120]. *Clostridium ljungdahlii* has shown similar capabilities in producing these compounds from CO_2 [120]. In another example, *Ralstonia eutropha* has been utilized for the generation of biofuels [121]. Microbial communities derived from freshwater bog sediments have enabled the production of chemicals such as 1-butanol, propionate, and butyrate. Likewise, microbial consortia from brewery wastewater have been harnessed to produce hydrogen gas and acetate [122]. Additional electroactive bacteria, such as Megasphaera and Desulfitobacterium species, contribute to MES performance by adhering to electrode surfaces, where they act as low-cost, self-sustaining biocatalysts that drive essential redox reactions.

Other types of BESs include microbial solar cells (MSCs), which utilize photosynthetic microorganisms, and microbial desalination cells (MDCs), which desalinate water, as well as bioelectrochemical treatment (BET) systems for pollutant removal. The sources do not provide extensive details on the specific microorganisms involved in all these less-discussed BES types within this context. However, BET systems utilize a combination of biological and electrochemical processes for effluent treatment, relying on the metabolic activities of diverse microorganisms to degrade pollutants. The efficiency of all types of BES is strongly influenced by the growth of microbial communities involved, their ability to form biofilms, and their capacity for extracellular electron transfer (Table 4).

Table 4 Microbial communities involved in BESs for organic waste processing

BES type	Types of substrates	Microorganisms involved	Primary role	References
MFC	Organic matter, domestic effluent, leachate from food waste, activated sludge, glucose, acetate, glucuronic acid, livestock manure, brewery wastewater, OFMSW, MFC spent substrate, rice hull, bean residue, ground coffee waste, rice straw, vegetable waste, diesel range organics, synthetic wastewater	Ex-electrogenic microbes (e.g., *Geobacter sulfurreducens*, *Shewanella oneidensis*, *clostridium* spp., *Enterobacter cloacae*, *Escherichia coli*), fungi (*aspergillus awamori*, *Phanerochaete chrysosporium*), anodophilic bacteria, mixed microbial consortia, *Citrobacter* sp., *pseudomonas* sp., and *Stenotrophomonas* sp., *Klebsiella pneumoniae* L17, *Rhodopseudomonas palustris* DX-1, *Geobacter* spp., *Bacteroidetes*, *b-proteobacteria*, anaerobically cultured waste-activated sludge (WAS), *Lactobacillaceae*, *Bacillaceae*, *clostridia*, *Pseudomonadaceae*, & *Pseudomonas aeruginosa*, native microbial populations	Oxidation of organic material at the anode, facilitating extracellular electron transfer for bioelectricity generation	[113, 123–125]

(continued)

Table 4 (continued)

BES type	Types of substrates	Microorganisms involved	Primary role	References
MEC	Organic substrates, polysaccharides, monosaccharides, proteins, alcohols, acetate, carboxylic acid, municipal effluent, animal manure, biomass hydrolysate, domestic and industrial effluent, beverage effluent), ethanol, cis-DCE, succinate, cl^-, H_2O, P-glycerol, food waste leachate, glucose, sludge fermentation liquid, swine manure, sewage sludge + food waste, sugar industry wastewater, algal biomass.	Exoelectrogenic bacteria, anaerobic sludge microorganisms, hydrogenophilic dechlorinating cultures, *Acinetobacter johsonii*, *Geobacter metallireducens*, *Actinobacillus succinogenes*, *Dechloromonas agitata*, *Geobacter lovleyi*, hydrogenotrophic methanogens, *Geobacter* spp., sulfate-reducing bacteria, methanogens	Oxidation of organic material at the anode, extracellular electron transfer, and protons reduction to hydrogen at the cathode (with external voltage)	[5, 120, 126–128]
MES	Reducing power generated from anodic oxidation, CO_2, acetate	Cathodic biocatalysts (e.g., *Ralstonia eutropha*, *Sporomusa ovata*, *clostridium ljungdahlii*, freshwater sediments microorganisms, brewery wastewater microorganisms, *Megasphaera* sp., *Desulfitobacterium* sp.), homo-acetogenic bacterial cultures, phototrophic mixed culture	Uptake of electrons at the cathode to reduce electron acceptors (e.g., CO_2) into value-added products	[5, 51, 120, 129, 130]
MSC	Light energy	Photosynthetic microorganisms	Conversion of light energy into electrical energy	[10]
BET	Pollutants in wastewater	Diverse microbial communities	Degradation and removal of pollutants from wastewater through combined biological and electrochemical processes	[54]

8 Bioelectricity Generation and Wastewater Treatment Using BES

The primary outcome of MFC technology has been the generation of bioelectricity. Additionally, BES technology offers several other benefits. One of the most significant uses of BES is wastewater treatment. Current biological treatment methods for low-concentration wastewater rely on conventional, energy-intensive technologies that are environmentally detrimental. Utilizing innovative procedures that enable substantial energy recovery is the most effective way to offset the substantial amount of energy consumed in conventional wastewater treatment. BES links the advantages of treating wastewater from various sources with the recovery of energy and resources by producing bioelectricity or other useful goods [8]. Aeration in the conventional activated sludge method for treating domestic wastewater consumes 0.3 kWh/m^3 while pumping and other uses require almost twice as much energy [131]. By anaerobically converting organic material to methane, energy can be recovered from household wastewater. However, only a fraction of the potentially available energy is recovered when anaerobic digestion and traditional aerobic treatment of wastewater are combined. During aerobic oxidation processes, a significant portion of the energy contained in dissolved organic matter is converted into heat. In contrast, utilizing anaerobic digestion enables energy recovery in the form of methane, providing a more energy-efficient alternative to conventional aerobic methods, such as activated sludge treatment. However, anaerobic digestion has limitations—it requires a waste stream with a high organic loading rate (greater than 3 kg/m^3) and temperatures consistently above 20 to operate effectively [88].

Additionally, to make this method economically viable, the use of large-scale digesters is necessary. Consequently, anaerobic digestion is typically restricted to treating wastewater sludge and high-strength wastewater and is not suitable for low-strength wastewater. In comparison, BESs, such as MFCs, generate significantly lower volumes of residual solids or sludge during wastewater treatment. This minimizes or eliminates the need for further sludge handling and disposal. Moreover, unlike combustion-based systems, MFCs are not constrained by Carnot cycle efficiency limits, allowing for potentially higher energy conversion efficiency from wastewater.

Nevertheless, the actual energy recovery from BESs remains modest. For instance, MFCs treating domestic wastewater have achieved power densities of up to 12 W/m^3, equivalent to approximately 0.07 kWh/m^3 over a 6-hour treatment period—comparable to the typical retention time in activated sludge processes [132]. However, this remains significantly lower than the potential energy yield from anaerobic digestion, where methane recovery from household wastewater can reach approximately 2 kWh/m^3.

While microbial fuel cells (MFCs) may not yet offer a compelling case for widespread implementation based solely on energy production, their use in organic waste removal presents a promising opportunity from a sustainability and economic perspective. Economic projections suggest that expanded deployment of MFCs for wastewater treatment could be cost-effective, even without relying on electricity generation as the primary justification. Instead, a more practical approach involves harnessing wastewater not only as a waste stream but also as a resource for in-situ production of valuable chemicals through MECs and MESs [88]. In these systems, the intrinsic energy within wastewater is harnessed more efficiently—particularly in MECs, where chemicals such as hydrogen and hydrogen peroxide are produced using electrons derived from wastewater. This approach offers energetic advantages over conventional methods such as standard electrolysis [88]. However, comparative studies reveal inevitable trade-offs. For example, an experimental analysis by Cusick et al. [133] demonstrated that MFCs surpassed MECs in both chemical oxygen demand (COD) removal efficiency and energy recovery (expressed in kWh per kg of COD removed) when treating domestic and winery wastewater. Despite this, the economic potential of BES technologies—especially in terms of recovering chemicals and energy from low-value waste—remains significant. The estimated cost of hydrogen production using MECs was calculated at \$4.51/kg-H_2 for winery wastewater and \$3.01/kg-H_2 for domestic wastewater, both of which are below the market price of hydrogen, estimated at \$6/kg-H_2 [133]. These findings highlight the dual environmental and economic benefits of employing BES technologies for sustainable waste treatment and resource recovery.

9 Challenges and Limitations in Bioelectricity Generation from Organic Waste

Bioelectrochemical systems (BESs) offer a promising pathway for simultaneously treating organic waste and recovering energy; however, their widespread adoption faces several significant challenges and limitations. Overcoming these technical and economic hurdles is crucial for transitioning BES technology from laboratory-scale demonstrations to commercial implementation.

- Challenges in achieving commercially viable power densities: One of the primary challenges of BES, particularly MFCs, is the relatively low power density achieved in practice compared to theoretical values. While MFCs can theoretically produce a potential of 1.2 V with oxygen as the terminal electron acceptor, real-world voltages are typically much lower, ranging from 500 to 650 mV, because of numerous electrochemical losses. Achieving the theoretical value of power density remains a major challenge. Low power densities impede the economic viability of BES for large-scale electricity generation. Operating settings, electrode materials, feedstock compositions, and inoculum sources all significantly impact the catalytic performance of BES.
- Strategies to overcome low power densities: Recent investigations focus on various methods to enhance power output and treatment efficiency. Model microorganisms that are either electrochemically active or inactive are screened, electrode materials are chosen and optimized, and electrochemical devices are designed. To improve the overall efficacy of BES, researchers are trying to lower the cathode potential by employing novel catalysts or composites. Modifying the engineering and individual components of the reactors has extraordinarily improved BES performance over the past decade. Studies have explored a variety of cost-effective and efficient cathode and anode types, as well as the materials and catalysts used in them. Integrating different biomass and/or biowaste could also enhance substrate accessibility for microbes, increasing bioelectricity production.
- Techno-economic limitations to large-scale application: The transition of BES technologies from laboratory to industrial-scale applications faces several criti-

cal techno-economic barriers. One of the primary technical challenges is the low power output typically observed in BESs, which limits their practicality at larger scales. Additional issues include underperforming cathodes, limited efficiency in pollutant removal or degradation, and reliance on costly membrane materials that significantly drive up system costs. From an economic standpoint, the scalability of BESs is hindered by substantial capital investment requirements, as well as ongoing operational and maintenance (O&M) expenses. As BES systems are scaled up, internal resistance tends to increase, which can further compromise performance.

- Furthermore, the use of high-cost materials for components such as electrodes and membranes becomes more burdensome at larger scales. Long-term operation also introduces challenges such as membrane clogging and electrode fouling, which can negatively affect both efficiency and maintenance demands. A significant step toward industrial viability involves the successful demonstration of large-scale or pilot-scale Battery Energy Storage (BES) systems. However, achieving this remains a formidable obstacle. Despite these challenges, overcoming them is essential to establish the feasibility of BES technologies for full-scale wastewater treatment and resource recovery applications.
- Ensuring long-term operational stability: Maintaining the performance of BES over extended periods is critical for practical application. Operational stability is influenced by various parameters, including system dynamics, fluid mechanics, and overall system complexity. Long-duration operations can lead to reduced performance.
- Biofouling, electrode degradation, and microbial community shifts: One of the key operational limitations in bioelectrochemical systems (BESs) is membrane fouling, which significantly affects system efficiency and presents a significant barrier to broader adoption. In particular, biofouling of membranes becomes more pronounced during prolonged operation of microbial fuel cells (MFCs) and microbial electrolysis cells (MECs), leading to performance degradation over time. Similarly, electrode fouling is another persistent issue that hampers system output during long-term use. To ensure sustained performance, it is essential to implement effective biofouling management strategies. These may include physical cleaning methods, chemical treatments, or the development and application of antimicrobial membrane materials. Such interventions help to maintain membrane integrity and reduce performance loss. The formation of microbial biofilms on electrodes is also a critical factor, as these biofilms facilitate electron transfer—a fundamental process in BES function. The efficiency of this biofilm formation is highly dependent on the interactions between microbial communities and electrode surfaces. Although microbial community structure and activity are vital to BES performance, the literature does not highlight microbial community shifts as a major operational limitation. Nonetheless, variations in substrate characteristics can affect microbial processes and contribute to fluctuations in performance, indicating the importance of maintaining consistent feedstock quality for stable operation.

- Cost of BES systems and potential economic incentives: The economic feasibility of BES is significantly impacted by high initial setup costs and ongoing operational and maintenance (O&M) costs. The cost of materials, particularly for electrodes, catalysts, and membranes, has historically been high (e.g., platinum electrodes, Nafion membranes). The cost of materials and maintenance can account for between 50 and 60% of the total cost of developing a reactor. A strategy to address scaling-up costs involves selecting inexpensive materials for electrodes and choosing low-cost membrane materials. Potential economic incentives stem from BES's ability to convert waste into valuable products, thereby adding revenue through resource recovery. These recovered products can include minerals, nutrients, methane, volatile fatty acids (VFAs), and hydrogen. Decentralized power production using BES can lead to additional revenue in the form of credits for the release of greenhouse gases and the production of renewable energy.
- Economic feasibility compared to conventional technologies: Although BESs are emerging as promising sustainable technologies, their economic competitiveness compared to established waste-to-energy methods, such as anaerobic digestion, remains to be evaluated. AD is a mature and widely adopted process for biogas generation with proven economic viability. In contrast, the electricity produced by MFCs is currently insufficient to offset their high capital investment, making BESs less economically attractive in direct comparison. However, BESs offer distinct operational advantages. They can effectively treat wastewaters with lower COD or higher VFA concentrations, operate efficiently at lower temperatures (around 20 °C), produce minimal sludge, and may reduce aeration costs. These factors contribute to lower operational burdens and environmental impact. Moreover, BESs enable simultaneous wastewater treatment and renewable energy generation, supporting resource recovery and advancing sustainability goals.
- Optimization of reactor design, microbial inoculation, and substrate handling: Enhancing the performance and scalability of bioelectrochemical systems (BESs) relies heavily on optimizing reactor components and operational conditions. Key areas of focus include the selection and structural refinement of electrode materials to improve electron transfer and system efficiency. The configuration of electrochemical reactors and the overall device design also play a critical role in maximizing performance. Another vital aspect is the identification and cultivation of electrochemically active microorganisms. Optimizing microbial inocula—whether pure cultures or mixed consortia—can significantly influence electron flow and substrate degradation. Operational parameters, such as pH, temperature, applied voltage, organic loading rate, conductivity, and salinity, must also be carefully regulated to maintain ideal conditions for microbial activity and optimal system output. Efficient substrate utilization by microorganisms is equally important, particularly when processing complex organic matter such as lignocellulosic biomass, which often requires pretreatment to enhance biodegradability. Additionally, real wastewater streams can vary in composition, posing challenges for consistent performance. Strategies such as co-digestion or blending different biomass types can improve substrate availabil-

ity and overall system efficiency, supporting more robust and adaptable BES operation.

- Compliance with environmental regulations: Effective management of agro-food waste and wastewater is crucial for achieving sustainability and mitigating climate change. BESs contribute to these goals by enabling waste reduction and wastewater treatment. Their implementation aligns with several United Nations Sustainable Development Goals (SDGs), including clean water and sanitation (Goal 6), affordable and clean energy (Goal 7), and sustainable cities and communities (Goal 11). Although BESs offer clear environmental benefits, current literature does not provide detailed insights into specific regulatory compliance challenges associated with their operation.
- Government incentives and support: The potential of BES to support renewable energy and sustainable waste management aligns with government initiatives aimed at achieving these goals. The possibility of generating renewable energy credits suggests a potential area for policy support. Policymakers are identified as a key target audience for comprehensive information on BES, highlighting the importance of governmental awareness and potential incentives for the adoption of this technology. However, specific details on existing government incentives for integrating BES with waste management and renewable energy programs are not provided in the source material.

10 Conclusion

Bioelectrochemical systems (BESs) represent a promising approach to simultaneously tackle the dual challenges of renewable energy generation and organic waste management. This chapter provides an overview of the core principles of BES technology, with an emphasis on microbial electron transfer processes and various system configurations. Different organic waste streams, including agricultural by-products, municipal wastewater, and food processing residues, have been identified as viable substrates for electricity generation. Critical factors influencing BES performance, such as substrate biodegradability, electrode material selection, microbial community structure, and operational parameters, were discussed to identify areas for performance enhancement.

Looking forward, BES holds substantial potential as a sustainable and scalable solution for waste valorization and energy recovery. The future integration of BES into decentralized wastewater treatment systems, agricultural waste management frameworks, and renewable energy networks could be enabled by advances in reactor engineering, material innovation, and microbial optimization. Nonetheless, current limitations, including low power densities, stability issues, and economic constraints, require ongoing interdisciplinary research and technological progress. As the global focus sharpens on clean energy and sustainable development, BESs are positioned as key technologies in the transition toward a circular economy. By transforming organic waste into usable energy and minimizing environmental impacts, BESs support carbon reduction goals and environmental sustainability. Continued research, development, and implementation will be essential for unlocking the full potential of BES in building a cleaner, more resilient future.

References

1. Fazal T, Saif Ur Rehman M, Mushtaq A, Hafeez A, Javed F, Aslam M, Fatima M, Faisal A, Iqbal J, Rehman F, Farooq R (2019) Simultaneous production of bioelectricity and biogas from chicken droppings and dairy industry wastewater employing bioelectrochemical system. Fuel 256:115902. https://doi.org/10.1016/j.fuel.2019.115902
2. Rojas Flores S, Nazario-Naveda R, Delfín-Narciso D, Gallozo Cardenas M, Diaz Diaz N, Valverde Ravelo K (2021) Generation of bioelectricity from organic fruit waste. Environ Res Eng Manag 77(3):6–14. https://doi.org/10.5755/j01.erem.77.3.28493
3. Gautam R, Nayak JK, Talapatra KN, Amit, Ghosh UK (2023) Assessment of different organic substrates for bio-electricity and bio-hydrogen generation in an integrated bio-electrochemical system. Mater Today Proc 80:2255–2259. https://doi.org/10.1016/j.matpr.2021.06.223
4. Maddirala S, Bhadra S, Mozumder MSI, Garlapati VK, Sevda S (2024) Advancements in bioelectrochemical systems for solid organic waste valorization: a comprehensive review. PRO 12(4):805. https://doi.org/10.3390/pr12040805
5. Naradasu D, Long X, Okamoto A, Miran W (2020) Bioelectrochemical systems, vol 1. Springer, Singapore. https://doi.org/10.1007/978-981-15-6872-5
6. Ramanaiah SV, Chandrasekhar K, Cordas CM, Potoroko I (2023) Bioelectrochemical systems (BESs) for agro-food waste and wastewater treatment, and sustainable bioenergy-A review. Environ Pollut 325:121432. https://doi.org/10.1016/j.envpol.2023.121432
7. Guo X, Liu J, Xiao B (2013) Bioelectrochemical enhancement of hydrogen and methane production from the anaerobic digestion of sewage sludge in single-chamber membrane-free microbial electrolysis cells. Int J Hydrog Energy 38(3):1342–1347. https://doi.org/10.1016/j.ijhydene.2012.11.087
8. Pandey P, Shinde VN, Deopurkar RL, Kale SP, Patil SA, Pant D (2016) Recent advances in the use of different substrates in microbial fuel cells toward wastewater treatment and simultaneous energy recovery. Appl Energy 168:706–723. https://doi.org/10.1016/j.apenergy.2016.01.056
9. Virdis B, Freguia S, Rozendal RA, Rabaey K, Yuan Z, Keller J (2011) Microbial fuel cells. In: Treatise on water science. Elsevier, Amsterdam, pp 641–665. https://doi.org/10.1016/B978-0-444-53199-5.00098-1
10. Khan N, Khan MD, Sultana S, Khan MZ, Ahmad A (2018) Bioelectrochemical systems for transforming waste to energy. In: Modern age environmental problems and their remediation. Springer, Cham, pp 111–128. https://doi.org/10.1007/978-3-319-64501-8_7
11. Pant D, Singh A, Van Bogaert G, Irving Olsen S, Singh Nigam P, Diels L, Vanbroekhoven K (2012) Bioelectrochemical systems (BES) for sustainable energy production and product recovery from organic wastes and industrial wastewaters. RSC Adv 2(4):1248–1263. https://doi.org/10.1039/C1RA00839K
12. Sharma M, Salama E-S, Thakur N, Alghamdi H, Jeon B-H, Li X (2023) Advances in the biomass valorization in bioelectrochemical systems: a sustainable approach for microbial-aided electricity and hydrogen production. Chem Eng J 465:142546. https://doi.org/10.1016/j.cej.2023.142546
13. Xu L, Zhao Y, Doherty L, Hu Y, Hao X (2016) The integrated processes for wastewater treatment based on the principle of microbial fuel cells: a review. Crit Rev Environ Sci Technol 46(1):60–91. https://doi.org/10.1080/10643389.2015.1061884

14. Arun S, Sinharoy A, Pakshirajan K, Lens PNL (2020) Algae based microbial fuel cells for wastewater treatment and recovery of value-added products. Renew Sust Energ Rev 132:110041. https://doi.org/10.1016/j.rser.2020.110041
15. Sevda S, Jyoti Sarma P, Mohanty K, Sreekrishnan TR, Pant D (2018) Microbial fuel cell Technology for bioelectricity generation from wastewaters. Springer, Cham, pp 237–258. https://doi.org/10.1007/978-981-10-7431-8_11
16. Saran C, Purchase D, Saratale GD, Saratale RG, Romanholo Ferreira LF, Bilal M, Iqbal HMN, Hussain CM, Mulla SI, Bharagava RN (2023) Microbial fuel cell: A green eco-friendly agent for tannery wastewater treatment and simultaneous bioelectricity/power generation. Chemosphere 312:137072. https://doi.org/10.1016/j.chemosphere.2022.137072
17. Mukherjee A, Patel V, Shah MT, Jadhav DA, Munshi NS, Chendake AD, Pant D (2022) Effective power management system in stacked microbial fuel cells for onsite applications. J Power Sources 517:230684. https://doi.org/10.1016/j.jpowsour.2021.230684
18. Singh NK, Mathuriya AS, Mehrotra S, Pandit S, Singh A, Jadhav D (2024) Advances in bioelectrochemical systems for bio-products recovery. Environ Technol 45(19):3853–3876. https://doi.org/10.1080/09593330.2023.2234676
19. Sevda S, Sreekishnan TR, Pous N, Puig S, Pant D (2018) Bioelectroremediation of perchlorate and nitrate contaminated water: a review. Bioresour Technol 255:331–339. https://doi.org/10.1016/j.biortech.2018.02.005
20. Sevda S, Sreekrishnan TR (2014) Removal of organic matters and nitrogenous pollutants simultaneously from two different wastewaters using biocathode microbial fuel cell. J Environ Sci Health A 49(11):1265–1275. https://doi.org/10.1080/10934529.2014.910064
21. Tommasi T, Salvador GP, Quaglio M (2016) New insights in microbial fuel cells: novel solid phase anolyte. Sci Rep 6(1):29091. https://doi.org/10.1038/srep29091
22. Dominguez-Benetton X, Sevda S, Vanbroekhoven K, Pant D (2012) The accurate use of impedance analysis for the study of microbial electrochemical systems. Chem Soc Rev 41(21):7228. https://doi.org/10.1039/c2cs35026b
23. Yu H, Jiang J, Zhao Q, Wang K, Zhang Y, Zheng Z, Hao X (2015) Bioelectrochemically-assisted anaerobic composting process enhancing compost maturity of dewatered sludge with synchronous electricity generation. Bioresour Technol 193:1–7. https://doi.org/10.1016/j.biortech.2015.06.057
24. Zhao Q, Yu H, Zhang W, Kabutey FT, Jiang J, Zhang Y, Wang K, Ding J (2017) Microbial fuel cell with high content solid wastes as substrates: a review. Front Environ Sci Eng 11(2):13. https://doi.org/10.1007/s11783-017-0918-6
25. Fantozzi F, Buratti C (2009) Biogas production from different substrates in an experimental continuously stirred tank reactor anaerobic digester. Bioresour Technol 100(23):5783–5789. https://doi.org/10.1016/j.biortech.2009.06.013
26. Abu-Reesh IM, Kunju A, Sevda S (2022) Performance of microbial fuel cells in treating petroleum refinery wastewater. J Water Process Eng 49:103029. https://doi.org/10.1016/j.jwpe.2022.103029
27. Zhu Y, Zhang Y, Luo D, Chong Z, Li E, Kong X (2021) A review of municipal solid waste in China: characteristics, compositions, influential factors and treatment technologies. Environ Dev Sustain 23(5):6603–6622. https://doi.org/10.1007/s10668-020-00959-9
28. Du H, Shao Z (2022) Synergistic effects between solid potato waste and waste activated sludge for waste-to-power conversion in microbial fuel cells. Appl Energy 314:118994. https://doi.org/10.1016/j.apenergy.2022.118994
29. Callegari A, Bolognesi S, Cecconet D (2019) Operation of a 2-stage bioelectrochemical system for groundwater denitrification. Water 11(5):959. https://doi.org/10.3390/w11050959
30. Li Z, Yao L, Kong L, Liu H (2008) Electricity generation using a baffled microbial fuel cell convenient for stacking. Bioresour Technol 99(6):1650–1655. https://doi.org/10.1016/j.biortech.2007.04.003

31. Richard EN, Hilonga A, Machunda RL, Njau KN (2019) A review on strategies to optimize metabolic stages of anaerobic digestion of municipal solid wastes towards enhanced resources recovery. Sustain Environ Res 29(1):36. https://doi.org/10.1186/s42834-019-0037-0
32. Hou H, Li L, Cho Y, de Figueiredo P, Han A (2009) Microfabricated microbial fuel cell arrays reveal electrochemically active microbes. PLoS One 4(8):e6570. https://doi.org/10.1371/journal.pone.0006570
33. Kim M, Hyun MS, Gadd GM, Kim GT, Lee S, Kim HJ (2009) Membrane-electrode assembly enhances performance of a microbial fuel cell type biological oxygen demand sensor. Environ Technol 30(4):329–336. https://doi.org/10.1080/09593330902732077
34. Liu Z, Li H, Liu J, Su Z (2008) Effects of inoculation strategy and cultivation approach on the performance of microbial fuel cell using marine sediment as bio-matrix. J Appl Microbiol 104(4):1163–1170. https://doi.org/10.1111/j.1365-2672.2007.03643.x
35. Pons L, Délia M-L, Basséguy R, Bergel A (2011) Effect of the semi-conductive properties of the passive layer on the current provided by stainless steel microbial cathodes. Electrochim Acta 56(6):2682–2688. https://doi.org/10.1016/j.electacta.2010.12.039
36. Ward AJ, Hobbs PJ, Holliman PJ, Jones DL (2008) Optimisation of the anaerobic digestion of agricultural resources. Bioresour Technol 99(17):7928–7940. https://doi.org/10.1016/j.biortech.2008.02.044
37. Ai C, Hou S, Yan Z, Zheng X, Amanze C, Chai L, Qiu G, Zeng W (2019) Recovery of metals from acid mine drainage by bioelectrochemical system inoculated with a novel Exoelectrogen, pseudomonas sp. E8 Microorg 8(1):41. https://doi.org/10.3390/microorganisms8010041
38. Rahimnejad M, Adhami A, Darvari S, Zirepour A, Oh S-E (2015) Microbial fuel cell as new technology for bioelectricity generation: A review. Alex Eng J 54(3):745–756. https://doi.org/10.1016/j.aej.2015.03.031
39. Srikanth S, Kumar M, Singh MP, Das BP (2016) Bioelectro chemical systems: a sustainable and potential platform for treating waste. Procedia Environ Sci 35:853–859. https://doi.org/10.1016/j.proenv.2016.07.102
40. Venkata Mohan S, Velvizhi G, Annie Modestra J, Srikanth S (2014) Microbial fuel cell: critical factors regulating bio-catalyzed electrochemical process and recent advancements. Renew Sust Energ Rev 40:779–797. https://doi.org/10.1016/j.rser.2014.07.109
41. Gupta SK, Rachna, Singh B, Mungray AK, Bharti R, Nema AK, Pant KK, Mulla SI (2022) Bioelectrochemical technologies for removal of xenobiotics from wastewater. Sustain Energy Technol Assess 49:101652. https://doi.org/10.1016/j.seta.2021.101652
42. Obileke K, Onyeaka H, Meyer EL, Nwokolo N (2021) Microbial fuel cells, a renewable energy technology for bio-electricity generation: a mini-review. Electrochem Commun 125:107003. https://doi.org/10.1016/j.elecom.2021.107003
43. Kumar SS, Kumar V, Malyan SK, Sharma J, Mathimani T, Maskarenj MS, Ghosh PC, Pugazhendhi A (2019) Microbial fuel cells (MFCs) for bioelectrochemical treatment of different wastewater streams. Fuel 254:115526. https://doi.org/10.1016/j.fuel.2019.05.109
44. Rezaei F, Richard TL, Logan BE (2009) Analysis of chitin particle size on maximum power generation, power longevity, and coulombic efficiency in solid–substrate microbial fuel cells. J Power Sources 192(2):304–309. https://doi.org/10.1016/j.jpowsour.2009.03.023
45. Khoo KS, Chia WY, Tang DYY, Show PL, Chew KW, Chen W-H (2020) Nanomaterials utilization in biomass for biofuel and bioenergy production. Energies 13(4):892. https://doi.org/10.3390/en13040892
46. Santoro C, Arbizzani C, Erable B, Ieropoulos I (2017) Microbial fuel cells: from fundamentals to applications. A review. J Power Sour 356:225–244. https://doi.org/10.1016/j.jpowsour.2017.03.109
47. Logan BE, Hamelers B, Rozendal R, Schröder U, Keller J, Freguia S, Aelterman P, Verstraete W, Rabaey K (2006) Microbial fuel cells: methodology and technology. Environ Sci Technol 40(17):5181–5192. https://doi.org/10.1021/es0605016

48. Dessie Y, Tadesse S (2022) Advancements in bioelectricity generation through nanomaterial-modified anode electrodes in microbial fuel cells. Front Nanotechnol 4. https://doi.org/10.3389/fnano.2022.876014
49. Arun J, SundarRajan P, Grace Pavithra K, Priyadharsini P, Shyam S, Goutham R, Hoang Le Q, Pugazhendhi A (2024) New insights into microbial electrolysis cells (MEC) and microbial fuel cells (MFC) for simultaneous wastewater treatment and green fuel (hydrogen) generation. Fuel 355:129530. https://doi.org/10.1016/j.fuel.2023.129530
50. Wilberforce T, Sayed ET, Abdelkareem MA, Elsaid K, Olabi AG (2021) Value added products from wastewater using bioelectrochemical systems: current trends and perspectives. J Water Process Eng 39:101737. https://doi.org/10.1016/j.jwpe.2020.101737
51. Bajracharya S, Sharma M, Mohanakrishna G, Dominguez Benneton X, Strik DPBTB, Sarma PM, Pant D (2016) An overview on emerging bioelectrochemical systems (BESs): technology for sustainable electricity, waste remediation, resource recovery, chemical production and beyond. Renew Energy 98:153–170. https://doi.org/10.1016/j.renene.2016.03.002
52. Katuri KP, Ali M, Saikaly PE (2019) The role of microbial electrolysis cell in urban wastewater treatment: integration options, challenges, and prospects. Curr Opin Biotechnol 57:101–110. https://doi.org/10.1016/j.copbio.2019.03.007
53. Jacobson KS, Drew DM, He Z (2011) Use of a liter-scale microbial desalination cell as a platform to study bioelectrochemical desalination with salt solution or artificial seawater. Environ Sci Technol 45(10):4652–4657. https://doi.org/10.1021/es200127p
54. Choudhary M, Verma P, Ray S (2024) A comprehensive review on bio-electrochemical systems for wastewater treatment: process, electricity generation and future aspect. Environ Dev Sustain. https://doi.org/10.1007/s10668-024-05866-x
55. Tschörtner J, Lai B, Krömer JO (2019) Biophotovoltaics: green power generation from sunlight and water. Front Microbiol 10. https://doi.org/10.3389/fmicb.2019.00866
56. Zheng T, Li J, Ji Y, Zhang W, Fang Y, Xin F, Dong W, Wei P, Ma J, Jiang M (2020) Progress and prospects of bioelectrochemical systems: electron transfer and its applications in the microbial metabolism. Front Bioeng Biotechnol 8. https://doi.org/10.3389/fbioe.2020.00010
57. Surti P, Kailasa SK, Mungray AK (2021) Genetic engineering strategies for performance enhancement of bioelectrochemical systems: a review. Sustain Energy Technol Assess 47:101332. https://doi.org/10.1016/j.seta.2021.101332
58. Prévoteau A, Carvajal-Arroyo JM, Ganigué R, Rabaey K (2020) Microbial electrosynthesis from CO2: forever a promise? Curr Opin Biotechnol 62:48–57. https://doi.org/10.1016/j.copbio.2019.08.014
59. Bhattacharyya KG, Gupta SS (2008) Adsorption of a few heavy metals on natural and modified kaolinite and montmorillonite: a review. Adv Colloid Interf Sci 140(2):114–131. https://doi.org/10.1016/j.cis.2007.12.008
60. Kondaveeti S, Govindarajan D, Mohanakrishna G, Thatikayala D, Abu-Reesh IM, Min B, Nambi IM, Al-Raoush RI, Aminabhavi TM (2023) Sustainable bioelectrochemical systems for bioenergy generation via waste treatment from petroleum industries. Fuel 331:125632. https://doi.org/10.1016/j.fuel.2022.125632
61. ElMekawy A, Srikanth S, Bajracharya S, Hegab HM, Nigam PS, Singh A, Mohan SV, Pant D (2015) Food and agricultural wastes as substrates for bioelectrochemical system (BES): the synchronized recovery of sustainable energy and waste treatment. Food Res Int 73:213–225. https://doi.org/10.1016/j.foodres.2014.11.045
62. Budihardjo MA, Syafrudin, Effendi AJ, Hidayat S, Purnawan C, Lantasi AID, Muhammad FI, Ramadan BS (2021) Waste valorization using solid-phase microbial fuel cells (SMFCs): recent trends and status. J Environ Manag 277:111417. https://doi.org/10.1016/j.jenvman.2020.111417
63. Florio C, Nastro RA, Flagiello F, Minutillo M, Pirozzi D, Pasquale V, Ausiello A, Toscano G, Jannelli E, Dumontet S (2019) Biohydrogen production from solid phase-microbial fuel cell spent substrate: a preliminary study. J Clean Prod 227:506–511. https://doi.org/10.1016/j.jclepro.2019.03.316

64. Wang C-T, Liao F-Y, Liu K-S (2013) Electrical analysis of compost solid phase microbial fuel cell. Int J Hydrog Energy 38(25):11124–11130. https://doi.org/10.1016/j.ijhydene.2013.02.120
65. Gurjar R, Behera M (2020) Treatment of organic fraction of municipal solid waste in bioelectrochemical systems: a review. J Hazard Toxic Radioactive Waste 24(3). https://doi.org/10.1061/(ASCE)HZ.2153-5515.0000505
66. Lay C-H, Kumar G, Mudhoo A, Lin C-Y, Leu H-J, Shobana S, Thi Nguyen M-L (2020) Recent trends and prospects in biohythane research: an overview. Int J Hydrog Energy 45(10):5864–5873. https://doi.org/10.1016/j.ijhydene.2019.07.209
67. Li Y, Hua D, Mu H, Xu H, Jin F, Zhang X (2017) Conversion of vegetable wastes to organic acids in leaching bed reactor: performance and bacterial community analysis. J Biosci Bioeng 124(2):195–203. https://doi.org/10.1016/j.jbiosc.2017.02.020
68. Xu SY, Lam HP, Karthikeyan OP, Wong JWC (2011) Optimization of food waste hydrolysis in leach bed coupled with methanogenic reactor: effect of pH and bulking agent. Bioresour Technol 102(4):3702–3708. https://doi.org/10.1016/j.biortech.2010.11.095
69. Pathy A, Ray J, Paramasivan B (2021) Challenges and opportunities of nutrient recovery from human urine using biochar for fertilizer applications. J Clean Prod 304:127019. https://doi.org/10.1016/j.jclepro.2021.127019
70. Sonawane JM, Mahadevan R, Pandey A, Greener J (2022) Recent progress in microbial fuel cells using substrates from diverse sources. Heliyon 8(12):e12353. https://doi.org/10.1016/j.heliyon.2022.e12353
71. Sundberg C (2004) Low pH as an inhibiting factor in the transition from mesophilic to thermophilic phase in composting. Bioresour Technol 95(2):145–150. https://doi.org/10.1016/j.biortech.2004.01.016
72. Digman B, Kim D (2008) Review: alternative energy from food processing wastes. Environ Prog 27(4):524–537. https://doi.org/10.1002/ep.10312
73. Li W-W, Sheng G-P, Yu H-Q (2013) Electricity generation from food industry wastewater using microbial fuel cell technology. In: Food industry wastes. Elsevier, Amsterdam, pp 249–261. https://doi.org/10.1016/B978-0-12-391921-2.00014-7
74. Venkata Mohan S, Srikanth S (2011) Enhanced wastewater treatment efficiency through microbially catalyzed oxidation and reduction: synergistic effect of biocathode microenvironment. Bioresour Technol 102(22):10210–10220. https://doi.org/10.1016/j.biortech.2011.08.034
75. Lin C-Y, Hung C-H, Chen C-H, Chung W-T, Cheng L-H (2006) Effects of initial cultivation pH on fermentative hydrogen production from xylose using natural mixed cultures. Process Biochem 41(6):1383–1390. https://doi.org/10.1016/j.procbio.2006.01.021
76. Tyagi VK, Fdez-Güelfo LA, Zhou Y, Álvarez-Gallego CJ, Garcia LIR, Ng WJ (2018) Anaerobic co-digestion of organic fraction of municipal solid waste (OFMSW): progress and challenges. Renew Sust Energ Rev 93:380–399. https://doi.org/10.1016/j.rser.2018.05.051
77. Addagada L, Goel M, Shahid MK, Prabhu SV, Chand S, Sahoo NK, Rout PR (2023) Tricks and tracks in resource recovery from wastewater using bio-electrochemical systems (BES): A systematic review on recent advancements and future directions. J Water Process Eng 56:104580. https://doi.org/10.1016/j.jwpe.2023.104580
78. Muduli M, Chanchpara A, Choudhary M, Saravaia H, Haldar S, Ray S (2022) Critical review on sustainable bioreactors for wastewater treatment and water reuse. Sustain Water Resour Manage 8(5):159. https://doi.org/10.1007/s40899-022-00747-5
79. Rumky J, Deb A, Shim MJ, Laakso E, Repo E (2023) A review on the recent advances in electrochemical treatment technologies for sludge dewatering and alternative uses. J Hazard Mater Adv 11:100341. https://doi.org/10.1016/j.hazadv.2023.100341
80. Lawan J, Wichai S, Chuaypen C, Nuiyen A, Phenrat T (2022) Constructed sediment microbial fuel cell for treatment of fat, oil, grease (FOG) trap effluent: role of anode and cathode chamber amendment, electrode selection, and scalability. Chemosphere 286:131619. https://doi.org/10.1016/j.chemosphere.2021.131619

81. Du Y, Ju T, Meng Y, Lan T, Han S, Jiang J (2021) A review on municipal solid waste pyrolysis of different composition for gas production. Fuel Process Technol 224:107026. https://doi.org/10.1016/j.fuproc.2021.107026
82. Panigrahi S, Dubey BK (2019) A critical review on operating parameters and strategies to improve the biogas yield from anaerobic digestion of organic fraction of municipal solid waste. Renew Energy 143:779–797. https://doi.org/10.1016/j.renene.2019.05.040
83. Uddin MN, Siddiki SK, Mofijur M, Djavanroodi F, Hazrat MA, Show PL, Ahmed SF, Chu Y-M (2021) RETRACTED: prospects of bioenergy production from organic waste using anaerobic digestion technology: a mini review. Front Energy Res 9. https://doi.org/10.3389/fenrg.2021.627093
84. Lee DH, Behera SK, Kim JW, Park H-S (2009) Methane production potential of leachate generated from Korean food waste recycling facilities: a lab-scale study. Waste Manag 29(2):876–882. https://doi.org/10.1016/j.wasman.2008.06.033
85. Zeller MA, Hunt R, Jones A, Sharma S (2013) Bioplastics and their thermoplastic blends from *spirulina* and *chlorella* microalgae. J Appl Polym Sci 130(5):3263–3275. https://doi.org/10.1002/app.39559
86. Zhang F, Ahn Y, Logan BE (2014) Treating refinery wastewaters in microbial fuel cells using separator electrode assembly or spaced electrode configurations. Bioresour Technol 152:46–52. https://doi.org/10.1016/j.biortech.2013.10.103
87. Huang D, Du Y, Xu Q, Ko JH (2022) Quantification and control of gaseous emissions from solid waste landfill surfaces. J Environ Manag 302:114001. https://doi.org/10.1016/j.jenvman.2021.114001
88. Logan BE, Rabaey K (2012) Conversion of wastes into bioelectricity and chemicals by using microbial electrochemical technologies. Science 337(6095):686–690. https://doi.org/10.1126/science.1217412
89. Jenol MA, Ibrahim MF, Kamal Bahrin E, Kim SW, Abd-Aziz S (2019) Direct bioelectricity generation from sago Hampas by Clostridium beijerinckii SR1 using microbial fuel cell. Molecules 24(13):2397. https://doi.org/10.3390/molecules24132397
90. Kalathil S, Patil SA, Pant D (2018) Microbial fuel cells: electrode materials. In: Encyclopedia of interfacial chemistry. Elsevier, Amsterdam, pp 309–318. https://doi.org/10.1016/B978-0-12-409547-2.13459-6
91. Mohanakrishna G, Abu-Reesh IM, Pant D (2020) Enhanced bioelectrochemical treatment of petroleum refinery wastewater with Labaneh whey as co-substrate. Sci Rep 10(1):19665. https://doi.org/10.1038/s41598-020-76668-0
92. Behera M, Jana PS, More TT, Ghangrekar MM (2010) Rice mill wastewater treatment in microbial fuel cells fabricated using proton exchange membrane and earthen pot at different pH. Bioelectrochemistry 79(2):228–233. https://doi.org/10.1016/j.bioelechem.2010.06.002
93. Cui W, Liu G, Zeng C, Lu Y, Luo H, Zhang R (2019) Improved hydrogen production in the single-chamber microbial electrolysis cell with inhibition of methanogenesis under alkaline conditions. RSC Adv 9(52):30207–30215. https://doi.org/10.1039/C9RA05483A
94. Jadhav DA, Ghosh Ray S, Ghangrekar MM (2017) Third generation in bio-electrochemical system research—A systematic review on mechanisms for recovery of valuable by-products from wastewater. Renew Sust Energ Rev 76:1022–1031. https://doi.org/10.1016/j.rser.2017.03.096
95. Luo H, Jin S, Fallgren PH, Park HJ, Johnson PA (2010) A novel laccase-catalyzed cathode for microbial fuel cells. Chem Eng J 165(2):524–528. https://doi.org/10.1016/j.cej.2010.09.061
96. Cercado-Quezada B, Delia M-L, Bergel A (2010) Testing various food-industry wastes for electricity production in microbial fuel cell. Bioresour Technol 101(8):2748–2754. https://doi.org/10.1016/j.biortech.2009.11.076
97. Chandrasekhar K, Amulya K, Venkata Mohan S (2015) Solid phase bio-electrofermentation of food waste to harvest value-added products associated with waste remediation. Waste Manag 45:57–65. https://doi.org/10.1016/j.wasman.2015.06.001

98. Goud RK, Babu PS, Mohan SV (2011) Canteen based composite food waste as potential anodic fuel for bioelectricity generation in single chambered microbial fuel cell (MFC): bio-electrochemical evaluation under increasing substrate loading condition. Int J Hydrog Energy 36(10):6210–6218. https://doi.org/10.1016/j.ijhydene.2011.02.056
99. Subha C, Kavitha S, Abisheka S, Tamilarasan K, Arulazhagan P, Rajesh Banu J (2019) Bioelectricity generation and effect studies from organic rich chocolaterie wastewater using continuous upflow anaerobic microbial fuel cell. Fuel 251:224–232. https://doi.org/10.1016/j.fuel.2019.04.052
100. Venkata Mohan S, Mohanakrishna G, Sarma PN (2010) Composite vegetable waste as renewable resource for bioelectricity generation through non-catalyzed open-air cathode microbial fuel cell. Bioresour Technol 101(3):970–976. https://doi.org/10.1016/j.biortech.2009.09.005
101. Song Y, Agrawal R, Wang C (2015) Micro enzymatic biofuel cells: from theoretical to experimental aspect. Proc SPIE 9493:949302. https://doi.org/10.1117/12.2180125
102. Thygesen A, Poulsen FW, Angelidaki I, Min B, Bjerre A-B (2011) Electricity generation by microbial fuel cells fuelled with wheat straw hydrolysate. Biomass Bioenergy 35(11):4732–4739. https://doi.org/10.1016/j.biombioe.2011.09.026
103. Wang H, Qian F, Li Y (2014) Solar-assisted microbial fuel cells for bioelectricity and chemical fuel generation. Nano Energy 8:264–273. https://doi.org/10.1016/j.nanoen.2014.06.004
104. Zuo Y, Maness P-C, Logan BE (2006) Electricity production from steam-exploded corn Stover biomass. Energy Fuel 20(4):1716–1721. https://doi.org/10.1021/ef060033l
105. Li X, Zhu N, Wang Y, Li P, Wu P, Wu J (2013) Animal carcass wastewater treatment and bioelectricity generation in up-flow tubular microbial fuel cells: effects of HRT and non-precious metallic catalyst. Bioresour Technol 128:454–460. https://doi.org/10.1016/j.biortech.2012.10.053
106. Lin J, Zuo J, Gan L, Li P, Liu F, Wang K, Chen L, Gan H (2011) Effects of mixture ratio on anaerobic co-digestion with fruit and vegetable waste and food waste of China. J Environ Sci 23(8):1403–1408. https://doi.org/10.1016/S1001-0742(10)60572-4
107. Wang X, Tang J, Cui J, Liu Q, Giesy JP, Hecker M (2014) Synergy of electricity generation and waste disposal in SolidState microbial fuel cell (MFC) of cow manure composting. Int J Electrochem Sci 9(6):3144–3157. https://doi.org/10.1016/S1452-3981(23)07998-1
108. Li X, Zhang R, Qian Y, Angelidaki I, Zhang Y (2017) The impact of anode acclimation strategy on microbial electrolysis cell treating hydrogen fermentation effluent. Bioresour Technol 236:37–43. https://doi.org/10.1016/j.biortech.2017.03.160
109. Lu L, Ren N, Xing D, Logan BE (2009) Hydrogen production with effluent from an ethanol–H2-coproducing fermentation reactor using a single-chamber microbial electrolysis cell. Biosens Bioelectron 24(10):3055–3060. https://doi.org/10.1016/j.bios.2009.03.024
110. Feng Y, Wang X, Logan BE, Lee H (2008) Brewery wastewater treatment using air-cathode microbial fuel cells. Appl Microbiol Biotechnol 78(5):873–880. https://doi.org/10.1007/s00253-008-1360-2
111. Pepe Sciarria T, Merlino G, Scaglia B, D'Epifanio A, Mecheri B, Borin S, Licoccia S, Adani F (2015) Electricity generation using white and red wine lees in air cathode microbial fuel cells. J Power Sources 274:393–399. https://doi.org/10.1016/j.jpowsour.2014.10.050
112. Bridier A, Desmond-Le Quemener E, Bureau C, Champigneux P, Renvoise L, Audic J-M, Blanchet E, Bergel A, Bouchez T (2015) Successive bioanode regenerations to maintain efficient current production from biowaste. Bioelectrochemistry 106:133–140. https://doi.org/10.1016/j.bioelechem.2015.05.007
113. Pant D, Van Bogaert G, Diels L, Vanbroekhoven K (2010) A review of the substrates used in microbial fuel cells (MFCs) for sustainable energy production. Bioresour Technol 101(6):1533–1543. https://doi.org/10.1016/j.biortech.2009.10.017
114. Strik DPBTB, Hamelers HVM, Snel JFH, Buisman CJN (2008) Green electricity production with living plants and bacteria in a fuel cell. Int J Energy Res 32(9):870–876. https://doi.org/10.1002/er.1397

115. Timmers RA, Strik DPBTB, Hamelers HVM, Buisman CJN (2010) Long-term performance of a plant microbial fuel cell with Spartina anglica. Appl Microbiol Biotechnol 86(3):973–981. https://doi.org/10.1007/s00253-010-2440-7
116. Yang Y, Xu M, Guo J, Sun G (2012) Bacterial extracellular electron transfer in bioelectrochemical systems. Process Biochem 47(12):1707–1714. https://doi.org/10.1016/j.procbio.2012.07.032
117. Sun R, Xing D, Jia J, Liu Q, Zhou A, Bai S, Ren N (2014) Optimization of high-solid waste activated sludge concentration for hydrogen production in microbial electrolysis cells and microbial community diversity analysis. Int J Hydrog Energy 39(35):19912–19920. https://doi.org/10.1016/j.ijhydene.2014.09.163
118. Rojas-Flores S, De La Cruz-Noriega M, Benites SM, Delfín-Narciso D, Luis A-S, Díaz F, Luis C-C, Moises GC (2022) Electric current generation by increasing sucrose in papaya waste in microbial fuel cells. Molecules 27(16):5198. https://doi.org/10.3390/molecules27165198
119. Cheng S, Logan BE (2007) Ammonia treatment of carbon cloth anodes to enhance power generation of microbial fuel cells. Electrochem Commun 9(3):492–496. https://doi.org/10.1016/j.elecom.2006.10.023
120. Nevin KP, Richter H, Covalla SF, Johnson JP, Woodard TL, Orloff AL, Jia H, Zhang M, Lovley DR (2008) Power output and columbic efficiencies from biofilms of *Geobacter sulfurreducens* comparable to mixed community microbial fuel cells. Environ Microbiol 10(10):2505–2514. https://doi.org/10.1111/j.1462-2920.2008.01675.x
121. Li H, Opgenorth PH, Wernick DG, Rogers S, Wu T-Y, Higashide W, Malati P, Huo Y-X, Cho KM, Liao JC (2012) Integrated Electromicrobial conversion of CO_2 to higher alcohols. Science 335(6076):1596–1596. https://doi.org/10.1126/science.1217643
122. LaBelle EV, Marshall CW, Gilbert JA, May HD (2014) Influence of acidic pH on hydrogen and acetate production by an Electrosynthetic microbiome. PLoS One 9(10):e109935. https://doi.org/10.1371/journal.pone.0109935
123. Bond DR, Lovley DR (2003) Electricity production by *Geobacter sulfurreducens* attached to electrodes. Appl Environ Microbiol 69(3):1548–1555. https://doi.org/10.1128/AEM.69.3.1548-1555.2003
124. Lee Y, Nirmalakhandan N (2011) Electricity production in membrane-less microbial fuel cell fed with livestock organic solid waste. Bioresour Technol 102(10):5831–5835. https://doi.org/10.1016/j.biortech.2011.02.090
125. Rezaei F, Xing D, Wagner R, Regan JM, Richard TL, Logan BE (2009) Simultaneous cellulose degradation and electricity production by *Enterobacter cloacae* in a microbial fuel cell. Appl Environ Microbiol 75(11):3673–3678. https://doi.org/10.1128/AEM.02600-08
126. Aulenta F, Reale P, Catervi A, Panero S, Majone M (2008) Kinetics of trichloroethene dechlorination and methane formation by a mixed anaerobic culture in a bio-electrochemical system. Electrochim Acta 53(16):5300–5305. https://doi.org/10.1016/j.electacta.2008.02.084
127. Gregory KB, Bond DR, Lovley DR (2004) Graphite electrodes as electron donors for anaerobic respiration. Environ Microbiol 6(6):596–604. https://doi.org/10.1111/j.1462-2920.2004.00593.x
128. Park DH, Zeikus JG (1999) Utilization of electrically reduced neutral red by *Actinobacillus succinogenes*: physiological function of neutral red in membrane-driven fumarate reduction and energy conservation. J Bacteriol 181(8):2403–2410. https://doi.org/10.1128/JB.181.8.2403-2410.1999
129. Villano M, Monaco G, Aulenta F, Majone M (2011) Electrochemically assisted methane production in a biofilm reactor. J Power Sources 196(22):9467–9472. https://doi.org/10.1016/j.jpowsour.2011.07.016
130. Zhang S, Jiang J, Wang H, Li F, Hua T, Wang W (2021) A review of microbial electrosynthesis applied to carbon dioxide capture and conversion: the basic principles, electrode materials, and bioproducts. J CO2 Util 51:101640. https://doi.org/10.1016/j.jcou.2021.101640

131. McCarty PL, Bae J, Kim J (2011) Domestic wastewater treatment as a net energy producer–can this be achieved? Environ Sci Technol 45(17):7100–7106. https://doi.org/10.1021/es2014264
132. Hays S, Zhang F, Logan BE (2011) Performance of two different types of anodes in membrane electrode assembly microbial fuel cells for power generation from domestic wastewater. J Power Sources 196(20):8293–8300. https://doi.org/10.1016/j.jpowsour.2011.06.027
133. Cusick RD, Kiely PD, Logan BE (2010) A monetary comparison of energy recovered from microbial fuel cells and microbial electrolysis cells fed winery or domestic wastewaters. Int J Hydrog Energy 35(17):8855–8861. https://doi.org/10.1016/j.ijhydene.2010.06.077

Algae-Derived Electrodes in Bioelectrochemical Systems

Gracia Saral Gladison, R. Dhandapani, R. Lavanya Priyadarshini, and Senthilkumar Rathnasamy

Abstract Bioelectrochemical systems (BES) represent a novel interface between electrochemical processes and microorganisms, facilitating electron transfer between donors and acceptors. The paper extensively elaborates on the pioneering application of algae in enhancing the performance of BES, particularly in microbial fuel cells (MFCs). Algae serve a vital function through multiple pathways: supplying oxygen via photosynthesis as a final electron acceptor, sequestering carbon dioxide from the atmosphere and anode chambers, cleaning wastewater by removing organic pollutants and nutrients and yielding valuable biomass for biofuel production. This chapter addresses specific BES configurations, including MFCs, Microbial Electrolysis Cells, Microbial Electrosynthesis, and Microbial Desalination Cells, emphasising the extracellular electron transfer pathways and photosynthesis processes that enable these setups. The experimentally observed evidence supports algae-integrated BESs achieving remarkable performance standards, including a maximum power density of 410 mW/m^2 and COD removal rates of 74%–86.6%. Moreover, algae-based electrodes are also adaptable as electron donors and acceptors, establishing efficient syntrophic relationships with bacteria that enhance overall system stability and efficiency. Successful utilisation in wastewater treatment, electricity generation, and CO_2 capture, as demonstrated through case studies, establishes algae-based BESs as a promising green bioengineering technology that is both environmentally beneficial and sustainable, generating renewable energy and valuable resources simultaneously.

Keywords Microalgae · Microbial fuel cells · Electrodes · Conductivity · Algal residues · COD · Biofuel

G. S. Gladison · R. Dhandapani · R. Lavanya Priyadarshini · S. Rathnasamy (✉)
Green Separation Engineering Laboratory, School of Chemical and Biotechnology, SASTRA Deemed to Be University, Thanjavur, India
e-mail: senthilrathna@sastra.ac.in

A. Sattar Jatoi, N. M. Mubarak (eds.), *Bio-Electrochemical Conversion of Waste to Fuel*, Green Energy and Technology,
https://doi.org/10.1007/978-981-95-4310-6_13

1 Introduction

Bioelectrochemical systems (BESs) are emerging bioengineering technologies that couple microorganisms or enzymes and electrochemical processes to enable electron transfer from electron donors to acceptors [1]. BES's bio-inspired and biocatalysed systems integrate biological and electrochemical processes to produce renewable solutions for energy, waste treatment, and resource recovery. The key elements of BESs are an anode and a cathode, with an external circuit connection as an option. The most crucial mechanism within such systems is the ability of exoelectrogens to transfer electrons beyond their cells to an external electron acceptor, such as an electrode. Extracellular electron transfer (EET) facilitates the conversion of chemical energy into electrical energy or vice versa. The electron transfer processes in BESs are as follows: first, short-range direct electron transfers occur via redox proteins, such as outer membrane cytochromes. The second one is indirect electron transfer via redox mediators excreted by microorganisms, and the last one is Long-range direct electron transfer via conductive appendages such as nanowires [2]. Various BES types have been engineered to carry out numerous functions, which are as follows.

- Microbial Fuel Cells (MFCs): Produce electricity by oxidising organic materials using microbial metabolism. Bacteria degrade organic matter at the anode, producing electrons that are transferred in an external circuit to the cathode, where they produce an electrical current [3].
- Microbial Electrolysis Cells (MECs): Employ a low external electric input to facilitate reactions that produce valuable products, such as hydrogen. MECs operate anaerobically and utilise bacteria to convert chemical energy into electrical energy, with the potential for electrolysis [4].
- Microbial Electrosynthesis (MES): Transfers the reducing power produced in anodic oxidation to produce value-added products on the cathode. Alternatively termed bioelectrosynthesis, MES involves the electroreduction of CO_2 or other organic feedstocks using microbes as biocatalysts.
- Microbial Desalination Cells (MDCs): Coupling the processes of desalination with the generation of electricity or wastewater treatment.

Through carbon sequestration, bioremediation, and green energy generation in the BES, algae are game-changing in bioelectrochemical systems, particularly in microbial fuel cells (MFCs). These systems integrate electrochemistry and algal photosynthesis to address energy and environmental challenges. Their uses are discussed in greater detail below:

- Carbon Sequestration and Oxygen Production: External aeration is not required in algal-based microbial fuel cells (MFCs), as algae in the cathode chamber produce oxygen through photosynthesis. By serving as a terminal electron acceptor, this oxygen is responsible for triggering electron flow and enhancing electricity generation. Algae utilise CO_2 to produce biomass by removing it from air supplies and the anode chamber, where it is produced by oxidising organic substances. Apart from reducing greenhouse emissions, this double benefit enhances the cultivation of algae for biofuel production.
- Wastewater Treatment and Bioremediation: Algae-MFC systems decompose organic matter, heavy metals, and nutrients (e.g., phosphorus and nitrogen) in wastewater. For example, Species of *Spirulina* and *Chlorella* remove contaminants in a highly efficient manner while building up biomass. The treatability of wastewater can become energy-neutral or even energy-positive, depending on the process used to produce bioelectricity.
- Value-added products and Bioenergy: Algal lipids (e.g., Chlorella vulgaris triglycerides) are converted by transesterification to FAME (fatty acid methyl esters). Bioethanol: Ethanol is generated from carbohydrate-based biomass in algae (e.g., brown algae) through fermentation. Biogas: Methane is generated through the anaerobic digestion of residue biomass after lipid extraction. Biochar: Pyrolysis is applied to treat algae biomass for application in pollutant adsorption and carbon sequestration. High-value products make economic viability possible, including proteins, polyunsaturated fatty acids (PUFAs), and antioxidants (carotenoids).
- BES Electrode Material: The pyrolysis of algae produces carbon as a viable, sustainable electrode material. Its intrinsic nitrogen functions improve conductivity and catalytic properties in BES.

One of the key achievements in green material science, algae-based electrodes have unparalleled benefits for energy storage, environmental remediation, and biological applications. Their renewability, biocompatibility, and flexible electrochemical properties make them worthy of consideration. Algae-based electrodes represent a green alternative to addressing two key challenges in waste valorisation and resource efficiency. The integration of algae into bioelectrochemical systems represents a significant advancement in green bioengineering technology, offering improved performance, economic benefits, and environmental applications that enhance the acceptability of these systems for practical use in wastewater treatment, power generation, and resource recovery. These benefits are detailed below (Fig. 1).

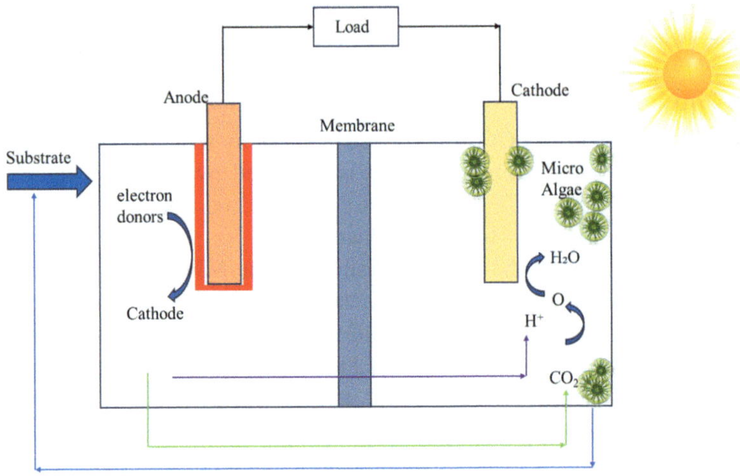

Fig. 1 Algae-based electrodes

2 Algae as Resources for Bioelectrochemical Systems

2.1 Composition and Properties of Algal Biomass

Algae have a large COH structure and plasticity, which render algae biomass a viable fuel for bioelectrochemical systems (BES) [5]. Recovery COH, sewage treatment, and green energy production are feasible through integration in microbial COH fuel cells (MFCs) and related COH technologies. Make-up, nature, and COH performance in BES are discussed in detail below: Make-Up of Algal Biomass—Two significant components of algae render them ideal for application in BES:

- Lipids (2.4–62% dry weight): Transesterified to fatty acid methyl esters (FAMEs) for biodiesel. While high concentrations of cold-flow improvable unsaturated fatty acids are desirable, hydroperoxide formation can clog the filter.
- Carbohydrates (up to 51%): Polymers like cellulose, alginate, and laminarin, which are oxidised directly in MFCs or hydrolysed to fermentable sugars for bioethanol. Electrogenic bacteria break down proteins (up to 32%) in MFCs to release electrons—species such as Scenedesmus obliquus and Chlorella vulgaris exhibit varying compositions under different growth conditions.

Algae enhance BES performance in electricity generation. Acetate and co-substrates enhance the performance of raw algae to 410 mW/m^2. Treat wastewater: Algal MFCs reduce COD by 74% to 86.6% while producing bioelectricity. Sequestration of CO2: Utilised CO_2 from anode reactions, cathodic chambers hold algae, reducing greenhouse gases in third-generation biofuels and bioelectrochemical systems (BES) [6]. Algae are prominent sources of green and clean energy production, as well as remediation capabilities. These systems are superior to previous generations of biofuels, as they utilise algae photosynthesis and microbial fuel cells (MFCs) for carbon capture facilitation, wastewater treatment, and renewable energy generation. In the cathode compartment of MFCs, algae utilise photosynthesis to produce oxygen, which serves as a substitute for external aeration and acts as a terminal electron acceptor, thereby enhancing energy output. For example, Spirulina and Chlorella species optimise this process in wastewater treatment. CO_2 Sequestration: Algae absorb CO_2 from air sources and the anode (formed during organic oxidation) to generate biomass for biofuel. Thus, the transportation costs of greenhouse gases and CO2 are reduced. Bioremediation: Nitrogen, phosphorus, and heavy metals are biodegraded through algae-MFC systems. *Chaetoceros* sp. enhances Coulombic efficiency by 77% with a power density of 159 mW/m^2 in pharmaceutical wastewater-treated MFCs.

2.2 Algae as Third-Generation Biofuel Sources

Due to their richness in lipids and metabolic diversity, algae have been utilised as multifunctional resources for the generation of biodiesel and bioelectrochemical systems (BES). High-value biodiesel microalgae include *Chlorella vulgaris* (11–43% lipids), *Botryococcus* (55%), and *Neochloris oleoabundans* (42%). ASTM/EN biodiesel specifications are met by lipid profiles comprising largely palmitic, oleic, and stearic acids. Cultivation strategies: Two-stage growth systems (biomass production, followed by induction of lipids) are commonly needed to provide balance in productivity, but nutrient stress (such as nitrogen starvation) promotes the accumulation of lipids. Enhancing cathode reactions and oxygen release by means of photosynthesis and treating wastewater by utilising CO_2 and nutrients (phosphorus and nitrogen).In co-product synergy, methane (0.2–0.4 m^3/kg biomass) or bioelectricity can be generated through the anaerobic digestion of waste from lipid-extracted algae (LEA) during biodiesel production in microbial fuel cells (MFCs) [7, 8].

Lipid extraction is a prevalent solvent-based process, typically using hexane; however, environmental expenses can be reduced by employing bio-based solvents, such as ethanol and hydrothermal liquefaction (HTL).Through the elimination of energy-consuming drying, wet biomass processing by HTL yields bio-crude yields of 34%–46%. Transesterification: Using methanol and TiO_2 or NaOH catalysts, triglycerides are catalytically converted into fatty acid methyl esters (FAME), which can be used to produce biodiesel with a purity of up to 98%. Methane yields per pretreatment and species are 0.024 to 0.6 L/g VS (volatile solids) with 0.2–0.4 m^3 CH_4/kg biomass; whole-cell digestion is superior to lipid-extracted biomass. Methane yields are boosted by 303% via hydrothermal-alkaline processes (e.g., 0.2 mol/L NaOH, 150 °C, 30 min). Co-digestion of food waste or bacterial biomass is advantageous for improving biogas production (e.g., 0.43 L/g VS).

Algae are critical components in bioelectrochemical systems (BES), also known as microbial fuel cells (MFCs), as they facilitate carbon sequestration, wastewater treatment, and the production of renewable energy while creating higher-value products. Unlike traditional methods, these systems utilise algae photosynthesis to achieve increased efficiency and lower operational costs. Algae MFCs are internally aeration self-sustaining, as they emit oxygen through photosynthetic reactions in the cathode chamber. Principal designs include single-chamber MFCs, where the co-location of bacteria and algae facilitates the sequestration of CO_2 and provides an oxygen supply. Dual-chamber MFCs: Cathode algae enhance biomass growth and power generation through CO_2 utilisation from the anode. Algae facilitate ion separation in three-chamber desalination units. These utilise the algal reduction of cathode oxygen and microbial oxidation of organics at the anode to achieve higher power densities. Wastewater treatment and nutrient removal. Carbon and nitrogen removal: Phosphates and nitrates are taken up by algae in the cathode, whereas heterotrophic bacteria in the anode decompose organic wastes [9]. Heavy metal removal: Algal biomass produces electricity while immobilising metal ions such as lead and cadmium. Such multi-attribute characteristics enhance the value of water reuse while reducing treatment costs. In BES, algae capture CO_2 from anode reactions or industrial wastewater and utilise it to produce biomass.

Thermal decomposition of algal biomass yields carbon-rich compounds that can serve as soil fertilisers or for energy storage. Biodiesel: Fatty acid methyl esters (FAMEs) are synthesised by transesterification of algal lipids; yield varies with the quantity of lipid and fatty acid profile. In biomethane production, biogas (50%–70% CH_4) was generated by anaerobic digestion of algal biomass. Ethanol is generated by fermenting algal carbohydrates (e.g., mannitol and laminarin), particularly from brown and red macroalgae. Through the integration of MFCs with photobioreactors or anaerobic digestion, the recovery of resources is maximised, and simultaneously, biofuels, electricity, and biopolymers are generated. In biomass pretreatment, the complex algal cell walls require significant energy to be treated, rendering the substrate more accessible. On a commercial scale, adoption will require optimising reactor design, membrane costs, and electrode materials. The increase in algal lipid and carbohydrate content, as well as enhanced electron transfer, may enhance production [10, 11]. A closed-loop model for a bioeconomy, algal-based BES minimises environmental impact and transforms waste into energy and by-products. Scale-up will primarily be a function of developments in genetic engineering, as well as systems integration.

3 Fundamentals of Algae-Derived Electrodes

3.1 Photosynthetic Mechanism in Algae

Algae electrodes are novel and sustainable elements in bioelectrochemical systems, featuring distinctive characteristics that enhance system performance and provide environmental benefits. Algae electrodes utilise the inherent properties of algae by undergoing several preparation processes to produce materials with superior electrochemical properties. Such fundamental mechanisms are as follows.

Algae are highly efficient photosynthetic organisms that contribute significantly to global oxygen production and carbon sequestration. Photosynthesis in algae occurs in chloroplasts, where light energy is converted into chemical energy. Their mode of photosynthesis is the capture of light by pigments such as chlorophyll and carotenoids, which fuel electron transport between Photosystems I and II. Photosystems I and II are responsible for generating ATP and NADPH, which are needed for the fixation of carbon in the Calvin cycle. Both cyanobacteria and green algae exhibit structural diversity in their photosystems, enabling them to adapt to various environmental conditions. There are two principal stages in the process: light-dependent reactions and light-independent (Calvin cycle) reactions. During the light-dependent reactions, pigments such as chlorophyll trap sunlight, exciting electrons that move through the photosynthetic electron transport chain. This process produces ATP and NADPH, which are utilised in the Calvin cycle to convert carbon dioxide into organic molecules [12–14]. Algae have carbon-concentrating mechanisms (CCMs) which maximise photosynthesis efficiency by accumulating CO_2 surrounding the enzyme RuBisCO [13, 14]. Moreover, oxygenic photosynthesis in algae dissociates water molecules, emitting oxygen and protons while transporting electrons to $NADP^+$ through Photosystems I and II [14, 15].

3.2 Electron Transfer Process

Electron transfer in algae occurs during respiration and photosynthesis. In BES, the electrons are harvested from metabolism and transferred to electrodes through direct electron transfer (DET) or mediated electron transfer (MET). DET is achieved through direct contact between algal cells and electrode surfaces, whereas MET involves the use of redox-active mediators for electron conduction. Environmental factors, including light intensity, substrate concentration, and electrode material, influence the efficiency of electron transfer. Electron transfer in algae is central to photosynthesis and bioelectrochemical systems (BES). Photosynthetic electrons are produced and pass through the electron transport chain involving Photosystem II (PSII), cytochrome b6f complex, plastocyanin, and Photosystem I (PSI) [14, 15]. The electrons are reduced from $NADP^+$ to NADPH. Algae are also responsible for extracellular electron transfer (EET) in BES applications, either through direct channels via membrane-bound proteins or indirect channels via secreted oxygen as an electron mediator [14, 16]. For example, microalgae such as Desmodesmus sp. have demonstrated direct EET capability through membrane proteins [16]. Moreover, auxiliary electron transport chains help in dissipating surplus energy under changing environmental conditions [14].

3.3 Types of Algae Used in BES

Certain algae species are utilised in bioelectrochemical systems because they are capable of photosynthesis and are adaptable, allowing them to be used either as living electrodes or microbial oxidation substrates. Such types are as follows.

- Green Algae: Green algae, such as Chlamydomonas reinhardtii and Chlorella vulgaris, are commonly used because they provide a good biomass yield and can generate oxygen at cathodes [17, 18].

- Blue-Green Algae (Cyanobacteria): Blue-green algae, such as Spirulina platensis, are most suitable for implementation in single-chamber microbial fuel cells (MFCs) since they can produce power even in dark environments [18].
- Other Microalgae: Additionally, microalgae such as Scenedesmus obliquus and Dunaliella tertiolecta have been studied for their activities in BES configurations. For instance, Chlorella vulgaris achieved optimal power densities of 15 mW/m^2 [18].

Specifically, the algae-derived electrodes in BES were classified into two types. One alga serves as an electron donor, and another alga serves as an electron acceptor. Raw algae serve as a direct source of electrons in BES systems. When introduced into MFCs, microbial oxidation of algal biomass generates electrons at the anode. This process is enhanced by co-substrates, such as acetate, which improve power density and current generation [19, 20]. For example, Raw algae blended with acetate produced a power density of 410 mW/m^2, while raw algae alone yielded a value of 230 mW/m^2 [19]. Dead biomass of microalgae can be utilised as a substrate for anaerobic respiration in double-chamber MFCs, providing twin sources of electrons from both organic substrates and microalgal biomass [21]. This method not only enables electricity production but also facilitates bioremediation through the breakdown of organic compounds. The algal metabolites, such as oxygen, evolved during photosynthesis and are natural electron acceptors in BES systems. Oxygen evolved in algal-assisted MFCs at the cathode is utilised to facilitate the reduction reactions necessary for power generation [18, 21].

Apart from these, residues extracted from lipids derived from algae were employed to induce sulfate reduction by acting as slow-release electron donors for SRB [22]. Oxygen generated by the living microalgae within the cathode chamber enhances the overall system's efficiency compared to mechanical aeration methods [21]. These metabolites enhance BES performance in addition to enabling sustainable energy production (Fig. 2).

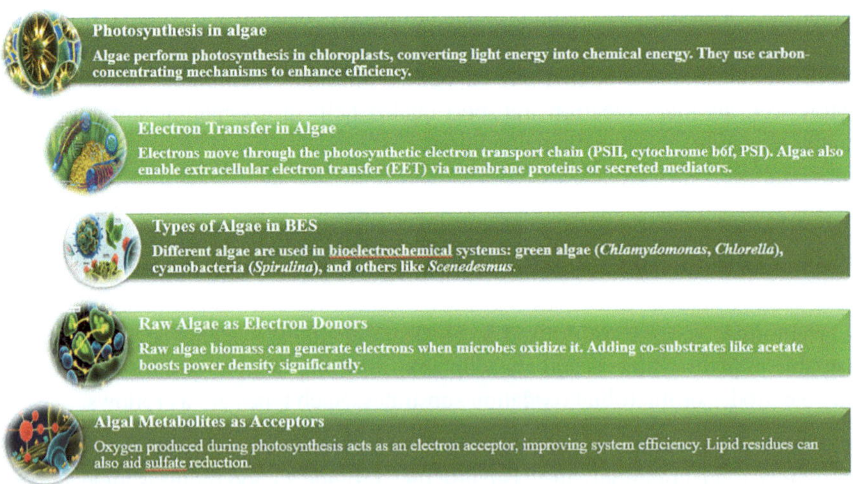

Fig. 2 Algae as electrodes

4 Integrated Algae-MFC Systems

Algae-MFC systems, when integrated, represent a promising and innovative advancement in green biotechnology, offering multifunctional technologies that simultaneously address environmental problems while generating renewable energy and helpful biomass resources. Those integrated systems are illustrated in Fig. 3.

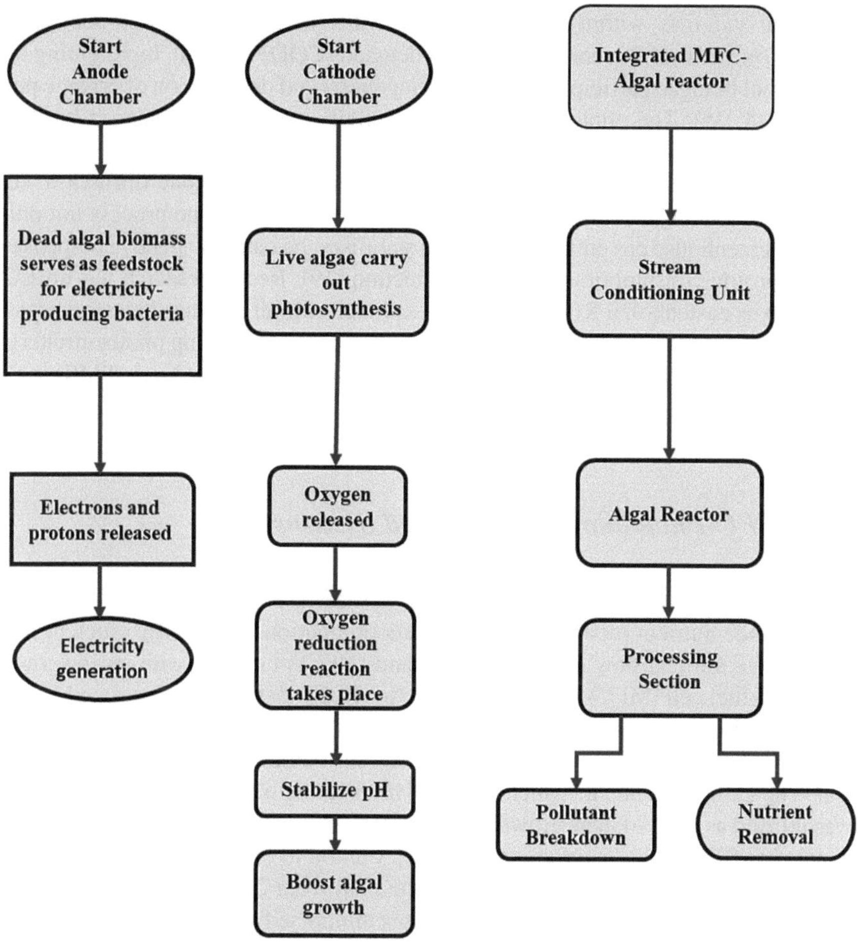

Fig. 3 Integrated algal-MFC systems

4.1 Photo-Bio Electrochemical Systems

Photo-bioelectrochemical systems (PBES) combine photosynthetic microorganisms, such as algae or cyanobacteria, within microbial fuel cells (MFCs) for wastewater treatment, carbon dioxide sequestration, and the concurrent generation of bioelectricity. Systems utilise algae's oxygenic photosynthesis in the cathode chamber to utilise dissolved oxygen (DO) as a terminal electron acceptor, thereby avoiding mechanical aeration and reducing operating costs [19, 23]. For instance, *Chlorella vulgaris* within a dual-chamber MFC achieved a power density of 250 mW/m^2 with 44.8% chemical oxygen demand (COD) removal, highlighting the dual aspect of algae participation in oxygenogenesis and degradation of organic pollutants [24, 25]. The photoelectrochemical activity of algae is enhanced by their ability to fix CO_2 via anode respiration. In a microbial carbon capture cell (MCC), CO_2 generated during the oxidation of the substrate in the anode diffuses to the cathode, where algae colonise it under illumination [23, 26]. The process not only reduces greenhouse gas emissions but also stabilises the cathodic pH by compensating for proton consumption in oxygen reduction [19]. Recent research has utilised Roseiflexus castenholzii RC-LH complexes, which exhibit enhanced electron transfer efficiency due to their tetraheme cytochrome structure, yielding photocurrents of 320 nA/cm^2 with 1-methoxy-5-methylphenazinium methyl sulfate (PMS) as a mediator [27].

4.2 MFC-Photobioreactor Coupled Systems

The combination of MFCs and photobioreactors (PBRs) creates synergistic systems that enhance nutrient recovery, algal biomass production, and energy efficiency. In a two-stage vertical flow constructed wetland (VFCW) system with an integrated microbial fuel cell (MFC) and photobioreactor (PBR), the system eliminated 95.8% of chemical oxygen demand (COD). It yielded 21,323 mg/L of algal biomass (wet weight) by utilising anode-generated electrons and cathodic oxygen from *Chlorella* spp. The cathode effluent, which contains high levels of nitrates and phosphates, was utilised as a microalgae growth medium, thereby closing the nutrient loop while generating 0.205 kWh/m^3 of energy [28, 29]. Central to this integration is hydraulic retention time (HRT). Increasing HRT in the PBR from 2 to 3 days increased algal biomass yield by 200%, as reduced flow rates minimised shear stress and enhanced light penetration [28]. Yet, recirculation rates in MFCs need to be optimised to keep energy consumption low; lowering pump speeds by 50% resulted in a 30% reduction in operating costs without sacrificing treatment efficiency [28]. Membrane photobioreactors (MPBRs) also optimise performance by utilising ultrafiltration membranes to isolate algal biomass, resulting in four times greater productivity compared to conventional systems [28, 30]. Recent studies investigate hybrid configurations, such as the integration of MFCs into algal raceway ponds. A 12-month pilot experiment demonstrated 98% ammonium removal and a power density of 2.2 W/m^3, with algal photosynthesis accounting for 60% of the energy consumption of the MFC [19]. Such systems are typical of the circular economy model, where waste streams are converted into biofuels (e.g., biodiesel from harvested algae), and electricity is produced [26, 30].

4.3 Syntrophic Interactions Between Bacteria and Algae

Syntrophy between bacteria and algae in microbial fuel cells (MFCs) enhances system stability, facilitates electron transfer, and promotes the degradation of substrates. Fermentative bacteria, such as Clostridium, break down complex organics into volatile fatty acids (VFAs) in anode chambers, which are further oxidised by exoelectrogens, including Geobacter and Shewanella, to yield electrons [31, 32]. Shewanella algae strains isolated from coastal sediment MFCs, for instance, yielded 68 mA/m^2 when using acetate as a substrate through metabolism, while syntrophic symbionts, such as Pelotomaculum, scavenged intermediates to prevent metabolic inhibition [20, 33]. Algae repay by providing oxygen and organic metabolites to cathode-related bacteria. Leptolyngbya cyanobacteria and Acutodesmus green algae biofilms, in integrated photo-bioelectrochemical (IPB) systems, enhance DO by 300% and allow Nitrosomonas and Nitrobacter to carry out nitrification-denitrification with 82% total nitrogen removal [17, 34]. Quorum-sensing molecules, such as acyl-homoserine lactones, enable cross-kingdom communication, which optimises substrate consumption and biofilm formation [34]. Dead algae are also syntrophic as photosensitisers. Light-irradiated Raphidocelis subcapitata biomass doubled the Geobacter sulfurreducens reduction rate by transferring photoelectrons through extracellular polymeric substances (EPS) [35]. Such interaction demonstrates the robustness of algal-bacterial consortia in bioelectrochemical systems to operate continuously, even in the presence of variable organic loads. Current studies examine hybrid systems, i.e., the incorporation of MFCs into algal raceway ponds. The 12-month pilot operation achieved 98% ammonium removal and a power density of 2.2 W/m^3, with algal photosynthesis accounting for 60% of the energy required by the MFC. Systems are traditional examples of the circular economy paradigm, where waste streams are cycled back to biofuels (e.g., harvested algae to biodiesel) and electricity production [31, 32].

4.4 Dead Vs. Live Microalgae Applications

The selection of dead or alive microalgae in MFCs is based on system applications: dead algae provide a cheap substrate, while live algae facilitate cathodic activities. Dead Microalgae: Thermal-acid pretreated Chlorella regularis biomass generated 15.6 g/L soluble COD, 40% higher than raw biomass and 0.98 W/m^2 power density

[26, 31]. Similarly, dried *Dunaliella tertiolecta* yielded 5.3 mW/m^2 by providing carbohydrates to *Shewanella* spp., though rigid cell walls needed pretreatment to achieve successful hydrolysis [30]. Dead algae are also biophotocatalysts; UV-irradiated Scenedesmus residues catalysed Cr(VI) reduction in MFCs by 150% through photo-induced electron shuttling [29]. Live Microalgae: Cathode-cultivated live Tetraselmis indica produced 560.8 mW/m^2 utilising photosynthetic oxygen addition and 78% phosphate utilisation from pharmaceutical wastewater effluent [19, 30]. Darkness, however, reduces efficiency; a Chlorella-MFC exhibited an 80% drop in voltage under nighttime conditions, and energy-storing buffers are necessary [23, 25]. This is alleviated by hybrid systems that combine live algae with electroactive biofilms (Table 1). A Nitzschia palea-Bacillus co-culture, for instance, removed 90% COD and 27.4 mW/m^2 under daytime, and Bacillus retained 40% activity during night [10, 16]. Comparative research indicates that dead algae provide more initial energy (e.g., 277 W/m^3 for Chlorella powder), but live systems can offer long-term stability because of the ongoing production of oxygen and biomass [24, 25].

Table 1 Integration strategy for algal-MFC systems

S. No	Integration strategy	Main features and components	Key benefits and outputs	References
1.	MFC–algal bioreactor integration	MFC is combined with an algal reactor in three parts: Water treatment, algal growth, and processing.	– Reduces chemical oxygen demand – Improves nutrient absorption by algae	[24]
2.	Mutualistic algae–bacteria interactions	Algae supply oxygen and food; bacteria offer CO_2 and nutrients.	– Better nutrient use – Stronger, more stable biofilm growth	[25]
3.	Photo-bio hybrid systems	It uses light-sensitive electrodes to break down waste to produce hydrogen. Includes microbial electrochemical cells for energy generation.	– Solar-powered water cleaning – Produces renewable hydrogen energy	[30]

5 Performance Metrics and Analysis

Performance metrics and analysis are crucial in determining the efficiency of electrodes made from algae in bioelectrochemical systems (BESs). By these metrics, researchers can approximate system efficiency, contrast different configurations, and optimise design parameters to achieve maximum performance.

5.1 Power Density and Conductivity

Conductivity is a measure of the capacity of an aqueous solution to conduct an electrical current. Ions transport the current, and hence, the conductivity will rise with an increasing concentration of ions in the solution, their mobility, and water temperature. Conductivity readings are proportional to ionic strength, and Thermal Conductivity will quantify how effectively a substance transports heat. It is significant in applications such as thermoelectric devices, where minimising thermal conductivity can improve efficiency by reducing heat loss. Thermal conductivity is often determined by methods such as the laser flash technique, which is based on computing it from thermal diffusivity, specific heat capacity, and mass density. A material with high thermal conductivity can conduct a large amount of heat quickly over a long distance.

In contrast, a material with low thermal conductivity can be employed as an insulating barrier to heat transfer. Thermal conductivity is a crucial factor in thermal interface materials, as it enables the transfer of heat between two surfaces. Electrical Conductivity: measures the degree to which a material facilitates the passage of electricity. It is crucial for applications that require efficient electrical current flow. Electrical conductivity can be measured using methods such as the Van der Pauw method [36].

Power density is a crucial consideration when selecting a location for renewable energy. This generally refers to the power output in terms of the area occupied by the system. High-power density implies that there are small devices, consequently resulting in low space and cost. Power density is measured in watts per cubic metre (W/m^3) or watts per kilogram (W/kg). It is a crucial measurement in a range of industries, from energy storage and power electronics to thermophotovoltaics (TPVs), as it specifies the efficiency and miniaturisation of these systems. Power density in energy storage systems measures the capacity of the system, by size or weight, to deliver instantaneous power supply. This differs from energy density, which focuses on the sustainment time of the power supply [37]. Power density varies in TPV devices due to emissive temperature, view factors, and material emissivity. High-power densities in near-field TPVs are achieved by closing the emitter-cell gap to the nanometer scale, thereby improving radial heat transfer even beyond the limits of a blackbody.

Power density is applied to estimate the performance of converters and power supplies. It is used for optimising design parameters, such as topology, heat management, and volume minimisation, to enhance performance within limited space [38, 39]. For example, next-generation TPV cells utilising spectrum-matched bandgaps can achieve power densities up to 3.5 times higher than mismatched designs, and power density plays a crucial role in constructing compact yet high-performance energy systems across various industries. Inverters, converters, and other electronic devices are based on high-power density structures to reduce size while preserving performance, which is critical in compact consumer products and industrial equipment. Materials with optimised critical power density are utilised in high-end optical systems, including lasers and imaging systems.

5.2 COD Removal Efficiency

COD removal efficiency is a crucial measure of wastewater treatment performance, indicating the capacity of treatment systems to minimise organic contaminants. Several studies have investigated the optimisation of COD removal in different technologies, reactors, and methods. Chemical oxygen demand (COD) is a measure of the organic and inorganic compounds present in industrial wastewater. Total organic carbon (TOC) is employed as a measure of the concentration of organic carbon within wastewater. COD and TOC are positively correlated and utilised to ascertain the quality of wastewater and the need for treatment. High levels of COD and TOC are present in industrial wastewater, posing a significant threat to human health, wildlife, and the environment. SPTFs employing PE and PU media have reported impressive COD removal efficiencies of up to 97%. The efficiency was uniform over a range of hydraulic loading rates (HLRs) from 2 to 5 m/day, with effluent COD concentrations persistently below 10 mg/L. At increased HLRs of 6 m/day, the COD removal efficiency decreased slightly due to elevated effluent concentrations [38]. The inherent properties of randomly packed reactors, including enhanced influent distribution and biomass growth, were responsible for increased contact opportunities between sewage and biomass, improving COD removal. Activated sludge systems are commonly used in wastewater treatment plants (WWTPs) for removing aerobic chemical oxygen demand (COD). The operational parameters of pH, food-to-microorganism (F/M) ratio, mixed liquor suspended solids (MLSS), and sludge volume index (SVI) have a significant impact on COD removal efficiency. A decrease in COD removal from 79% to 4% within 12 days was reported, resulting from high SVI values that led to biomass washout and a decrease in bacterial catabolic capacity. The growth of protozoa and rotifers inhibits flock-forming bacteria that are crucial for COD biodegradation. Electrocoagulation has proven to be a promising option for enhancing chemical oxygen demand (COD) removal in wastewater treatment plants (WWTPs), particularly for industrial wastewater. At optimum conditions—pH 3, current density 29.79 mA/cm^2, 5 min electrolysis time—the COD removal efficiency was 79%. Electrocoagulation accelerates the oxidation of organic pollutants by generating hydroxyl radicals at the anode.

TiO_2–chitosan nanocomposites were used for the treatment of industrial wastewater, with COD removal efficiencies as high as 93.67% at favourable conditions (pH 7, contact time 180 min, adsorbent mass 2.5 g). The surface charge and pore structure of the nanocomposite were tuned for ion exchange and adsorption processes. Nevertheless, COD removal was adversely affected at elevated temperatures due to the detachment of pollutants from the adsorbent surface. COD is a good indicator of organic pollution in water bodies and WWTP effluents. Measurement techniques employing potassium dichromate ($K_2Cr_2O_7$) or potassium permanganate ($KMnO_4$) have been optimised for precision but are subject to secondary pollution by chemical reagents. Improved electrochemical and photocatalytic oxidation methods offer promising alternatives for large-scale applications.

The concentration of Mixed Liquor Suspended Solids directly influences microbial activity, as well as the availability of substrate. Higher MLSS enhances COD removal up to a certain level, after which oxygen transfer limitations take over, and Operations with MLSS levels in the 2000–4000 mg/L range typically achieve stable COD removal (~79–96%). Very high MLSS (>5000 mg/L) reduces settlement efficiency, resulting in increased effluent COD. Food-to-Microorganism (F/M) Ratio is a parameter of the ratio of organic loading, in the form of Chemical Oxygen Demand (COD), to the biomass in the shape of Mixed Liquor Suspended Solids (MLSS). The optimal F/M range is 0.2–0.5 kg COD per kg MLSS per day, as it has been proven to improve COD removal efficiency at rates of up to 96%. However, a low F/M ratio, i.e., less than 0.1, can lead to excessive endogenous respiration, which reduces both microbial activity and COD removal efficiency. Alternatively, a high F/M ratio greater than 0.8 can lead to oxygen demand spikes, resulting in problems such as sludge bulking and incomplete oxidation [39].

COD destruction performance relies on treatment technology, operation, and reactor design. Efficient technologies, such as nanocomposites, electrocoagulation, and innovative reactor configurations, enhance performance while addressing environmental sustainability concerns. These technologies are crucial for meeting the stringent requirements of wastewater treatment in both industrial and municipal applications.

5.3 Biogas Yield from Algal Residues

Algal residue biogas has also been recognised as a potential renewable energy source, leveraging the high organic content and rapid growth rate of microalgae and macroalgae. Technical feasibility, optimal conditions, and process complexity are discussed in certain research studies, providing valuable insights for enhancing yield and efficiency. Organic Loading Rate (OLR): Maintaining OLR at 2.87 kg VS/m^3·day achieves maximum biogas production, as seen in co-digestion experiments using Cladophora glomerata macroalgae, cattle manure, and sewage sludge. The

Table 2 Performance metrics and its application

Sl No	Performance metrics	Key findings	Application	References
1.	Power density and conductivity	Achieved 22 W cm^2 in type half-heusler thermoelectric material. High-power factor of 106 μW cm^{-1} k^{-2} at room temperature	High-efficiency waste heat recovery systems, compact power electronics	[36]
2.	COD removal efficiency	97% removal in sponge-packed tower reactors (SPTFs) at 2–5 m/day hydraulic loading	Energy-efficient wastewater treatment for industries and municipalities	[38]
3.	Biogas yield from algal residues	Theoretical potential linked to algal biomass composition	Bioenergy production via anaerobic digestion	[39]
4.	High-power density batteries	High-power density lithium-ion batteries enhance energy storage efficiency	Electric vehicles, renewable energy systems	[37]
5.	Enhanced conductivity materials	Additive manufacturing enhances the electrical conductivity of metals like copper and silver	High-performance electronics and energy transmission lines	[36]

design produced 439 L biogas/kg VS and 306.5 L CH_4/kg VS, and the higher OLRs decreased the yield by 1.55 times compared to the initial.

Co-digestion strategies involve combining algal biomass with high-nutrient feedstocks, such as wheat straw or cattle manure, to stimulate biogas production. Co-digestion of microalgae and wheat straw, for example, boosted CH_4 yield by 77% over mono-digestion. Researchers are exploring ways to streamline this process. The use of waste from liquid biogas plants as a nutrient source for algae cultivation enables waste recycling without incurring additional expenses [37]. Incorporating technical glycerol into the cultivation process can enhance algal biomass yield by up to 20% by increasing the availability of feedstock. Additionally, treating agricultural algal wastewater not only contributes to biomass production for biogas but also offers several advantages (Table 2).

6 Applications and Case Studies

6.1 Wastewater Treatment

Algae growth wastewater treatment is a promising method to create bioelectric electrodes in bioelectrochemical systems. Algae-based wastewater treatment systems offer unique advantages, including high-nutrient removal rates, CO_2 capture, and the production of desirable biomass that can be reused for other purposes [40]. High-quality nutrient sources, such as aquaculture facilities and dairy farms, provide optimal cultivation media for microalgae strains that not only facilitate high growth but also concurrently remove impurities, including organic matter, nitrogen, and phosphorus. Experiments have demonstrated that some species, such as *Chlorella pyrenoidosa,* can grow well in unpasteurised wastewater [41], with high biomass productivity (5.03 g m^{-2} day^{-1}) and successful removal of NH4 + –N (75.9%) total phosphorus (68.4%), COD (74.8%), and even heavy metals such as zinc, copper, and iron. The resulting algal biomass, under its distinct biochemical properties and conductive capability, can be used to synthesise electrode materials for bioelectrochemical systems, promoting a circular economy model where waste treatment leads to the generation of resources for renewable energy technology. Biofilm-attached cultivation methods are especially promising for this strategy, as they can increase both treatment efficacy and biomass concentration, making them possible for electrode synthesis [42].

6.2 Electricity Generation

Algae play a crucial role in generating electricity through the utilisation of microbial fuel cells (MFCs) and photosynthetic power cells. In the context of MFCs, the combination of raw algae with an acetate co-substrate has demonstrated a significant enhancement in power density, achieving 410 mW/m^2, which surpasses the performance of single algae systems that yield only 230 mW/m^2. Furthermore, when Chlorella-based cathodes are employed in MFCs for wastewater treatment, they can produce a power density of 34.2 mW/m^2, showcasing the potential of algae in energy generation while simultaneously addressing environmental concerns. In the realm of photosynthetic power cells, microphotosynthetic Chlorella cells have been demonstrated to generate a power density of 62.93 mW/m^2, accompanied by an impressive nitrogen removal rate of 96%. The performance indicators for these technologies highlight their effectiveness [43], with the MFC utilising algae and acetate achieving a power density of 410 mW/m^2 and a chemical oxygen demand (COD) removal rate of 97%. In comparison, algal fuel cells reach a power density of 124 mW/m^2 and a COD removal of 94.8%. However, challenges remain, particularly regarding the cost and scalability of proton exchange membranes, as well as the competition between electroactive bacteria and algae for available substrates, which could hinder the widespread adoption of these innovative technologies [44].

6.3 Success Implementation of Algae-Derived Electrodes

Algae biomass serves as a precursor for electrode materials, playing a significant role in catalysis and energy storage applications. In the realm of batteries, carbon derived from harmful algal blooms (HABs) has demonstrated impressive

Table 3 Algal-derived electrodes' environmental benefit and their description

S. No.	Environmental benefit	Description	References
1.	Nutrient removal	Algae absorb nutrients such as nitrogen and phosphorus, thereby preventing eutrophication and improving water quality.	[48]
2.	Carbon sequestration	Algae capture CO_2 during photosynthesis, contributing to greenhouse gas mitigation.	[41]
3.	Energy efficiency	Algae-based systems reduce energy consumption by utilising sunlight for growth and supplementing aeration needs.	[36]
4.	Pathogen reduction	Algae can inhibit the growth of harmful bacteria and viruses, thereby enhancing water quality.	[49]
5.	Biomass production	Algal biomass can be converted into biofuels, fertilisers, and animal feed, supporting a circular economy.	[41]
6.	Heavy metal removal	Algae effectively remove heavy metals from wastewater, contributing to environmental remediation.	[43]
7.	Low environmental impact	Algae-based systems minimise chemical additives and sludge production	[50]

performance, achieving a capacity of 230 mAh/g in sodium-ion batteries while maintaining a capacity of 135 mAh/g even under high current conditions [45]. Additionally, Chlorella hydrogels exhibit a conductivity of 4.9 mS/cm, making them suitable for integration into LED circuits, thereby enhancing the efficiency and functionality of electronic devices. In the field of catalysis, catalysts derived from algae have proven effective in activating peroxymonosulfate [46, 47], resulting in the removal of over 90% of organic contaminants and demonstrating their potential for environmental remediation. A notable case study involves the use of Chlorella vulgaris embedded in alginate hydrogels, which functioned as living bioelectrodes. This innovative approach enabled sustained growth for 28 days, during which the bioelectrodes maintained consistent electrical activity [48], highlighting the promising applications of algal biomass in both the energy and environmental sectors (Table 3).

6.4 CO_2 Sequestration and Biofuel Production

Carbon Concentration Mechanisms (CCMs) play a crucial role in enhancing the efficiency of Rubisco in *Chlorella* species by concentrating CO_2 around pyrenoids, achieving an impressive annual rate of 12 tons of CO_2 per ton of biomass [51]. Additionally, certain algal strains from the Qatari desert utilise bicarbonate through HCO_3 Conversion, demonstrating their ability to thrive in high-salinity and intense light conditions. In the realm of biofuels, various pathways are being explored.

Biodiesel production involves the transesterification of lipids from Chlorella and Scenedesmus into fatty acid methyl esters (FAMEs). Furthermore, algal biomass can be subjected to dark fermentation, yielding 135 mL of hydrogen per gram of biomass under nutrient-stressed conditions. Bioethanol production is also viable, as algal polysaccharides, such as cellulose, can be hydrolysed to produce 0.45 g of ethanol/g of sugar [50]. A notable case study from Qatar highlights the achievement of a biomass yield of 25 $g/m^2/day$ using local algal strains and flue gas CO_2 capture techniques by researchers.

The biomass produced by algae-mediated BES can be harvested and utilised to produce biofuels, such as biodiesel, biogas, and bioethanol. Hydrothermal liquefaction (HTL) is a preferred process for treating wet algal biomass, as it can convert organic matter into bio-crude at moderate pressure and temperature. Pyrolysis and transesterification are also standard methods, particularly for lipid-rich algae species such as Nannochloropsis gaditana [49]. Genetic modification has also increased lipid production, with up to 50% of the dry weight consisting of lipids in modified strains.

The combination of BES and biofuel production is beneficial from a resource perspective. A report indicated, for instance, that algae-MFC systems can produce power outputs of 0.58 W/m^3 while simultaneously generating high-quality biomass for biodiesel production. Economic challenges remain, though, since the EROI of algal biofuels is currently lower than that of fossil fuels [52, 53], illustrating the comparison of energy yields between various biofuel production processes.

6.5 Removal of Heavy Metal Batteries

Algae-based electrodes in bioelectrochemical systems have been demonstrated to be a green technology for heavy metal removal, leveraging the dual features of energy harvesting and environmental detoxification. Chlorella vulgaris and Spirulina platensis are inexpensive, renewable algae materials used in synthesising highly conductive and biocompatible electrodes. These systems rely on the application of photosynthetic oxygen evolution at the cathode, enhancing efficiency in electron transfer and the immobilisation of heavy metals in adsorptive, reducible, or precipitating forms simultaneously. For example, studies indicate that algae microbial fuel cells (MFCs) can achieve a maximum iron (Fe^{3+}) removal of 99% by reducing it to Fe^{2+} at the cathode, aided by electroactive biofilms. Likewise, selenium (Se(IV)) is entirely reduced to non-toxic selenium (Se(0)) within 72 h on algae-engineered cathodes, as oxygenic photosynthesis provides favourable redox conditions.

Conductive polymer composite material (e.g., polyaniline) or graphene oxide further improves performance. For instance, a chitosan-alginate hydrogel with immobilised Chlorella cells showed a 40% improvement in lead (Pb^{2+}) adsorption capacity compared to conventional activated carbon electrodes [54]. The porous nature of algae-based biochar also facilitates the simultaneous adsorption of multiple metals, such as cadmium (Cd^{2+}) and copper (Cu^{2+}), through ion exchange and

Table 4 Heavy metal removal efficiencies using algae-derived electrodes

S. No	Heavy metal	Initial concentration	Removal efficiency (%)	Time (h)	Electrode material	References
1.	Fe^{3+}	100 mg/L	99	48	Chlorella-graphene anode	[50]
2.	Se(IV)	200 mg/L	100	72	Alginate-algae cathode	[48]
3.	Pb^{2+}	50 mg/L	85	96	Algae-biochar composite	[45]
4.	Cd^{2+}	30 mg/L	90	120	Spirulina-polyaniline	[48]

surface complexation. Algae-based BES for heavy metal removal demonstrates the dual function of algae in oxygen evolution and electron transfer at the cathode, as outlined in Table 1, which highlights recent developments in removal efficiencies.

Economic studies indicate that algae electrodes can lower operating costs by 20%–30% compared to platinum-based systems through the utilisation of waste biomass (e.g., aquaculture seaweed) as feedstock. Moreover, life-cycle analysis supports the fact that algae cultivation has a lower carbon footprint, as it sequesters CO_2 through the process of photosynthesis. Despite these advantages, long-term sustainability remains a challenge due to electrode fouling and biofilm detachment, resulting in a loss of performance after 6–8 months of operation.

Upcoming research emphasises hybrid configurations that pair algae with nanotechnology. As such, the iron nanoparticle doping of an algae electrode increases arsenic (As^{3+}) removal efficiency to 95% due to enhanced surface redox reactions (Table 4). Likewise, genetic manipulation of algae strains to overexpress metal-binding proteins such as metallothioneins still improves performance [55].

6.6 *Bioremediation Applications*

Algal-based BESs are effective in decontaminating wastewater with toxic metals and organic contaminants. Microalgae produce oxygen as a byproduct of photosynthesis, which acts as an electron acceptor when reduced toxic metals such as Pb^{2+}, Fe^{3+}, and Se(IV) are present. For instance, algae-encapsulated hydrogel electrodes were capable of achieving maximum efficiency of iron ion removal of 99%, thereby reducing it to less toxic species, such as Fe^{2+}. In the same way, selenium is reduced from Se(IV) to non-toxic Se(0) within 72 h by using algae-amended cathodes. Extracellular polymeric substances (EPS) that bind heavy metals and are secreted by algae also immobilise these metals.

Composite electrodes with embedded algae in materials such as graphene oxide or polyaniline are constantly improving their performance. The composites enhance the electron transfer rate and are biocompatible. For example, algae-polyaniline electrodes demonstrated a 40% improvement in lead adsorption compared to conventional activated carbon electrodes. Algal biomass biochar has also been utilised for the removal of cadmium and copper through ion exchange mechanisms.

6.7 Application in Desalination Systems

Algae electrodes are increasingly being utilised in microbial desalination cells (MDCs), where they increase salinity reduction and power generation. Algal biofilms increase the electrochemical gradient for ion separation. MDCs constructed using algae-based and biochar electrodes are up to 85% efficient in removing salt. Oxygen generated by photosynthesis at the cathode initiates the desalination process through proton exchange between the membranes.

Technologies like 3D-printed algae scaffolds have enhanced the conductivity and toughness of electrodes, which was an issue due to the detachment of biofilms in cases of a shift in salinity. Moreover, hybrid technologies, such as nanoparticles used in algae electrodes, have enhanced the efficiency of the desalination process by improving surface redox reactions.

7 Challenges and Future Perspectives

7.1 Technical and Operational Challenges

Technical and functional issues with algae-based electrodes for bioelectrochemical systems (BES) span material science, microbiology, and engineering integration. Algae-based electrodes have great potential for green BES applications [10] but are hindered by inherent issues that limit their generalisation. One of the primary issues is optimising electron transfer across the bio-abiotic interface, where limited knowledge of electron transfer kinetics between electrode interfaces and microorganisms hinders improvement through rational design. This lack of understanding makes it challenging to increase the performance of algae-based materials in real applications. Algae-based electrode development is also made challenging by the lingering issues of low efficiency and energy yield that most BES technologies face [55]. The systems are plagued by mass transfer constraints that limit the mobility of substrates, products, and charge carriers in the electrode framework, and this has a direct impact on reaction kinetics and system performance.

Material properties are another major challenge, as electrode materials from algae will tend to have less-than-ideal porosity and proton conductivity—the factors that determine ideal electrochemical processes [21]. The physical stability of such biologically derived materials raises concerns about mechanical strength and long-term reliability under operating conditions, particularly for large-scale systems.

Chemical stability remains a persistent challenge, as algae-derived materials can degrade or change over time, thereby modifying their electrochemical properties and catalytic activity. The inherent heterogeneity of the algae feedstocks introduces compositional and performance variability into the electrodes, making quality control and batch-to-batch standardisation challenging.

Scalability remains a recalcitrant problem, with laboratory success often not translating to commercially valuable larger systems due to challenges in maintaining uniform properties and performance at scale. Biologically, biocompatibility problems are conceivable between microbial communities necessary for BES operation and algal-derived materials, resulting in suppressed biofilm growth or diminished microbial activity [53]. Optimising the nitrogen content in carbons derived from algae is a technical challenge that requires precise control of pyrolysis and activation processes to provide optimal active sites for electrochemical reactions.

Existing characterisation methods are generally unable to deconvolute the intricate surface chemistry and structural details of such materials, thereby preventing controlled optimisation. Economic feasibility requires attempts to be directed towards avoiding noble metals in catalyst preparation without compromising comparable performance. This delicate task has only been partially achieved by substitutes based on algae. Integration problems arise when different algae-based electrodes are integrated into other renewable power systems, where compatibility and synergistic interaction must be carefully engineered. Ecological problems also arise from the provision of sustainable cultivation and processing conditions to create electrode materials, which raises questions regarding the environmental friendliness of the final technology. All these integrated problems require integrated approaches founded on multiscale principles, combining high-flying materials science, microbial biochemistry, and electrochemical design to create the final value of bioelectrochemical electrodes formed from algae for application in future-generation bioelectrochemical systems.

7.2 *Strategies to Overcome Limitations*

Algae-based bioelectrochemical systems (BES) [56, 57] hold great potential for generating renewable energy, treating wastewater, and sequestering carbon. Nevertheless, their effective use is limited by various constraints. Some of the most significant approaches to overcoming the problems associated with the use of algae-derived electrodes in bioelectrochemical systems are:

7.2.1 Understanding and Improving Electron Transfer Mechanisms

Mediatorless Electron Transfer

The need for electron mediators in algal microbial fuel cells (MFCs) poses significant challenges to sustainability, cost, and toxicity. Researchers have found some microalgae that can directly transfer electrons [11].

- *Desmodesmus* sp. showed direct electron transport through membrane-bound proteins.
- *Cyanobacterium Synechocystis* sp. PCC6803 showed direct electron transfer by conductive nanowires under excess light conditions and CO_2-limited conditions.

Biofilm Development

Biofilm formation on electrode surfaces can strongly enhance electron transfer efficiency:

- Biofilms facilitate direct long-range electron transfer, increased catalytic rates, and increased power outputs.
- The porosity and three-dimensionality of electrode surfaces enhance system performance and biofilm growth.

7.2.2 Optimisation of Algal Biomass Processing

Biomass Pretreatment

The structured cell wall of algae presents a strong obstacle to electron extraction [21].

- Pretreatment of algal biomass enhances the bio-electrokinetics of MFCs, thereby increasing the power output.
- Physical and chemical pretreatment methods lyse cell structures, increase surface area, and hydrolyze polymers.
- Pre-digested algae biomass can enable increased power output than undigested biomass.

Strategic Algae Selection

Different algae species don't react similarly in bioelectrochemical systems:

- Power output can be optimised by using the right algae species with favourable features.
- Chlorella vulgaris is found to be superior compared to Dunaliella tertiolecta, and the maximum power densities of both are 15 $mW{\cdot}m^{-2}$ and 5.3 $mW{\cdot}m^{-2}$, respectively [36].
- Macroalgae have certain benefits compared to microalgae for waste treatment and biomass recovery.

7.2.3 Improvement in Electrode and System Design

Advanced Materials

Electrode material selection has significant performance implications:

- Synthetic porous conductive materials enable enhanced electron conduction and facilitate biofilm growth.
- There needs to be an ideal balance between conductivity and porosity, as increased porosity may be at the expense of conductivity.
- Inexpensive and scalable electrode material formation remains a research focus area.

System Configuration Optimisation

Various system configurations are employed based on application and performance:

- Single-chamber, dual-chamber, H-shaped, and three-chamber configurations have distinct advantages.
- The incorporation of external photobioreactors can enhance algae growth and electricity generation.
- Miniaturisation by stacking multiple small reactors instead of scaling up solitary reactors can reduce internal resistance problems [55].

7.2.4 Integrated Strategies

Combinational Systems

Employment of multiple technologies results in synergistic effects:

- The incorporation of MFCs into algae-based bioreactors forms integrated photo-bioelectrochemical systems for simultaneous electricity generation and biomass production [58].
- Employing algae in cathode chambers eliminates the need for a constant supply of terminal electron acceptors, as algae release oxygen through the process of photosynthesis.
- CO_2 generated in anodic chambers can trigger algal growth in cathodic chambers, thereby reducing the costs and energy required for CO_2 transport.

Dual Functionality

Strategic deployment of algae at various points of the system

- Dead microalgae are utilised for the production of electrons in the anode chamber, while active microalgae in the cathode chamber produce oxygen and serve as a substrate.
- This approach introduces diversity into system operations, thereby increasing overall system efficiency.

7.2.5 Future Research Directions

Genetic Engineering

Gene modification of algae introduces high promise:

- Gene modification can improve cell potential and advance electron transfer functionality.
- More work needs to be done to investigate gene-modified algae, primarily for applications in biorefineries and energy systems (BES).

Omics Strategies

Improved knowledge of metabolic pathways by omics technologies can inform the optimisation strategy.

- An improved understanding of electron transfer processes between microbes and electrodes can inform more informed strain and material selection.
- Overcoming these limitations using multidisciplinary techniques will promote algae-derived electrodes in bioelectrochemical systems, potentially transforming sustainable energy production while delivering environmental benefits.

8 Conclusion

In summary, bioelectrochemical systems (BESs) are a cutting-edge frontier in green biotechnology, and the integration of algae represents a significant advancement in their efficiency and capabilities. The incorporation of algae in BESs, and especially in microbial fuel cells (MFCs), is a synergetic system that resolves several environmental issues on the one hand while producing green energy on the other. These systems efficiently utilise algal photosynthetic potential for carbon sequestration, oxygen generation, and wastewater treatment, regardless of external aeration, and facilitate the electron transfer process. Algae application in BESs extends beyond operational benefits, as they serve as renewable electrode materials with enhanced electrochemical performance and contribute to the synthesis of value-added products, including biofuels, biochar, and high-value compounds. Advanced configurations, such as photo-bioelectrochemical systems and combined MFC-photobioreactor systems, exhibit very good performance profiles, with maximum power densities of 410 mW/m^2 and COD removal efficiencies exceeding 90%. Algae-BES coupling represents the pinnacle of circular bioeconomy, where environmental resources are converted into valuable resources while addressing energy needs, waste disposal issues, and carbon emissions simultaneously, making such technology a viable candidate for sustainable development in the twenty-first century.

References

1. Corona-Martínez DA, Martínez-Amador SY, Rodríguez-De la Garza JA, Laredo-Alcalá EI, Pérez-Rodríguez P (2025) Recent advances in scaling up bioelectrochemical systems: a review. 3 Biotech 14(1):8
2. Adeloju S (2024) Bioelectrochemical systems and applications—taking A broader perspective and highlighting relevance to the circular economy. Front Plant Sci 2023(14):1081108
3. Rabaey K, Angenent L, Schroder U, Keller J (2009) Bioelectrochemical systems. IWA Publishing, London
4. Zheng T, Li J, Ji Y, Zhang W, Fang Y, Xin F, Jiang M (2020) Progress and prospects of bioelectrochemical systems: electron transfer and its applications in the microbial metabolism. Front Bioeng Biotechnol 8:10
5. Nayak JK, Gautam R, Guan CY, Ghosh UK (2024) Algae-based bio-electrochemical systems for carbon sequestration, bioremediation, and cogeneration of valuable chemicals: challenges and future prospects. In: Algae based bioelectrochemical systems for carbon sequestration, carbon storage, bioremediation and bioproduct generation. Elsevier, Amsterdam, pp 77–93
6. Mishra A, Chhabra M (2023) Microalgae cultivation in bio-electrochemical systems. Algal Syst Resour Recov 15:129
7. Bajwa K, Kumar S, Baloch MY, Kumar SS, Bishnoi NR (2024) Renewable sustainable biocatalyzed electricity production: challenges and prospects of algal-based bio-electrochemical systems. In: Algae based bioelectrochemical systems for carbon sequestration, carbon storage, bioremediation and bioproduct generation. Elsevier, Amsterdam, pp 51–68
8. Jadhav DA, Neethu B, Ghangrekar MM (2019) Microbial carbon capture cell: advanced bioelectrochemical system for wastewater treatment, electricity generation and algal biomass production. In: Application of microalgae in wastewater treatment: volume 2: biorefinery approaches of wastewater treatment. Springer, Cham, pp 317–338
9. Mahmoud RH, Abd El-Raheem H, Hassan RY (2022) Algae-mediated bioelectrochemical system: the future of algae in the electrochemical operations. In: Bio-electrochemical systems. CRC Press, Boca Raton, pp 93–123
10. Yang Q, Lin Y, Liu L, Yang F (2017) A bio-electrochemical membrane system for more sustainable wastewater treatment with MnO2/PANI modified stainless steel cathode and photosynthetic provision of dissolved oxygen by algae. Water Sci Technol 76(7):1907–1914
11. Saratale RG, Kuppam C, Mudhoo A, Saratale GD, Periyasamy S, Zhen G, Koók L, Bakonyi P, Nemestóthy N, Kumar G (2017) Bioelectrochemical systems using microalgae–a concise research update. Chemosphere 177:35–43
12. Larom S, Salama F, Schuster G, Adir N (2010) Engineering of an alternative electron transfer path in photosystem II. Proc Natl Acad Sci 107(21):9650–9655
13. Zhou Q, Yamamoto H, Shikanai T (2023) Distinct contribution of two cyclic electron transport pathways to P700 oxidation. Plant Physiol 192(1):326–341
14. Xu C, Poon K, Choi MM, Wang R (2015) Using live algae at the anode of a microbial fuel cell to generate electricity. Environ Sci Pollut Res 22:15621–15635

15. Kondaveeti S, Mohanakrishna G, Pagolu R, Kim IW, Kalia VC, Lee JK (2019) Bioelectrogenesis from raw algal biomass through microbial fuel cells: effect of acetate as co-substrate. Indian J Microbiol 59:22–26
16. Ali SS, Mastropetros SG, Schagerl M, Sakarika M, Elsamahy T, El-Sheekh M, Sun J, Kornaros M (2022) Recent advances in wastewater microalgae-based biofuels production: A state-of-the-art review. Energy Rep 8:13253–13280
17. Osman AM, Abdurahman NH, Noormazlinah A (2024) Revolutionizing bioenergy: the microalgae-microbial fuel cell frontier. Sci Technol Energy Trans 79:91
18. He R, Kraemer D, Mao J, Zeng L, Jie Q, Lan Y, Li C, Shuai J, Kim HS, Liu Y, Broido D (2016) Achieving high power factor and output power density in p-type half-Heuslers Nb1-xTixFeSb. Proc Natl Acad Sci 113(48):13576–13581
19. Tay ZH, Ng FL, Ling TC, Iwamoto M, Phang SM (2022) The use of marine microalgae in microbial fuel cells, photosynthetic microbial fuel cells and biophotovoltaic platforms for bio-electricity generation. 3 Biotech 12(7):148
20. Koltysheva D, Shchurska K, Kuzminskyi Y (2021) Microalgae and cyanobacteria as biological agents of biocathodes in biofuel cells. Biotechnologia 102(4):437
21. Ayala-Parra P, Sierra-Alvarez R, Field JA (2016) Algae as an electron donor promoting sulfate reduction for the bioremediation of acid rock drainage. J Hazard Mater 317:335–343
22. Du M, Zhang Y, Kang S, Guo X, Ma Y, Xing M, Zhu Y, Chai Y, Qiu B (2022) Trash to treasure: photoreforming of plastic waste into commodity chemicals and hydrogen over MoS2-tipped CdS nanorods. ACS Catal 12(20):12823–12832
23. Buceti G (2014) Sustainable power density in electricity generation. Manage Environ Qual 25(1):5–18
24. Morris BE, Henneberger R, Huber H, Moissl-Eichinger C (2013) Microbial syntrophy: interaction for the common good. FEMS Microbiol Rev 37(3):384–406
25. Yi G, Wang B, Feng Y, Fang D, Yang L, Liu W, Zhang Y, Shao P, Pavlostathis SG, Luo S, Luo X (2022) The ins and outs of photo-assisted microbial electrochemical systems for synchronous wastewater treatment and bioenergy recovery. Resour Conserv Recycl 181:106230
26. Tse HT, Luo S, Li J, He Z (2016) Coupling microbial fuel cells with a membrane photobioreactor for wastewater treatment and bioenergy production. Bioprocess Biosyst Eng 39:1703–1710
27. Fan X, Wang C, Zhu Z, Meng H (2022) Design and analysis of a high power density permanent magnet linear generator for direct-drive wave power generation. Inactuators 11(11):327
28. Liu T, Rao L, Yuan Y, Zhuang L (2015) Bioelectricity generation in a microbial fuel cell with a self-sustainable photocathode. Sci World J 2015(1):864568
29. Mahmoud RH, Samhan FA, Ibrahim MK, Ali GH, Hassan RY (2022) Waste to energy conversion utilizing nanostructured algal-based microbial fuel cells. Electrochem Sci Adv 2(6):e2100071
30. Gogoi J, Nalwad A, Ganguly A, Mutnuri S (2024) Effect of integrated microbial fuel cell in a two-stage vertical flow constructed wetland for piggery wastewater treatment and resource recovery by photobioreactor. Water Pract Technol 19(11):4556–4571
31. Du J, Xin J, Liu M, Zhang X, He H, Wu J, Xu X (2022) Preparation of photo-bioelectrochemical cells with the RC-LH complex from Roseiflexus castenholzii. Front Microbiol 13:928046
32. Wang Z, He Z, Young EB (2020) Toward enhanced performance of integrated photo-bioelectrochemical systems: taxa and functions in bacteria–algae communities. Curr Opin Chem Biol 59:130–139
33. Xiao L, Young EB, Berges JA, He Z (2012) Integrated photo-bioelectrochemical system for contaminants removal and bioenergy production. Environ Sci Technol 46(20):11459–11466
34. Chen S, Chen J, Zhang L, Huang S, Liu X, Yang Y, Luan T, Zhou S, Nealson KH, Rensing C (2023) Biophotoelectrochemical process co-driven by dead microalgae and live bacteria. ISME J 17(5):712–719
35. Angelov A, Bratkova S, Velichkova P (2021) Integration of microbial fuel cells in A system for biomethanation and photosynthesis. J Chem Technol Metall 56(4):796

36. Florez JZ, Camus C, Hengst MB, Buschmann AH (2017) A functional perspective analysis of macroalgae and epiphytic bacterial community interaction. Front Microbiol 8:2561
37. Zhang H, Yan Q, An Z, Wen Z (2022) A revolving algae biofilm based photosynthetic microbial fuel cell for simultaneous energy recovery, pollutants removal, and algae production. Front Microbiol 13:990807
38. Jayaraman J, Kumaraswamy J, Rao YK, Karthick M, Baskar S, Anish M, Sharma A, Yadav AS, Alam T, Ammarullah MI (2024) Wastewater treatment by algae-based membrane bioreactors: a review of the arrangement of a membrane reactor, physico-chemical properties, advantages and challenges. RSC Adv 14(47):34769–34790
39. Arun S, Sinharoy A, Pakshirajan K, Lens PN (2020) Algae based microbial fuel cells for wastewater treatment and recovery of value-added products. Renew Sust Energ Rev 132:110041
40. Wang H, Man S, Wang H, Presser V, Yan Q (2024) A sustainable approach: repurposing harmful algal biomass as carbon-based catalysts for nitrogen fertilizer electrosynthesis from nitrate and CO2. Chem Eng J 497:154455
41. Lakatos GE, Ranglová K, Manoel JC, Grivalský T, Kopecký J, Masojídek J (2019) Bioethanol production from microalgae polysaccharides. Folia Microbiol 64:627–644
42. Shahparnia M, Packirisamy M, Juneau P, Zazubovich V (2015) Micro photosynthetic power cell for power generation from photosynthesis of algae. Technology 3(2):119–126
43. Riveros A, Asefaw BK, Wang Q, Maqbool T, Tang Y, Jiang D (2025) Selenium treatment via integrating flow electrode capacitive deionization (FCDI) and bio-electrochemical systems (BES). Water Res 271:122844
44. Raj R, Dhanda A, Kumar S, Das S, Ghangrekar MM (2024) Repurposing waste-iron electrocoagulated algal biomass as effective heterogenous (bio) electro-Fenton catalyst for phthalate removal from wastewater. Sci Rep 14(1):23932
45. Sun H, Gong Q, Fan Y, Wang Y, Wang J, Zhu C, Mou H, Yang S, Liu J (2024) Unlocking 3D printing technology for microalgal production and application. Adv Biotechnol 2(4):36
46. Alsulaiti F. Identifying Qatar's potential for carbon sequestration (Doctoral dissertation, Swansea University)
47. Khan SA, Hussain MZ, Prasad S, Banerjee UC (2009) Prospects of biodiesel production from microalgae in India. Renew Sust Energ Rev 13(9):2361–2372
48. Kumar S, Yadav S, Ghangrekar MM (2025) Wastewater treatment employing microbial carbon-capture cells with concomitant bioenergy production and valuable product recovery. In: Algal biorefinery. Elsevier, Amsterdam, pp 263–271
49. Szkoda M, Skorupska M, Łukaszewicz JP, Ilnicka A (2023) Enhanced supercapacitor materials from pyrolyzed algae and graphene composites. Sci Rep 13(1):21238
50. Khera HK, Srivastava AK (2022) Genetic engineering of algae. In: An integration of phycoremediation processes in wastewater treatment. Elsevier, Amsterdam, pp 487–502
51. Singh G, Samuchiwal S, Hariprasad P, Sharma S (2022) Melioration of paddy straw to produce cellulase-free xylanase and bioactives under solid state fermentation and deciphering its impact by life cycle assessment. Bioresour Technol 360:127493
52. Yadav P, Azeem PA, Patel S, Mahar G, Yadav R, Borkar H (2024) Novel industrial biomass derived materials for super capacitor application in powering up electronic gadgets. J Energy Storage 97:112653
53. Singh UB, Ahluwalia AS (2013) Microalgae: a promising tool for carbon sequestration. Mitig Adapt Strateg Glob Chang 18(1):73–95
54. Rani A, Saini KC (2023) Algal metal remediation for contaminated source. In: Algae refinery. CRC Press, Boca Raton, pp 126–147
55. Fang Z, Chen H, Wei YQ, Fan Q, Zhu MW, Zhang Y, Liu J, Yong YC (2024) Bioelectricity and CO2-to-butyrate production using photobioelectrochemical cells with bio-hydrogel. Bioresour Technol 398:130530
56. Sharma R, Jasrotia T, Kumar R, Kumar R, Alothman AA, Mana Al-Anazy M, Alqahtani KN, Umar A (2021) Multi-biological combined system: a mechanistic approach for removal of multiple heavy metals. Chemosphere 276:130018

57. Zhang L, Wang Y (2023) Decoupled artificial photosynthesis. Angew Chem Int Ed 62(23):e202219076
58. Thorne R. Bio-photo-voltaic cells (photosynthetic-microbial fuel cells)
59. Arora K, Kumar P, Bose D, Li X, Kulshrestha S (2021) Potential applications of algae in biochemical and bioenergy sector. 3 Biotech 11(6):296
60. Mekuto L, Olowolafe AV, Huberts R, Dyantyi N, Pandit S, Nomngongo P (2020) Microalgae as a biocathode and feedstock in anode chamber for a selfsustainable microbial fuel cell technology: a review. S Afr J Chem Eng 31(1):7–16
61. Inohana Y, Katsuya S, Koga R, Kouzuma A, Watanabe K (2020) Shewanella algae relatives capable of generating electricity from acetate contribute to coastal-sediment microbial fuel cells treating complex organic matter. Microbes Environ 35(2):19161

New Trends and Advancements in the Field of BES

Abdul Sattar Jatoi, Nabisab Mujawar Mubarak, and Arshad Iqbal

Abstract Amid growing efforts to combat climate change, bioelectrochemical systems (BES) have gained attention as a sustainable technology for converting carbon dioxide (CO_2) into value-added products, particularly methanol. This chapter reviews recent advancements in BES for CO_2-to-methanol conversion, covering core components, microbial and electrochemical innovations, and integrative strategies that enhance system performance. Key topics include the evolution of reactor design, novel electrocatalysts, electron transfer mechanisms, and the integration of renewable energy. Broader BES applications in hydrogen production, wastewater treatment, and bioplastics synthesis are also discussed. The chapter highlights the role of digital technologies, alongside techno-economic and environmental assessments, in advancing BES towards scalable deployment in future circular bioeconomy models.

Keywords Trends · Advancement · BES

A. S. Jatoi (✉) · A. Iqbal
Department of Chemical Engineering, Dawood University of Engineering and Technology, Karachi, Pakistan
e-mail: abdul.sattar@duet.edu.pk

N. M. Mubarak
Department of Chemical Engineering, College of Engineering, Imam Mohammad Ibn Saud Islamic University (IMSIU), Riyadh, Saudi Arabia

Department of Biosciences, Saveetha School of Engineering, Saveetha Institute of Medical and Technical Sciences, Chennai, India

A. Sattar Jatoi, N. M. Mubarak (eds.), *Bio-Electrochemical Conversion of Waste to Fuel*, Green Energy and Technology,
https://doi.org/10.1007/978-981-95-4310-6_14

1 Introduction

Bioelectrochemical systems (BES) have emerged as a multidisciplinary technological platform that merges microbiology, electrochemistry, materials science, and environmental engineering to address some of the most pressing global challenges related to sustainable energy, wastewater treatment, resource recovery, and environmental remediation [1]. The foundational principle of BES lies in harnessing the metabolic activities of electroactive microorganisms that can catalyse redox reactions at electrode interfaces, facilitating the conversion of chemical energy from organic or inorganic substrates into electrical energy or other value-added products. Over the past two decades, significant research has propelled BES from laboratory curiosities into versatile tools with the potential to revolutionise decentralised wastewater treatment, renewable electricity generation, biosensing, desalination, and carbon capture [1, 2]. The early focus on microbial fuel cells (MFCs) as low-power generators for wastewater-powered applications has evolved into a broader exploration of various BES configurations, including microbial electrolysis cells (MECs), microbial desalination cells (MDCs), and microbial electrosynthesis (MES) systems—each tailored for specific applications ranging from hydrogen production to the bioconversion of carbon dioxide into fuels and chemicals [3]. Recent trends in BES research are marked by innovations across multiple dimensions: advanced electrode materials with nanostructured architectures to enhance electron transfer; genetically engineered microbial strains and synthetic consortia designed for superior performance and product selectivity; novel reactor designs that promote scalability, efficiency, and system robustness; and integration of artificial intelligence, modelling tools, and omics-driven insights to optimise system-level performance [4]. In tandem with material and biological advancements, BES platforms are being increasingly coupled with renewable energy sources and incorporated into circular economy frameworks, enabling resource-efficient and low-carbon solutions for diverse sectors.

Furthermore, the development of miniaturised and portable BES-based biosensors has unlocked new possibilities for real-time environmental monitoring. Meanwhile, emerging hybrid systems are exploring the synergistic benefits of combining phototrophic and electroactive organisms to harness solar energy in bioelectrochemical reactions. As BES technology matures, significant challenges remain, including low volumetric power densities, high initial capital costs, limited long-term operational stability, and the need for standardisation in reactor design and performance benchmarking [2, 5]. However, the rapid pace of interdisciplinary research, fueled by advancements in synthetic biology, nanotechnology, materials engineering, and computational modelling, is steadily addressing these bottlenecks [3]. Integrating cutting-edge fabrication techniques, such as 3D printing, utilising low-cost biomass-derived electrode materials, and developing innovative control systems for autonomous operation are propelling BES closer to commercial feasibility. This chapter aims to provide a comprehensive overview of the recent trends and cutting-edge advancements in bioelectrochemical systems. It begins by exploring the latest breakthroughs in BES applications, followed by in-depth analyses of electrode material innovations, microbial engineering strategies, and enhancements to reactor design. Subsequent sections will explore the synergistic integration of BES with renewable technologies, novel bioproduct synthesis through microbial electrosynthesis, and the role of artificial intelligence and data-driven tools in enhancing BES performance. Finally, the chapter will highlight the key challenges, regulatory considerations, and future directions needed to realise the full potential of BES as a transformative tool for sustainable development. By capturing the dynamic and evolving landscape of BES research, this chapter seeks to serve as a critical resource for researchers, engineers, policymakers, and practitioners aiming to harness bioelectrochemical platforms for environmental resilience, energy security, and industrial innovation in the decades ahead.

2 Advances in Electrode Materials

Advances in electrode materials have been pivotal in driving the performance, scalability, and versatility of bioelectrochemical systems (BES), as the electrode serves as the critical interface for microbial electron transfer and energy conversion [6, 7]. Traditional carbon-based electrodes, such as graphite rods, carbon cloth, and carbon felt, laid the foundation for early BES configurations due to their chemical stability, conductivity, and biocompatibility. However, limitations such as a low surface area, poor microbial adhesion, and limited catalytic properties spurred the exploration of next-generation materials [8]. Among the most promising innovations are nanostructured carbon materials, including carbon nanotubes (CNTs), graphene, and carbon nanofibers. These materials offer an ultra-high surface area, tunable conductivity, and superior microbial colonisation surfaces, as highlighted in Table 1. These materials enhance direct extracellular electron transfer (DEET)

Table 1 Overview of different electrode materials, their properties, and their applications

Electrode material type	Sub-type/example	Key properties	Main benefits	Recent applications
Carbon-based materials	Graphite, carbon felt, carbon cloth	Chemically stable, conductive	Cost-effective, biocompatible	Traditional MFC anodes, wastewater treatment
Nanostructured carbons	CNTs, graphene, rGO	High surface area, high conductivity	Enhanced microbial adhesion, faster EET	High-performance MFCs and MECs
3D porous structures	Carbon foam, RVC, hierarchical carbon	Porous, large surface area	Supports thick biofilms, efficient mass transport	Scalable reactor designs, stacked MFCs

(continued)

Table 1 (continued)

Electrode material type	Sub-type/example	Key properties	Main benefits	Recent applications
Metal electrodes	Stainless steel, nickel, titanium	Strong, conductive, catalytically active	Durable cathodes, good for oxygen reduction	MEC cathodes, ORR enhancement
Noble metal catalysts	Platinum (Pt), palladium (Pd), ruthenium	High catalytic efficiency	Fast ORR kinetics, improved power output	High-efficiency cathodes in lab-scale BES
Transition metal catalysts	MnO_2, Fe–N–C, co–N–C, metal Phthalocyanines	Affordable, modifiable	Lower-cost ORR catalysts	Sustainable cathodes in wastewater MFCs
Conductive polymers	Polypyrrole (PPy), polyaniline (PANI)	Redox-active, flexible	Enhanced EET, biocompatible surfaces	Anode modifications, dynamic electrodes
Functionalized surfaces	Amino groups, redox mediators	Customisable chemistry	Improved electron transfer, microbial adhesion	BES optimisation with engineered microbes
Hybrid composites	Graphene + MnO_2, CNTs + PANI	Synergistic conductivity + catalysis	Enhanced performance, stability	Cathode composites, MES systems
Bio-derived carbons	Activated biochar, coconut shell carbon	Renewable, porous	Low-cost, sustainable alternative	Rural/low-cost BES designs
Anti-biofouling coatings	Ag-doped materials, self-cleaning polymers	Biocidal, hydrophilic	Reduced clogging, longer electrode life	Real wastewater treatment reactors
Additively manufactured	3D-printed carbon or metal composites	Custom geometry, scalable	Tailored performance, reproducibility	BES prototypes, wearable biosensors
Photoactive materials	TiO_2, CdS with conductive supports	Light-sensitive, semiconductor	Solar-driven reactions, improved sustainability	Solar-assisted MFCs and MES
Living electrodes	Microbe-infused conductive hydrogels	Biohybrid, flexible	Self-regulating EET, high integration potential	Emerging BES applications, biosensing

by increasing the density of active sites and facilitating tighter biofilm formation. CNTs and graphene derivatives, such as reduced graphene oxide (rGO), have been utilised to fabricate composite electrodes incorporating metal oxides (e.g., MnO_2, TiO_2) or conductive polymers (e.g., polyaniline, polypyrrole), resulting in synergistic effects that enhance electrochemical reactivity and microbial affinity [9]. Additionally, three-dimensional (3D) structured electrodes, such as reticulated vitreous (RVC), carbon foam, and hierarchical porous carbon, have emerged as leading platforms for accommodating thick biofilms and facilitating mass transport. These architectures provide mechanical robustness and promote the convective flow of nutrients and metabolites, making them suitable for scale-up in modular BES systems [10]. Moreover, metal-based electrodes have garnered attention for their catalytic capabilities, especially at the cathode where oxygen or CO_2 reduction occurs. Materials such as stainless steel, titanium, nickel, and copper have been investigated, with surface treatments like anodization, doping, or nanoparticle deposition used to enhance conductivity and biocompatibility. For instance, cathodes modified with platinum-group metals (Pt, Pd, Ru) or transition metal catalysts (e.g., Co–N–C, Fe–N–C complexes) exhibit high activity for oxygen reduction reactions (ORR), improving BES efficiency. However, due to the high cost of noble metals, alternative catalysts such as manganese oxides, iron phthalocyanines, and metal-organic frameworks (MOFs) are increasingly favoured for their lower cost and tunable properties. Meanwhile, advances in conductive polymers have enabled the development of bioinspired and flexible electrode materials. Polymers like polypyrrole (PPy) and polyaniline (PANI) can be synthesised in situ on electrode surfaces to improve electron shuttling, water wettability, and microbial adhesion. Some BES designs utilise dynamic or adaptive electrodes that respond to pH or redox changes, enabling self-regulation of microbial activity and charge transfer. In addition, surface functionalization techniques—such as the attachment of redox mediators (e.g., neutral red, anthraquinone derivatives), amino or carboxyl groups, and microbial adhesion peptides—have been employed to bridge the gap between electrode surfaces and electroactive microbial membranes, especially in systems relying on mediated electron transfer (MET). Another critical area of innovation is electrode durability and anti-biofouling properties, particularly for real-world wastewater treatment applications. Coatings that resist corrosion and inhibit non-electrogenic microbial growth, as well as those that enable self-cleaning, are being actively researched. Hydrogels embedded with conductive nanoparticles are one such approach, offering a hydrated matrix that supports microbial activity while maintaining low impedance and high proton conductivity. Emerging technologies, such as additive manufacturing (3D printing) and laser scribing, have introduced new possibilities for customising electrode geometries, pore structures, and surface features to optimise BES performance across scales—from microscale sensors to pilot-scale reactors. Furthermore, the integration of advanced characterisation tools, such as atomic force microscopy (AFM), electrochemical impedance spectroscopy (EIS), and confocal microscopy, has allowed researchers to gain a better understanding of microbe-electrode interactions, biofilm development, and charge transfer pathways at the nanoscale. These insights inform the rational design of electrode materials tailored to specific microbial strains or reaction pathways. Importantly, sustainability and cost-effectiveness are gaining prominence as BES progresses to commercialisation. Biomass-derived carbons from agricultural residues, activated carbon from coconut shells, and biochar, as well as recycled graphite from electronic waste, have been explored as low-cost, eco-friendly electrode materials. These alternatives align well with circular economy principles and reduce the environmental footprint of BES installations. Ultimately, the future of electrode materials in BES is likely to be shaped by the convergence of interdisciplinary insights—combining knowledge from materials science, synthetic biology, electrochemistry, and environmental engineering. Innovations such as living electrodes (biohybrid constructs of microbes embedded in conductive matrices), photoactive electrode materials for solar-assisted battery energy storage (BES), and dynamically tunable interfaces driven by stimuli-responsive materials suggest a move towards more intelligent, efficient, and multifunctional electrodes. As these material advancements unfold, they are expected to significantly enhance the energy efficiency, product selectivity, operational lifespan, and economic feasibility of BES technologies deployed across various sectors, including wastewater treatment, resource recovery, carbon capture, and sustainable chemical production [11].

3 Microbial Engineering and Synthetic Biology

The heart of BES lies in the microbial catalysts that drive electron transfer and metabolic conversions. While native electroactive microorganisms, such as Geobacter sulfurreducens, Shewanella oneidensis, and Clostridium ljungdahlii, have been extensively studied for their ability to interact with electrodes, the potential of microbial engineering has only recently begun to be fully realised. Advances in synthetic biology and metabolic engineering enable the customisation of microbial strains for enhanced electron transfer efficiency, increased tolerance to environmental stressors, and expanded substrate utilisation [12]. For instance, genetically modifying *Escherichia coli* or *Corynebacterium glutamicum* to express cytochrome complexes can render these industrially robust strains electroactive. Tools like CRISPR/Cas9 and synthetic promoters enable precise control over gene expression, allowing for the development of tailored biosynthetic pathways that produce specific chemicals, such as methanol, from carbon dioxide (CO_2). Synthetic consortia offer modularity and flexibility in BES design, where multiple engineered microbes perform sequential tasks (e.g., one reduces CO_2 to formate, another converts formate to methanol). Metabolic flux analysis and systems biology approaches optimise carbon and electron flow through microbial networks, increasing product yields and energy efficiency. Moreover, omics technologies—genomics, transcriptomics, proteomics, and metabolomics—provide deep insights into microbial behaviour under electrochemical stress, aiding in the rational design of high-performance bioelectrochemical consortia [13]. Integrating quorum sensing circuits and programmable logic gates into microbes opens avenues for autonomous control of BES operation, enabling self-regulation based on substrate availability or redox state. Despite these advances, challenges remain in maintaining genetic stability and preventing plasmid loss during long-term operation. Nevertheless, microbial engineering is a powerful lever in advancing BES, offering unprecedented control over biological and electrochemical functions essential for CO_2 conversion and other high-value applications [14].

4 Innovations in Reactor Design

Reactor configuration is crucial for the efficiency, scalability, and economic feasibility of bioelectrochemical systems (BES). Traditional BES designs often have limitations, including high internal resistance, poor mass transfer, and inefficient electron recovery. Addressing these issues has led to the development of innovative reactor architectures that enhance performance while simplifying operations [15]. One of the most notable trends is the shift towards modular and scalable designs, facilitating easy upscaling for industrial applications. These systems comprise smaller, independently operable units that can be interconnected in series or parallel, allowing flexible capacity adjustments and redundancy. Another major innovation is the emergence of membranelles BES, which eliminates costly and fouling-prone ion-exchange membranes, thereby reducing capital and operational costs [16]. These systems use controlled flow patterns and selective biofilm formation to separate anodic and cathodic processes. Researchers are also exploring stacked BES configurations, where multiple cells are layered to increase volumetric productivity and energy density—a concept particularly promising for compact installations in urban or space-constrained environments [17].

Additionally, microfluidic BES platforms have become a staple in laboratory research, enabling precise control over flow rates, gradients, and microbial interactions at the microscale. These platforms enable high-throughput screening of microbial strains and electrode materials with minimal resource consumption [18]. Once the standard, dual-chamber reactors are replaced or enhanced by multi-compartment reactors that incorporate intermediate chambers for buffering, nutrient supply, or intermediate chemical recovery. Innovations also include air-cathode BES, which utilises ambient oxygen as a terminal electron acceptor, thereby eliminating the need for external aeration systems and reducing energy input [19]. The integration of fluidised bed electrodes, in which conductive particles are suspended in an electrolyte, has shown promise in enhancing electron transfer and mass transport. Reactors are now being designed with automated control systems that regulate voltage, pH, temperature, and substrate feed in real time, ensuring optimal microbial activity and consistent performance. Including gas diffusion electrodes (GDEs) in cathodic chambers for CO_2 conversion enables higher gas-liquid contact efficiency, which is crucial for improving the availability of gaseous substrates to microbes or catalysts. Moreover, 3D-printed reactor components and custom housing made from biocompatible polymers or conductive composites allow for the rapid prototyping of novel geometries tailored to specific BES applications. Collectively, these advancements in reactor design are breaking traditional performance barriers, enabling longer operational cycles, higher yields, and more practical applications of BES technology for CO_2 reduction and beyond [20].

5 Enhanced Electron Transfer Mechanisms

Efficient electron transfer is the cornerstone of bioelectrochemical system (BES) performance, dictating the speed and effectiveness of biochemical reactions, particularly in processes such as CO_2 conversion [21]. Traditionally, electron transfer in BES was understood to occur directly, through physical contact between microbial cells and electrodes, or indirectly, via soluble redox mediators. Recent discoveries have expanded this understanding, uncovering sophisticated mechanisms and strategies that greatly enhance electron flow between electrodes and microbial cells. One breakthrough is the identification and engineering of conductive nanowires or pili, particularly in species such as Geobacter sulfurreducens and Shewanella oneidensis [22]. These structures enable direct interspecies electron transfer (DIET), allowing electron flow across longer distances within biofilms or between different microbial species, facilitating complex syntrophic relationships. Genetic modification to overexpress or enhance these conductive structures has significantly improved current densities and substrate utilisation.

Additionally, researchers have explored the application of redox-active mediators, both naturally occurring (e.g., flavins, phenazines, humic substances) and synthetic (e.g., neutral red, methylene blue), which shuttle electrons between microbes and electrodes. These mediators bridge the gap in systems where direct contact is limited, or biofilm growth is uneven, thereby increasing the overall efficiency of electron transfer [23]. Further innovation arises from the chemical modification of electrode surfaces with redox-active functional groups, such as quinones, which facilitate faster extracellular electron transfer (EET) by reducing the activation energy barrier. Incorporating conductive nanomaterials, such as carbon nanotubes, graphene flakes, or metal nanoparticles, into biofilms has also been shown to create percolation networks that facilitate faster and more uniform electron flow [24]. Moreover, efforts to understand and manipulate microbial outer membrane cytochromes—key proteins involved in EET—have led to the development of bioengineered strains with superior electroactivity. Some advanced systems use co-cultures engineered for DIET, where one species oxidises organic matter and transfers electrons to a partner species that reduces CO_2, forming a closed-loop carbon cycle driven by electricity [25]. Researchers also employ electrochemical gating, where applied voltages are dynamically adjusted to optimise redox potentials for specific metabolic pathways, thereby tuning electron availability and guiding microbial behaviour [26]. Real-time monitoring of EET using electrochemical impedance spectroscopy (EIS) and scanning electrochemical microscopy (SECM) provides insights into the spatial and temporal dynamics of electron transfer, informing the design of better materials and biofilms. These developments transform BES into highly efficient and responsive platforms capable of driving complex biotransformations, such as CO_2 to methanol while minimising energy input and maximising yield [27].

6 Integration with Renewable Energy Sources

Integrating renewable energy sources with bioelectrochemical systems (BES) represents a strategic advancement to make the technology fully sustainable, energy-efficient, and compatible with future green infrastructure [28]. BES are inherently electricity-driven, particularly in microbial electrosynthesis (MES) for CO_2 reduction, where an external voltage is often required to drive thermodynamically unfavourable reactions. Leveraging electricity from intermittent renewable sources, such as solar, wind, geothermal, and hydropower, enables BES to align with global decarbonisation efforts [29]. For example, solar panels can directly power microbial electrolysis cells (MECs) and MES systems during peak daylight hours, converting solar energy into chemical energy stored in products like methanol, formate, or acetic acid. Recent developments in photoelectrochemical BES (photo-BES) have introduced a new layer of innovation, where light-absorbing materials such as TiO_2, $BiVO_4$, or perovskite films are incorporated into the cathode or anode, enabling light-driven microbial reactions. In these systems, photons directly stimulate charge separation, facilitating redox reactions without the need for an external power supply [30]. Wind energy is another promising input for BES, especially in regions with strong wind patterns but limited energy storage infrastructure [31]. Pairing BES with wind turbines efficiently converts excess wind energy into storable biofuels or chemicals. Similarly, geothermal heat can support thermophilic microbial consortia in BES, potentially increasing metabolic rates and enhancing gas solubility. Importantly, coupling BES with renewables addresses the issue of energy intermittency by acting as a biological energy storage system, converting fluctuating power into stable, long-lived chemical bonds. Innovative grid technologies and battery-buffered systems can also regulate power flow to BES during off-peak hours, ensuring steady operation [32].

Additionally, surplus electricity from industrial renewable installations can be diverted to BES units for on-site carbon capture and utilisation. Advanced control systems, including real-time power matching algorithms and predictive load balancing, are being developed to optimise the synergy between renewable energy sources and battery energy storage systems (BES) [33]. Moreover, distributed energy systems based on microgrids are facilitating the deployment of decentralised Battery Energy Storage Systems (BESS) in rural and off-grid areas, where both renewable energy sources and waste streams (e.g., organic waste, CO_2 emissions) are available. The ability to co-locate BES near renewable energy farms and point sources of CO_2 (such as biogas plants or fermentation industries) offers logistical and economic advantages [34]. In summary, integrating renewable energy with BES represents a technical enhancement and a strategic step towards achieving carbon neutrality, enabling the efficient and equitable transformation of excess renewable energy into sustainable fuels and value-added chemicals.

7 Advances in Applications

The bioelectrochemical systems (BES) field has witnessed significant strides in its applications, expanding its utility beyond traditional energy generation to a diverse array of environmental, industrial, and sensing technologies [35]. Among the most notable advances is the evolution of microbial fuel cells (MFCs), which harness the metabolic activity of electroactive microorganisms to convert organic matter directly into electricity. Recent innovations in MFCs have focused on enhancing power output, improving operational stability, and increasing substrate flexibility. Researchers have developed high-performance electrode materials and tailored reactor architectures, such as air-cathode single-chamber designs and stacked modules, to overcome limitations in internal resistance and scalability [36]. MFCs are increasingly integrated into decentralised wastewater treatment systems, where they treat sewage and generate electricity as a byproduct, offering a sustainable solution for off-grid or rural areas [37]. In parallel, microbial electrolysis cells (MECs) represent a significant advancement, wherein external voltage input drives the microbial reduction of protons to produce value-added products such as hydrogen, methane, and other biofuels [38]. Modern MECs incorporate hybrid catalysts, low-energy input designs, and tailored microbial consortia to achieve high yields with minimal energy losses. Some studies have demonstrated that MECs achieve hydrogen production efficiencies exceeding 90% using a range of feedstocks, including domestic wastewater, industrial effluents, and lignocellulosic hydrolysates [39]. Moreover, microbial desalination cells (MDCs) and microbial electrosynthesis (MES) platforms have expanded the landscape of BES applications. MDCs combine desalination with wastewater treatment and energy generation, positioning them as ideal candidates for water-stressed regions with salinity challenges. Conversely, MES utilises microbes on cathodes to reduce CO_2 into multicarbon compounds, such as acetate, ethanol, and butanol, offering a bioelectrochemical route for carbon capture and biotransformation [19]. In recent advancements, MES systems have been optimised through the genetic engineering of cathodic microbes and electrode modifications to enhance selectivity, coulombic efficiency, and product titers. Furthermore, bioelectrochemical remediation has emerged as a novel approach for treating environments contaminated with heavy metals, nitrates, hydrocarbons, and chlorinated compounds [40]. For instance, BES-based bioreactors have successfully reduced hexavalent chromium and perchlorate by utilising microbial electron transfer pathways coupled with external circuitry. These systems provide redox control in situ, allowing for the targeted degradation of pollutants under low-energy conditions. In agricultural and aquaculture applications, BES platforms are being tested to remove nitrogen and phosphorus from runoff and recirculated waters, helping mitigate eutrophication and support sustainable practices. Another fast-emerging domain is bioelectrochemical sensing, where BES-based biosensors are being developed for real-time monitoring of pollutants, biochemical oxygen demand (BOD), heavy metals, and even pathogens [41]. These sensors exploit the direct correlation between microbial metabolic activity and current generation to provide a rapid, low-cost, and field-deployable detection mechanism [42]. With advances in miniaturisation, signal amplification, and wireless communication technologies, BES biosensors are now being integrated into smart environmental monitoring networks. Additionally, hybrid BES systems that combine phototrophic organisms (e.g., algae or cyanobacteria) with electroactive bacteria are being explored to utilise solar energy in conjunction with microbial processes, thereby creating bio-photoelectrochemical cells capable of driving synthetic pathways or generating electricity with improved sustainability [43]. In industrial bioprocessing, BES technologies are being applied to biorefinery operations, where waste streams such as glycerol, molasses, and fermentation byproducts are valorised through electroactive microbial pathways into higher-value chemicals, including succinate, propionate, and polyhydroxyalkanoates (PHAs). The ability to integrate BES with existing fermentation units offers a promising avenue for process intensification and waste minimisation. Additionally, the concept of power-to-chemicals (P2C) has taken root, where surplus renewable electricity is used to drive electrochemical reactions in BES, allowing for flexible energy storage and chemical production. This approach synergizes well with the increasing penetration of intermittent renewable energy sources, thereby contributing to grid stability and decarbonisation strategies. On a global scale, the implementation of BES in humanitarian and disaster-relief settings has gained interest, particularly for the deployment of portable BES units that treat greywater while powering LED lighting or mobile devices—demonstrating the technology's versatility and adaptability. Despite these promising applications, challenges such as low

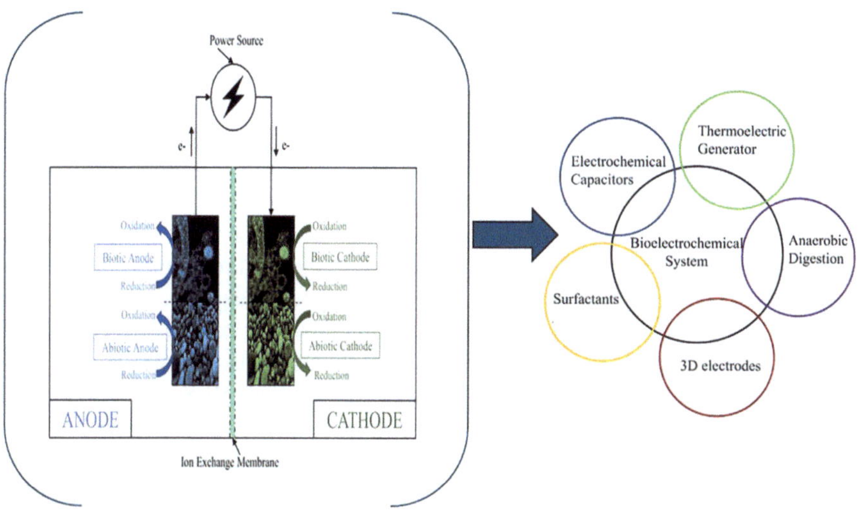

Fig. 1 Advances in applications [44]

volumetric power density, high capital costs, system longevity, and microbial-electrode compatibility remain to be fully addressed (Fig. 1). However, recent research indicates that with strategic use of synthetic biology, systems integration, and advanced materials science, many of these bottlenecks are gradually being overcome. Overall, the horizon of BES applications continues to expand, underpinned by a multidisciplinary convergence of microbiology, electrochemistry, environmental engineering, and systems design. As these systems evolve, their role in the circular economy, renewable energy production, and sustainable waste management is expected to become increasingly prominent, offering transformative solutions to some of the most pressing environmental and energy challenges of the twenty-first century [45].

8 Digitalization and Process Monitoring

As bioelectrochemical systems (BES) evolve towards more complex and large-scale applications, the role of digital technologies in monitoring, control, and optimisation has become increasingly crucial [46]. Modern BES research and development are leveraging tools from the Industrial Internet of Things (IIoT), artificial intelligence (AI), and data analytics to achieve precise, real-time management of biological and electrochemical processes [47]. For example, biosensors integrated into electrodes or reactor walls can continuously monitor parameters such as pH, redox potential, conductivity, substrate concentration, and product yield [48]. These sensors feed data into supervisory control and data acquisition (SCADA) systems, enabling automated feedback control loops that maintain optimal conditions for microbial activity [49]. Furthermore, machine learning algorithms enable the predictive modelling of BES performance based on historical and real-time data, allowing for the early detection of performance degradation, biofilm disruption, or reactor imbalance [50]. Digital twins—virtual replicas of BES reactors—are being developed to simulate various operating scenarios, test optimisation strategies, and conduct risk analyses without interfering with actual operations [51]. These models incorporate biological, chemical, and electrical dynamics to provide comprehensive insights into the system's behaviour. Big data analytics is also being employed to process the massive datasets generated by high-frequency sensors, uncovering hidden patterns and correlations that inform reactor design and microbial management [52]. The integration of blockchain technology is being explored to ensure data security and transparency, particularly in decentralised or community-based BES systems. Mobile applications and web-based dashboards also enable operators and researchers to access, remotely control, and troubleshoot BES systems, making management more efficient and accessible. Edge computing devices installed at BES sites can preprocess data locally, reducing latency and bandwidth needs [53]. Advanced visualisation tools, including augmented reality (AR), enable operators to understand and interact with BES components in real time. These digital solutions enhance system reliability and efficiency while also lowering maintenance

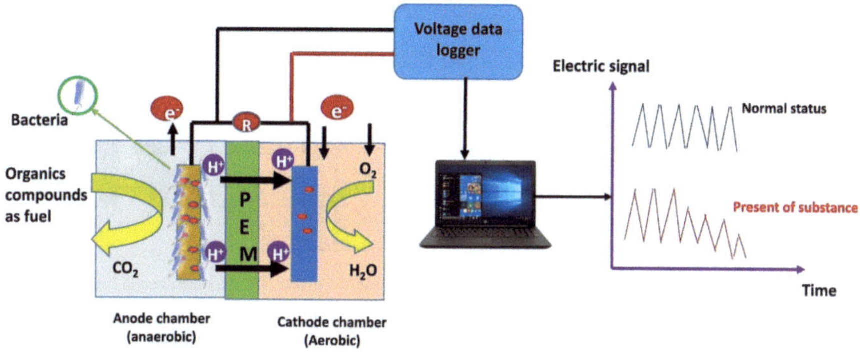

Fig. 2 Schematic representation of process control in microbial fuel cell [56]

costs and operational risks [54, 55]. As digitalization becomes a standard feature of modern BES, it will play a central role in scaling the technology for industrial deployment and integrating it into smart grids, circular bioeconomy systems, and automated waste management infrastructures (Fig. 2).

9 Techno-Economic Analysis and Scale-up Challenges

Despite the promising advances in bioelectrochemical systems (BES), transitioning from laboratory research to commercial-scale deployment remains fraught with both economic and technical challenges [57]. Techno-economic analysis (TEA) has become essential in evaluating the feasibility, cost-effectiveness, and scalability of BES technologies. Key cost drivers include the materials used for electrodes, membranes (if any), reactor construction, and operational expenses such as power input, nutrient supply, and maintenance. Although innovative materials, such as carbon nanotubes and metal catalysts, improve performance, they often remain prohibitively expensive for large-scale use [58]. TEA helps quantify trade-offs between performance and cost, guiding researchers towards cost-optimised solutions. Life cycle assessment (LCA) is another critical tool for assessing the environmental impacts associated with BES, including raw material extraction, energy use, emissions, and end-of-life disposal [59]. While BES offer clear sustainability benefits regarding carbon reduction and energy recovery, LCAs reveal hotspots that must be addressed for holistic impact reduction—such as the use of rare metals or synthetic polymers [60]. Scaling up BES introduces new complexities, including biofilm uniformity, mass transfer limitations, increased ohmic losses, and risks of microbial contamination. Maintaining stable microbial communities over extended periods in industrial environments presents a significant challenge, particularly when feedstocks vary or environmental conditions fluctuate [61]. Fouling of electrodes and clogging of flow channels increase downtime and reduce efficiency (Fig. 3). Furthermore, low product concentrations in large-scale BES necessitate

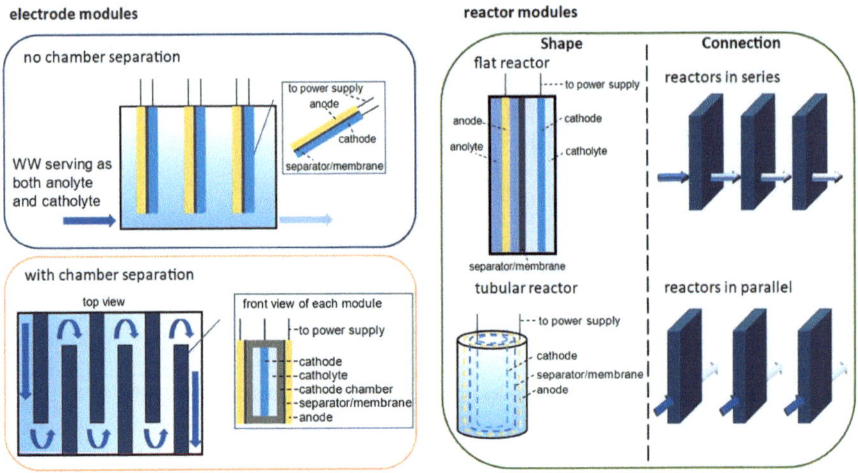

Fig. 3 Key approaches for scale-up of microbial fuel cell [57]

energy-intensive downstream processing, which can offset the environmental benefits of the process. Innovations in continuous operation, in situ product recovery, and process intensification are needed to overcome this bottleneck. Regulatory uncertainty also poses a barrier, especially for applications involving genetically modified organisms (GMOs) or waste-to-food pathways. Lack of standardised performance metrics, safety guidelines, and quality control procedures hampers industrial investment and public acceptance [62]. Nonetheless, partnerships between academia, industry, and government agencies are increasingly addressing these issues, leading to the development of pilot projects and demonstration units. For BES to reach commercial viability, it must demonstrate technical robustness and clear economic and environmental advantages over incumbent technologies. TEA and LCA are pivotal in making this case.

10 Future Perspectives and Research Directions

The future of bioelectrochemical systems (BES) holds great promise, driven by interdisciplinary innovation and the urgent global demand for sustainable solutions to energy, waste, and climate challenges. One key trend is the convergence of BES with other emerging technologies, including synthetic biology, artificial intelligence, and green chemistry, leading to hybrid platforms capable of complex biotransformations. Integrating artificial photosynthesis systems with BES is being explored for more efficient solar-to-fuel conversion, combining natural microbial metabolism with semiconductor-based light harvesting (Fig. 4). Soon, we can expect the rise of programmable BES, where synthetic genetic circuits enable microbes to respond autonomously to environmental cues, turning specific

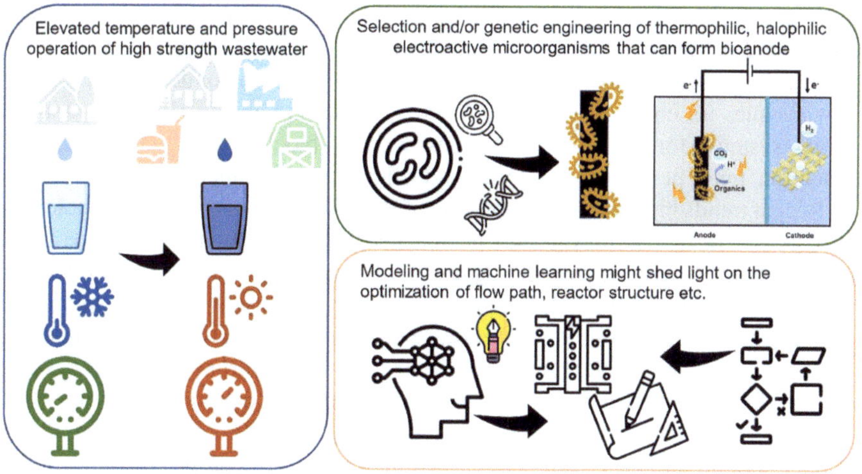

Fig. 4 Challenges and future direction [57]

metabolic pathways on or off as needed. There is also significant interest in developing closed-loop BES, capable of operating in circular systems where inputs and outputs are continuously recycled, such as in space habitats, military bases, or remote communities. Research is expanding into non-traditional microbial chassis with unique properties, such as acidophiles, halophiles, or extremophiles—that can operate under harsh conditions and broaden the range of possible feedstocks. Future work will also explore multi-functional BES reactors, which simultaneously purify water, recover nutrients, produce chemicals, and generate energy. From a materials perspective, there is a growing push to develop eco-friendly, biodegradable, and locally sourced materials for reactor construction, aiming to minimise environmental footprints. At the policy level, international frameworks promoting carbon credits, waste valorisation incentives, and renewable energy integration could help catalyse BES adoption. Educational and outreach efforts are needed to build a skilled workforce that operates and maintains BES infrastructure. Finally, the long-term vision includes the deployment of BES at a global scale, integration into smart cities, bio-refineries, and even climate engineering strategies. With ongoing research, strategic investment, and interdisciplinary collaboration, BES hold the potential to reshape how we produce energy, treat waste, and manage carbon in a post-fossil-fuel world.

11 Conclusion

Bioelectrochemical systems (BES) have witnessed remarkable advancements over the past decade, driven by the urgent global need for sustainable energy production, waste remediation, and resource recovery. Innovations in microbial engineering, electrode materials, and system configurations have significantly improved the performance and versatility of BES technologies. Cutting-edge research in genetically modified microbes, nanostructured electrodes, and novel reactor designs has opened new avenues for enhancing electron transfer, increasing energy efficiency, and tailoring systems for specific applications such as bioelectricity generation, CO_2 reduction, nutrient recovery, and value-added chemical production.

Recent trends highlight the integration of BES with other emerging technologies, such as microbial electrosynthesis, photocatalysis, and artificial intelligence-based process control, pushing the boundaries of what BES can achieve in both lab and real-world environments. Developing hybrid systems and modular platforms enables scalable solutions for decentralised wastewater treatment and renewable energy production. Furthermore, advancements in omics technologies and systems biology facilitate a deeper understanding of microbial communities within BES, enabling the design of more robust and efficient systems. The continued interdisciplinary collaboration and focus on real-world applications will be crucial for translating these innovations into commercially viable solutions, ultimately contributing to a more sustainable and circular bioeconomy.

References

1. Jatoi AS, Akhter F, Mazari SA, Sabzoi N, Aziz S, Soomro SA et al (2021) Advanced microbial fuel cell for waste water treatment—a review. Environ Sci Pollut Res 28:5005–5019
2. Jatoi AS, Ahmed J, Bhutto AA, Shaikh MS (2024) Sustainable bio-energy generation via the conversion of spent wash using dual chamber microbial fuel cell. Waste Disp Sustain Energy 6:219–231
3. Jatoi AS, Hashmi Z, Anjum A, Bhatti ZA, Siyal SH, Mazari S et al (2021) Overview of bioelectrochemical approaches for sulfur reduction: current and future perspectives. Biomass Convers Biorefinery 13:1–16
4. Jatoi AS, Unar IN, Aziz S, Soomro SA, Abro M, Khoso JA et al (2018) Microbial fuel cell for bio-electricity generation via bio-electrochemical conversion of fermentation industry sludge. Abasyn J Life Sci 2:42–53
5. Opoku PA, Shaowei X, Jingyu H (2025) Bioenergy generation, organic matter, and simultaneous nitrogen and phosphorus removal in comparative pyrite and graphite-based constructed wetland-microbial fuel cells at different anode scales. J Water Process Eng 69:106882
6. Wei J, Liang P, Huang X (2011) Recent progress in electrodes for microbial fuel cells. Bioresour Technol 102:9335–9344
7. Zhou M, Chi M, Luo J, He H, Jin T (2011) An overview of electrode materials in microbial fuel cells. J Power Sources 196:4427–4435
8. Agrahari R, Bayar B, Abubackar HN, Giri BS, Rene ER, Rani R (2022) Advances in the development of electrode materials for improving the reactor kinetics in microbial fuel cells. Chemosphere 290:133184
9. Mustakeem NM (2015) Electrode materials for microbial fuel cells: nanomaterial approach. Mater Renew Sustain Energy 4:1–11
10. Kalathil S, Patil S, Pant D (2018) Microbial fuel cells: electrode materials. Encycl Interfac Chem 309:318
11. Nitisoravut R, Thanh CN, Regmi R (2017) Microbial fuel cells: advances in electrode modifications for improvement of system performance. Int J Green Energy 14:712–723

12. Rabiço F, Pedrino M, Narcizo JP, De Andrade AR, Reginatto V, Guazzaroni M-E (2023) Synthetic biology toolkit for a new species of pseudomonas promissory for electricity generation in microbial fuel cells. Microorganisms 11:2044
13. Lee SK, Chou H, Ham TS, Lee TS, Keasling JD (2008) Metabolic engineering of microorganisms for biofuels production: from bugs to synthetic biology to fuels. Curr Opin Biotechnol 19:556–563
14. Ghosh D, Das S, Debnath S, Pradhan S, Karmakar M, Mondal S (2022) Chapter 13 - metabolic engineering and synthetic biology key players for improving efficacy of microbial fuel cell technology. In: Jadhav DA, Pandit S, Gajalakshmi S, Shah MP (eds) Scaling up of microbial electrochemical systems. Elsevier, Amsterdam, pp 217–238
15. Li F, Li Y, Sun L, Li X, Yin C, An X et al (2017) Engineering Shewanella oneidensis enables xylose-fed microbial fuel cell. Biotechnol Biofuels 10:196
16. Li F, Wang L, Liu C, Wu D, Song H (2018) Engineering exoelectrogens by synthetic biology strategies. Curr Opin Electrochem 10:37–45
17. Ding M-Z, Song H, Wang E-X, Liu Y, Yuan Y-J (2016) Design and construction of synthetic microbial consortia in China. Synth Syst Biotechnol 1:230–235
18. Clomburg JM, Gonzalez R (2010) Biofuel production in Escherichia coli: the role of metabolic engineering and synthetic biology. Appl Microbiol Biotechnol 86:419–434
19. Mohd Noor NN, Kim K (2025) Boosting bioelectricity performance in sediment microbial fuel cells with raw bamboo biochar as a sustainable energy source. Renew Energy 251:123457
20. Yamada S, Hashimoto K, Nedachi T, Takahashi E, Ando R, Kim M et al (2025) Identification and characterization of electrogens in tidal flat soils using microbial fuel cells and evaluation of the electrogenic potential of Zobellella denitrificans. Next Res 2:100264
21. Sreelekshmy BR, Basheer R, Shibli SMA (2022) Exploration of bifurcated electron transfer mechanism in Bacillus cereus for enhanced power generation in double-chambered microbial fuel cells. J Environ Chem Eng 10:107601
22. Chen H, Yu Y, Yu Y, Ye J, Zhang S, Chen J (2021) Exogenous electron transfer mediator enhancing gaseous toluene degradation in a microbial fuel cell: performance and electron transfer mechanism. Chemosphere 282:131028
23. Ai T, Zhan H, Zou L, Fu J, Fu Q, He Q et al (2020) Potential applications of endogenous sulfide for enhanced denitrification of low C/N domestic wastewater in anodic mixotrophic denitrification microbial fuel cell: the mechanism of electrons transfer and microbial community. Sci Total Environ 722:137830
24. Christwardana M, Prayoga BA, Kuntolaksono S, Joelianingsih J, Maulana AY (2025) Enhanced electron transfer in microbial fuel cells using calcined eggshell anode with optimized performance. Bioresour Technol Rep 29:102088
25. Min Z, Tian Q, Shi Z, Ding X, Guo C, Li CM et al (2025) Enhanced electron transfer efficiency in microbial fuel cells utilizing hybrid microorganisms integrated with conjugated polymer-TiO2 photocatalyst under light illumination. J Power Sources 630:236078
26. Christwardana M, Khoirunnisa K, Asy'ari M, Hadiyanto H (2025) Evaluating nitrogen sources for enhanced halophilic bacteria growth, electron transfer, and microbial fuel cell performance. Chemosphere 378:144397
27. Feng M, Meng L, Zhang Z, Zheng Q, Wang R, Yang C et al (2024) Hierarchical modulation of extracellular electron transfer processes in microbial fuel cell anodes for enhanced power output through improved Geobacter adhesion. Electrochim Acta 487:144165
28. Aftab Q, Wang X, Lu J, Tariq M, Liu Y (2025) Advancing soil microbial fuel cells: exploring bioelectrogenesis mechanisms for integration into environmental bioremediation. Renew Sust Energ Rev 214:115495
29. Baniam M, Gholamian E, Yari M, Mehr AS (2025) Innovative integration of DMFC in polygeneration energy systems for enhanced renewable fuel and power outputs. Process Saf Environ Prot 199:107263
30. Kanani B, Zahedi A (2024) Step toward sustainable development through the integration of renewable energy systems with fuel cells: a review. Sustain Energy Technol Assess 70:103935

31. Youssef YA, Abuarab ME, Mahrous A, Kurniawan TA, Abdellatif HRS, Mahmoud M (2024) A comprehensive overview of constructed wetland-microbial fuel cells for enhancing organic waste treatment and renewable energy production: from fundamentals to applications. J Water Process Eng 64:105592
32. Meena PK, Patane PM (2025) Biohydrogen: advancing a sustainable transition from fossil fuels to renewable energy. Int J Hydrog Energy 100:955–970
33. Srivastava S, Pandit C, Chen H-Y, Wang C-T (2025) Harnessing microbial fuel cells for mitigating metal corrosion in offshore wind power infrastructure: mechanisms, applications, and future prospects. Chem Eng J 510:161759
34. Genç N, Pískin ED, Turk M (2025) Microbial fuel cell as a green technology for sludge oxidation and abiotic nitrate reduction: integration of Taguchi-multi criteria decision method. Biomass Bioenergy 199:107955
35. Pal P, Wang C-T (2025) Radio wave-driven enhancement of microbial fuel cells: converting waste waves to power. Process Saf Environ Prot 196:106885
36. Ballestas ER, Bortoluzzi EC, Hamad Minervino AH, Palma HH, Neckel A, Ramos CG et al (2024) Power generation potential of plant microbial fuel cells as a renewable energy source. Renew Energy 221:119799
37. Brzęczek M, Kotowicz J (2024) Integration of alternative fuel production and combined cycle power plant using renewable energy sources. Appl Energy 371:123738
38. Qi S, Xu Y, Fan Z, Chen S, Zhang S, Zhang L et al (2025) Integrating MXene with carbon dots modified bacteria as hybrid biofilm for photo-assisted microbial fuel cells. J Colloid Interface Sci 687:65–73
39. Kanani B, Zahedi A (2025) Efficiency enhancement by integration of fuel cells in downstream of power plants: next step in energy generation systems. Next Energy 7:100226
40. Singh S, Kirrolia AS, Pannu D, Bishnoi NR (2025) Exploring the synergy between microbial fuel cells and constructed wetlands in dairy waste water treatment: a comprehensive review. J Water Process Eng 71:107211
41. Wang Y (2024) Carbon felt electrode modified with RGO/PANI composite material for enhancing renewable energy storage in microbial fuel cells. Renew Energy 232:121109
42. Murugan M, Rajendran T, Ramakrishnan G, Sahadevan R (2025) Chapter 15 - biomass-fueled microbial electrochemical systems: a renewable energy solution. In: Gurunathan B, Venkatesh G (eds) Biofuels and bioenergy. Elsevier Science, Amsterdam, pp 343–379
43. Zhou S, Wang G, Huang L, Li S, Li C, Zhao R et al (2025) Novel microbial fuel cell integrating anoxic/oxic-membrane bioreactor design for simultaneous energy recovery and carbon–nitrogen removal from shale oil and gas wastewater. Bioresour Technol 432:132675
44. Debnath K, Dutta S (2023) Bio-electrochemical system analysis and improvement: a technical review. Clean Circ Bioecon 6:100052
45. Javanmard A, Zuki FM, Patah MFA, Daud WMAW (2024) Revolutionizing microbial fuel cells: biochar's energy conversion odyssey. Process Saf Environ Prot 187:26–58
46. Souza L, Recouvreux D, Watzko E, Taufemback W, Carminatti C, Antônio R et al (2025) Bacterial nanocellulose/hydroxyapatite hybrid for proton exchange membrane applied in microbial fuel cells. Int J Biol Macromol 307:141569
47. Gu M, Guo CF, Song Y (2025) Multimodal bioelectronics: a pathway to digital health management. Matter 8:102048
48. Costa NL, Olorounto G, Lebègue E, Barrière F (2022) Electrografted anthraquinone to monitor pH at the biofilm-anode interface in a wastewater microbial fuel cell. Colloids Surf B: Biointerfaces 210:112274
49. Rusyn I, Medvediev O (2024) Stacking and design optimization of novel plant microbial fuel cell based on dwarf indoor decorative and culinary plants as a compact biobattery for a low energy consumption devices. Bioresour Technol Rep 26:101860
50. Salvian A, Saini S, Ramirez-Moreno M, Avignone-Rossa C, Varcoe JR, Kousi K et al (2025) Cost-effective and stable biosensing of biochemical oxygen demand in wastewater treat-

ment: Exsolved rhodium-titanate perovskite catalyst in microbial fuel cell-based biosensors. J Environ Chem Eng 13:115692
51. Fu L, Wang J, Fu X, Zhao G (2025) Finite-time Pade-based adaptive FNN controller implementation for microbial fuel cell with delay and multi-disturbance. Int J Hydrog Energy 98:1034–1043
52. Zhang X, Jia L, Wang Z, Qin J, Cui C, Lv C et al (2025) Efficient electricity generation and wastewater degradation simultaneously in microbial fuel cell: synergistic effect of amplifying localized charge density of carbon shell and confinement of hollow structures for oxygen reduction reaction. Sep Purif Technol 360:130890
53. Chandra S, Pandey KK, Pandit S, Roy A, Rajeev M, Sharma K et al (2025) Ultrasonication of microalgal biomass: acoustics study and enhanced bioelectricity generation in microbial fuel cells. Bioresour Technol 424:132289
54. Sreelakshmi C, Kiruthika S, Jeyalakshmi R (2025) Optimization modeling and economics assessment on simultaneous struvite and bioelectricity production from waste nutrient solution in the microbial fuel cell. Biochem Eng J 218:109705
55. Güneş S, Ayol A, Tezer O (2025) Biodegradation and power generation from olive oil mill wastewater using soil-based microbial fuel cells. Int J Hydrog Energy 116:410–419
56. Do MH, Ngo HH, Guo W, Chang SW, Nguyen DD, Liu Y et al (2020) Microbial fuel cell-based biosensor for online monitoring wastewater quality: a critical review. Sci Total Environ 712:135612
57. Jiang J, Lopez-Ruiz JA, Bian Y, Sun D, Yan Y, Chen X et al (2023) Scale-up and techno-economic analysis of microbial electrolysis cells for hydrogen production from wastewater. Water Res 241:120139
58. Deka R, Shreya S, Mourya M, Sirotiya V, Rai A, Khan MJ et al (2022) A techno-economic approach for eliminating dye pollutants from industrial effluent employing microalgae through microbial fuel cells: barriers and perspectives. Environ Res 212:113454
59. Sharma R, Choudhary P, Thakur G, Pathak A, Singh S, Kumar A et al (2025) Sustainable management of biowaste to bioenergy: a critical review on biogas production and techno-economic challenges. Biomass Bioenergy 196:107734
60. Srivastava N, Singh R, Lal B, Haque S (2025) Evaluation of bioprocess parameters for pilot scale fermentative biohydrogen production using organic waste: environmental remediation, techno-economic challenges & future solutions. Int J Hydrog Energy 107:396–409
61. Ragini YP, Palanivelu J, Hemavathy RV (2024) Critical analysis of enhanced microbial bioremediation strategies of PAHs contaminated sites: toxicity and techno-economic analysis. Groundwater Sustain Dev 27:101369
62. Jadhav DA, Mungray AK, Arkatkar A, Kumar SS (2021) Recent advancement in scaling-up applications of microbial fuel cells: from reality to practicability. Sustain Energy Technol Assess 45:101226

Energy-Efficient Bio-Electrochemical System for Treatment of Agricultural Wastes

Suman Pawar, Nabisab Mujawar Mubarak, and Ashagowda Karegowda

Abstract Bioelectrochemical systems (BESs) have emerged as a promising new technology for converting waste into energy. BESs are effective in curtailing organic contaminants and conserving nutrients. Agricultural waste and treated wastewater (TWW) can be utilised as a feedstock for microbes and as an irrigation source. The WW can also be used as a source for concurrently engendering power. BESs are energy-efficient systems that integrate with traditional approaches and contribute to the generation of value-added products. Agricultural production and subsequent food manufacturing are the primary sources of the global food supply. Still, they generate substantial amounts of organic waste, which hurts the atmosphere and climate. This demands the proper reuse of agro-food waste (AFW) and wastewater (WW). BES is a promising biotechnology that integrates microorganisms (or enzymes) to achieve sustainability in food production by converting polluting waste into biodegradable materials. This chapter presents the benefits and remedial potentials of BES in managing AFW and WW. It also provides a brief overview of BES's current and potential applications.

Keywords Bio-electrochemical systems · Agro-food waste · Wastewater · Biohydrogen

S. Pawar (✉)
Department of Chemical Engineering, Siddaganga Institute of Technology, Tumkur, India
e-mail: svp@sit.ac.in

N. M. Mubarak
Department of Chemical Engineering, College of Engineering, Imam Mohammad Ibn Saud Islamic University (IMSIU), Riyadh, Saudi Arabia

Department of Biosciences, Saveetha School of Engineering, Saveetha Institute of Medical and Technical Sciences, Chennai, India

A. Karegowda
Master of Computer Applications, Siddaganga Institute of Technology, Tumkur, India

A. Sattar Jatoi, N. M. Mubarak (eds.), *Bio-Electrochemical Conversion of Waste to Fuel*, Green Energy and Technology,
https://doi.org/10.1007/978-981-95-4310-6_15

1 Introduction

Bioelectrochemical systems are innovative technologies integrating biological and electrochemical processes, offering energy-efficient and environmental-friendly solutions. These systems can significantly reduce waste and treat wastewater to recover electricity, bio-H_2, and other valuable chemicals through specific redox reactions. Energy plays a vital role in modern agriculture, where crop and livestock production rely on direct energy inputs, such as fuel or electricity, to operate various machinery and equipment. Additionally, energy is indirectly consumed through the use of compost and insecticides. Integrating BESs in agricultural practices (AC) can enhance sustainability by minimising waste and optimising resource use, thereby contributing to more efficient farming operations [1]. The over-reliance on fossil fuels, manure, and insecticides in crop production, along with the improper use of antibiotics in livestock farming, can lead to significant environmental issues. These practices contribute to orangery gas emissions and the contamination of water bodies, posing sombre threats to ecosystems and public health. Addressing these challenges is crucial for promoting sustainable air conditioning (AC) practices and protecting the environment [2–4].

Furthermore, many AC systems generate waste materials that require proper treatment or disposal. Agricultural wastes, including crop residues such as stover and livestock manure, typically contain significant quantities of microbially degradable organic matter [5]. Effectively managing these waste products minimises environmental impact and promotes sustainability in AC practices [6]. AC's byproducts (Diesel, Gasoline, LPG, electricity) can be reused to power AC machinery or vehicles.

Among the innovations for enhanced sustainability in agriculture, BESs are promising solutions, offering significant potential for improving sustainability within AC practices [7]. BESs exhibit considerable variability in their reactor configurations, dimensions, substrates, and bacterial communities, depending on the specific end goals they aim to achieve. This adaptability allows BESs to be tailored for various applications, enhancing their effectiveness in energy production and resource recovery [7]. Despite the differences, BESs are usually classified according to their application purposes and can be divided into microbial fuel cells (MFCs), microbial electrolysis cells (MECs), microbial electrosynthesis (MESs), microbial desalination cells (MDCs), microbial solar cells (MSCs), and enzymatic biofuel cells (EFCs) [8]. The mechanisms and uses of these dissimilar types of BESs are elaborated in the next section.

Food waste (FW) is produced at each phase of the food cycle, starting from AC production and continuing through industrial manufacturing and processing, retail, and home ingesting. For example, 42% of food waste (FW) in the European Union occurs in households, while the remaining 58% is generated within the food manufacturing, distribution, and retail sectors [9]. Each year, approximately 1.3 billion tons of food intended for human consumption are lost globally, resulting in an estimated economic cost of around $940 billion, according to the United Nations Food

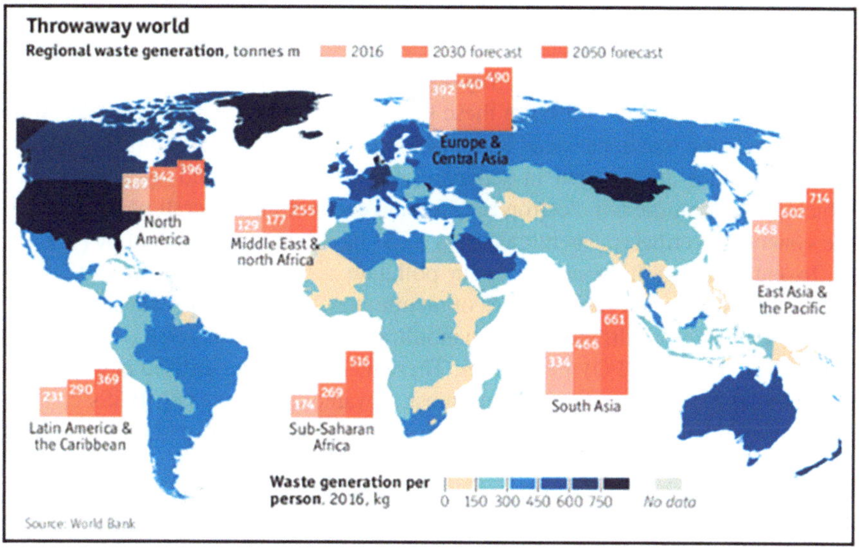

Fig. 1 Universal FW scattering and prediction [11]

and Agriculture Organisation (FAO). This staggering amount represents nearly one-third of all food produced, highlighting a significant disorganisation in the global food system [10]. According to forecasts by the World Bank, FW is growing alarmingly (Fig. 1).

By 2050, the scale of FW generated is expected to become a significant challenge [11, 12]. While some countries lack accessible public data on FW, the estimates presented in Fig. 1 provide the most complete global figures despite certain uncertainties [13, 14]. Research indicates that urbanised countries tend to engender additional food losses.

In Eco-tech, regenerative industrial design principles such as the Perpetual resource cycle and circular economy frameworks are gradually recognised as foundational concepts aimed at achieving a "zero waste economy." This necessitates efforts to convert waste into value-added products. Concurrently, advancements in biotechnology hold promise for benefiting ecosystems by transforming pollutants into recyclable materials and enhancing manufacturing and disposal processes to become more environmentally friendly.

BESs combine the microbes or other biobased catalysts with electrochemical systems powered by redox reactions. This integration enables various innovative applications, making BESs versatile in addressing environmental challenges and enhancing sustainability [15, 16].

These systems can effectively reduce waste and treat wastewater while simultaneously performing various additional utilities, contingent on the specific microbial systems and types of BESs employed. For instance, they can recover bioelectricity (BE) through microbial fuel cells (MFCs), produce biohydrogen using microbial electrolysis cells (MECs), generate value-added compounds via microbial electrolysis systems (MESs), and desalinate saltwater through microbial desalination cells (MDCs).

BESs are widely regarded as highly promising advancements in biotechnology. At least two electrodes are utilised in these systems: the anode and the cathode. At the anode, oxidation reactions occur, and reduction takes place at the cathode. Typically, these electrodes are situated in separate compartments, divided by a membrane, which facilitates efficient electron transfer and enhances the overall performance of the system. Nevertheless, membrane-less BESs have been corroborated by [17].

To enhance handling efficiency in pollutant removal and resource recovery, BESs may incorporate various types of membranes, including ion exchange, microfiltration (MF), ultrafiltration (UF), nanofiltration (NF), and reverse osmosis (RO) and forward osmosis (FO) membranes. These membranes are utilised in wastewater treatment systems and can eliminate most unwanted species [18]. BESs can also accompany other biological usage units, such as actuated slush vessels, contributing an additional 10%–20% in energy generation. The low biomass yield associated with BESs is another advantage, as it reduces costs related to sludge disposal. These systems can achieve high elimination efficacies for chemical oxygen demand, reaching 92% and total suspended solids exceeding 99.0% [19].

From the Science Citation Index Expanded (SCIE) database, an overview of research on BES technology can be obtained, highlighting its primary applications in the management of agricultural freshwater (FW) and wastewater (WW). Keyword analysis suggests that research on BESs can be categorised into three primary areas: various types of BES reactors, potential applications, and methods for performance enhancement. MFCs are the most widely studied BES, particularly for their dual role in direct power generation and simultaneous WW treatment. Despite significant advancements in BES performance over the past 10–15 years, ongoing research continues to explore and refine these technologies to address environmental challenges effectively [20]. However, research and reviews on BESs connected to the agro-food sector are extremely limited.

Despite the potential of BESs, their applications in AC have received significantly less attention. There is also a notable lack of comprehensive reviews addressing the use of BESs in AC systems. This article aims to fill that gap by critically reviewing past studies on BESs, with a specific focus on their applications in AC systems. Additionally, it discusses the current and future uses of various BES technologies in AC contexts to outline a framework for sustainable AC systems that address the interconnected challenges of food, energy, and water.

2 Types of BESs

2.1 MFCs

MFCs (Fig. 2a) are the supreme central BESs and have been widely studied for real-world applications [21]. The bacteria (i.e., exoelectrogens) such as *Geobacter sulfurreducens* and *Shewanelle putrefaciens* on the anode, MFCs, can generate electrical

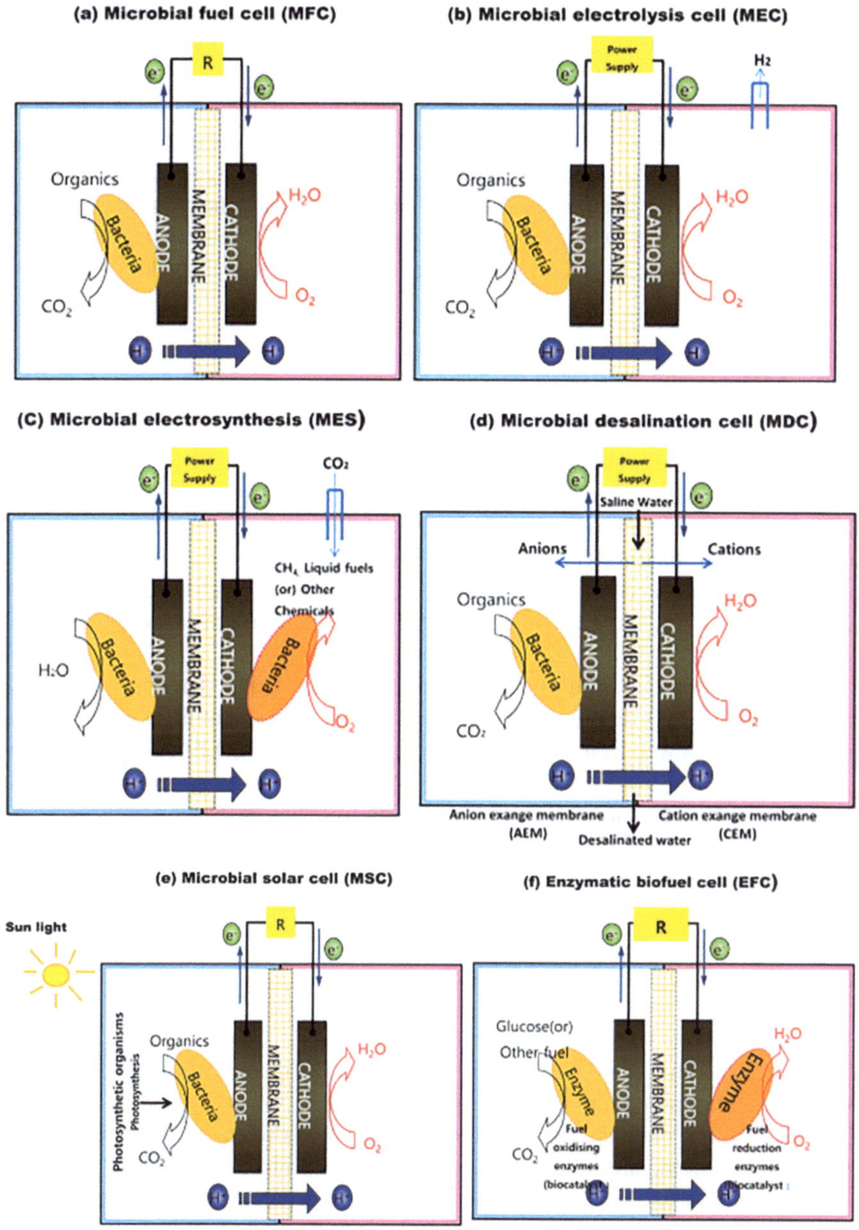

Fig. 2 Types of BESs: (**a**) MFC, (**b**) MEC, (**c**) MES, (**d**) MDC, (**e**) MSC, (**f**) EFC

energy directly from a wide range of recyclable organics [22]. In MFCs, organic matter in the anode (AE) compartment acts as electron (e^-) donors, which are metabolised by exoelectrogens. This process generates (e^-) that can be collected from the AE. The (e) then flows through an external circuit to the cathode (CE), creating an electric current that can be utilised immediately or stored for later use. The advancement of air-CE MFCs has significantly enhanced the feasibility of this technology for WW treatment and electricity generation [23]. Numerous studies, including some aviator works [24], have reported considerable electric power generation from domestic wastewater [25], industrial wastewater [26], landfill leachate [22], swine wastewater [27], and other sources. However, the widespread commercial application of large-scale MFCs is primarily limited by two main factors: the cost of electrodes and the reduced power output at greater scales. Overcoming these challenges is critical for making MFC technology economically viable and practically applicable in real-world settings [23].

2.2 MECs

Adapted from MFCs, MECs can produce H_2 gas, methane (CH_4), or other value-added chemicals for indirect energy use and storage [8]. Hypothetically, these have the potential to generate any additional worth of chemicals, irrespective of the electrodes [28]. The thermodynamic barriers that hinder required redox responses are overcomed by an energetic (e^-) flow with an outward electrical power source (Fig. 2b). The CE section needs to be anaerobic to maximise the efficiency of electrons in generating preferred products.

Among the various products generated in MECs, H_2 has demonstrated strong potential as a replenishable energy source, thanks to its suitability for supply and transportation, as well as the rapidly growing market demand [29]. The system efficiency of MECs is considerably (>BESs) [28]. Therefore, considerable effort has been devoted to generating H2 in MECs using a variety of waste and wastewater [30]. In recent years, significant advances have been reported in the production of H_2 using microbial electrochemical cells (MECs).

2.3 MESs

From an innovative standpoint within BESs, MES (Fig. 2c) represents a form of MEC capable of producing premium-quality products, such as CH_3CH_2-, $C_4H_8O_2$, CH_3CH_2OH, and biodiesel, through the electric power-driven reduction of CO_2 and other organics at the biocathode, utilising microbes as biocatalysts [31].

Depending on the feeding substrates, MES can generate innumerable industrially relevant products, including bioethanol, biofuels, and unpackaged chemicals, via oxidation reactions at the AE [31]. The substrates suitable for microbial electrolysis cells (MECs) are analogous to those used in microbial fuel cells (MFCs); however, unlike MFCs, MECs require external electric power to facilitate the generation of organics. Consequently, the processes in MES are essentially the reverse of those in MFCs, with energy being stored in covalent chemical bonds.

It is important to note that while MESs are related to microbial electrolysis cells (MECs), they differ in function; MECs utilise external electric power to provide the necessary electrical potential for decreased reactions at the CE, such as the mitigation of H^+ to H_2 [32]. These offer a diverse range of requests, producing several final goods that include various types of biofuels [33], industrially relevant chemicals [34], and pharmaceutical precursors [35]. Consequently, many scientists anticipate that MESs will become increasingly essential in upcoming bioproduction efforts [35].

2.4 *MDCs*

It evolved from MFCs to address the significant energy demands associated with conventional desalination technologies, such as reverse osmosis, electrodialysis, and mechanical vapour compression [36]. In this, a purification partition is positioned between the AE and CE compartments (Fig. 2d), integrating both a "Cation Exchange Membrane (CEM) and an Anion Exchange Membrane (AEM)". These will effectively desalinate saltwater by removing ions. The electrical energy generated from organic matter in the AE section can also be stored for other requests. MDCs have demonstrated a 90% desalination efficacy with a power density of 31 W/m^3 when using acetate as an (e^-) donor and $[Fe(CN)_6]^{3-}$ as an (e) acceptor [36]. Additional studies have revealed that residential WW can serve as an (e^-) donor, achieving substantial salinity removal of over 60% when combined with sparged air as an (e^-) acceptor [37]. Due to their capability to concurrently treat wastewater (WW) and desalinate saltwater while recovering electric power from these processes, MDCs are increasingly recognised as an auspicious low-energy technology. They offer solutions to pressing challenges related to salt and salted H_2O removal, speciality chemical synthesis, aquifer decontamination, wastewater treatment, and energy retrieval [38].

2.5 *MSCs*

2.5.1 Harnessing Bioelectricity Through Photosynthesis and Exoelectrogenesis

MSCs (Fig. 2e) represent a cutting-edge BES capable of recovering in situ BE or producing fine chemicals by integrating photosynthesis with *exoelectrogenesis* [39]. These systems utilise *photosynthetic organisms*, such as *photoautotrophic microorganisms* or higher plants, to convert solar energy into chemical energy stored in organic matter. Microbial fuel cell components then utilise this organic matter to generate power.

2.5.2 Key Processes in MSCs

- Photosynthesis begins with photosynthetic organisms converting solar energy into chemical energy.
- Organic Matter Transfer: There are three primary modes for transferring this organic matter to the anode compartment:
  - Rhizodeposition: High plants release organic compounds through their roots.
  - Diffusion: Phototrophic biofilms facilitate the diffusion of these compounds.
  - Pumping Mechanisms: Translocation from photobioreactors or coastal marine ecosystems can also occur.

2.5.3 Plant Microbial Fuel Cells (PMFCs)

Among various MSC types, those integrated with living higher plants—known as PMFCs—have demonstrated the most effective power generation capabilities, reaching approximately 1000 GJ/ha/year. The efficiency of PMFCs is directly linked to the obtainability of *rhizodeposits* electrooxidation reactions [36].

2.5.4 Advantages and Applications

Unlike conventional solar cells, MSCs offer several advantages:

- Diverse outputs: They can produce power and a wide range of premium chemicals.
- Self-sustaining systems: benefiting from continuously growing populations of microorganisms involved in photosynthetic and electrochemical reactions, MSCs are self-regenerating systems with longer lifetimes and reduced upkeep needs.
- Sustainability and compatibility: Their applications are broad and promising due to their compatibility as viable ecological vigour dealers [39].

Overall, MSCs offer a novel approach to harnessing renewable energy while promoting environmental sustainability through their waste management and chemical synthesis capabilities.

2.6 EFCs

EFAs utilise precise oxidoreductase enzymes as electrocatalysts to convert chemicals into usable energy. This process involves oxidation reactions at the anode (AE) and reduction at the cathode (CE) (Fig. 2f). Initially, enzymes were suspended in the electrolyte in early electrochemical fuel cells (EFCs). Still, these systems suffered low efficiency and stability [36]. Modern studies have established numerous techniques to immobilise and stabilise enzymes on the conductor surface. For instance, crosslinking redox hydrogels, sandwich techniques, and encapsulation have been used to entrap enzymes while allowing for the continuous conveyance of carbon-based materials within the sections [40].

3 Present Applications of BESs

3.1 On-Demand Electricity Generation

Among the various BESs, MFCs have been extensively studied across many countries for treating different types of AW and animal WW while producing viable BE that can be directly used as an on-farm power source [27]—the resulting power is diverse contingent on the type and focus of feed and the kind of MFC reactor. Hence, the upcoming study can use the 1-chamber air-CE to treat less focused swine WW [27]. Additionally, many scientists have utilised various types of feed materials for power generation using MFCs. Those are mentioned in Table 1.

3.2 Production of Biohydrogen

Various types of AC wastewater, including dairy compost slurry [47], swine wastewater [48], and fermentation discharge [49], have been utilised for biohydrogen production in MECs. Compared to other bio-H_2 technologies like "anaerobic digestion and photobiological processes", MECs have shown superior recital in generating hydrogen from a broader range of substrates. This is due to their ability to overcome thermodynamic constraints. However, the performance of MECs, measured by hydrogen production rate (HPR) and H_2 yield, can vary significantly depending on the substrate used for conversion [50].

Table 1 Electricity production in MFCs using diverse types of AW/WW

Feed	MFC	Electrical output (mW/m^2)	References
Swine WW	2-chamber (C); aqueous CE	45	[27]
	1-C; air-cathode	261	[27]
Raw corn Stover	1-C; air-cathode	296	[41]
Wheat straw	2-C: Aqueous cathode	123	[42]
Starch processing WW	1 C	239.4	[43]
Rice mill WW	Earthen pot -MFC	2.0	[44]
Chilled ready-meal food production	Tubular-MFC	3.34–5.86	[45]
MFC	Molasses WW	16.1	[46]

Applying a higher external voltage in the same MEC system using the same substrate generally increases hydrogen production rate and yield. This is due to enhanced electron transfer towards the CE, facilitated by the increased voltage. Nevertheless, an advanced exterior voltage typically reduces energy efficiency since it requires additional input. It has been observed that HPR and energy proficiency can be well-adjusted by using an ideal external voltage range of 0.6–0.8 V [51]. Anaerobic digesters, commonly employed for minimising AC waste and recovering valuable biogases like H_2 and CH_4, can be integrated with MECs to enhance HPR and produce more. The fermentation process in anaerobic digesters breaks down complex carbohydrates into smaller, more biodegradable metabolites, which are ideal for microbial electrolysis cells (MECs). The uppermost reported HPR of 189 m^3/m^3/day has been achieved using integrated systems [52]. Integrating MECs also facilitates the complete oxidation of organic compounds into CO_2, a step that fermentation alone cannot accomplish. These integrated systems have effectively produced hydrogen from lignocellulosic ingredients, which are typically considered recalcitrant. Such materials include corn stover, Beta vulgaris, plant greeneries, and corn stems [53].

3.3 *Treatment of AW and WW*

Most AW and WW, specifically those that are extremely focused on animal waste and WW, need to be treated before they can be discharged into national WW streams for further treatment [54]. Solid visceral trash is typically achieved through anaerobic digestion, but it can also be efficiently treated using BESs to recover energy as an added benefit. Chemical oxygen demand (COD) is removed using a needle to eliminate organic matter during wastewater treatment. Notably, substantial COD removals of 98% have been described Table 2. Additionally, BESs have shown usefulness in pretreating lignocellulosic biomass hydrolytes. For instance, Mohan et al. [59] achieved a 63% removal of 52 g/L COD from vegetable hydrolytes using a 1-C air-CE MFC with non-catalysed electrodes.

3.4 *Elimination and Retrieval of Nutrients*

Agricultural and animal wastewater is rich in decomposable nutrients. Animal WW with foci (>500 mg/L) of organic N and NH_3 is typically inhibitory to AE-respiring deeds and is unwanted for WW management [60]. Still, with suitable pre-treatment and modified reactor plans, the amputation and retrieval of N in the forms of ammonia (NH3), NH4+, and nitrite from N-rich wastewater streams can be successfully achieved, as presented by BES [61]. Recuperating nutrients from wastewater in aquaculture (AC) systems is more sustainable than eliminating them, as the recuperated nutrients can replenish nutrient-depleted fields. N is one of the most indispensable nutrients for plant growth. In BESs, nitrogen recovery primarily occurs through the extraction of ammonia (NH_3) driven by the generated electronic current. However, nitrogen can also be recuperated through integration and stored in algal cell systems [62]. Ammonia in the anolyte can be transported to the catholyte through diffusion and dispersion across the CEM, allowing NH_3-N to accumulate at concentrations of several grams per litre. For instance, MFCs integrated with algal photosynthesis have achieved high phosphorus recovery efficiencies, ranging from 58% to 92% [62].

4 Advantages and Disadvantages

The details of valuable resources are briefed in Table 2.

Table 2 Merits and demerits of BES

BESs	Beneficial byproducts	Merits	Demerits	References
MFC	BE	Transforms unused wastewater into energy, offers low treatment costs, enables nutrient recovery, and yields minimal sludge	High catalyst price, restricted power output, and moderate COD removal efficiency	[55]
MEC	Bio-H_2, CH_4, and (or) a diversity of biological compounds	It eliminates complex organic and inanimate pollutants and can be easily adapted to produce the desired end product	Exterior power source requires a low-priced manufacturing rate	[56]
MES	Valuable biochemicals	Carbon sequestration, reclamation of high-value C-C-C-C-compounds from industrial procedures, metal repossession, and sundry requests	Outward voltage supply needs, poorer manufacture rate of chemicals with lengthier C-chains	[31]
MDC	Salt-H_2O and briny water desalination, BE	Unlike other refining technologies, salt removal and concurrent treatment of unwanted waste and wastewater don't rely on an external power supply	Higher anolyte salinity reduces bacterial removal efficacy, promotes membrane fouling, and negatively impacts MDC performance due to the presence of inorganic elements in WW, leading to scaling	[57]
MSC	BE	Use sunshine, self-repairing systems, leftover and WW treatment, and wireless sensor network technology solutions in AC processes	Ineffective utilisation of "sun" energy sources is luxurious and extremely variable depending on the withstand conditions	[39]
EFC	BE	Used to power biosensors and transportable devices, support energy storage and transformation applications, and facilitate waste and WW treatment	Damage to enzyme action during immobilisation, high immobilisation costs, and mass transfer limitations	[58]

5 Future Potential Uses of BESs in Agro-Food Industries and Associated Defies

BESs hold significant potential for future applications in the agro-food sector, particularly in discarded management and sustainable energy production. These systems can efficiently treat AC runoff and wastewater, improving water quality and reducing environmental impacts [63, 64]. BESs can also convert organic waste into valuable resources such as biofertilisers, biopesticides, and bioenergy, contributing to a circular economy [65].

Key Applications

- Waste management and water quality improvement: BESs can break down pollutants in AC runoff, enhancing water quality and reducing the environmental footprint of farming practices [66].
- Sustainable energy production: BESs can generate electricity and produce biofuels from organic waste, providing renewable energy sources for AC operations [67].
- Nutrient recovery: By converting organic waste into biofertilisers, BESs can enhance soil fertility and reduce the need for synthetic fertilisers.
- Carbon Sequestration: Some BESs can capture CO_2, contributing to climate change mitigation efforts.

Challenges

Despite the promising applications, several challenges need to be addressed for the widespread adoption of BESs:

- Scalability and efficiency: Improving the scalability and efficiency of BESs is crucial for large-scale implementation.
- Cost and infrastructure: The initial setup costs and infrastructure requirements for BESs can be high, necessitating economic incentives and supportive policies.
- Public awareness and education: Raising awareness and educating stakeholders about the benefits and operation of BESs is essential for their adoption.
- Technological advancements: Continuous research and development are needed to enhance the performance and adaptability of BESs in diverse AC settings.

6 Conclusion

Despite the extensive study of various BESs, research on agriculture-driven applications remains limited, even though BESs can significantly impact food safety, energy sustainability, and river obtainability. BESs have the potential to recuperate significant amounts of power and chemicals from AW and WW, which can be used to decrease energy inputs in farming systems. Additionally, BESs can produce valuable resources such as biofuels, plant nutrients, and irrigation H_2O. They may also replace untenable batteries to power remote devices for AC monitoring. Furthermore, BESs have potential applications in producing food without sunlight and remediating dirty soils. AW can be transformed into building materials or C-rich electrodes to reduce the costs associated with BESs. BESs can be integrated into AC systems as viable technologies for energy generation, source retrieval, surplus reduction, wastewater treatment, irrigation, and remote monitoring. Moreover, BESs can be designed as self-sufficient biosensors for nourishment production and soil decontamination. AC waste can also be utilised to develop large-scale Battery Energy Storage Systems (BESs). Long-term studies are crucial for evaluating the durability and benefits of on-farm BES applications. With the rapid advancement of BES technology, particularly in green and alternative energy production, significant progress is predicted in the near future.

References

1. Heller MC, Keoleian GA (2003) Assessing the sustainability of the US food system: a life cycle perspective. Agric Syst 76(3):1007–1041
2. Lanzafame P, Abate S, Ampelli C, Genovese C, Passalacqua R, Centi G, Perathoner S (2017) Beyond solar fuels: renewable energy-driven chemistry. ChemSusChem 10(22):4409–4419
3. Srikanth S, Kumar M, Singh MP, Das BP (2016) Bioelectro chemical systems: a sustainable and potential platform for treating waste. Procedia Environ Sci 1(35):853–859
4. Li SiMeng LS, Barreto V, Li RunWei LR, Chen Gang CG, Hsieh YP (2018) Nitrogen retention of biochar derived from different feedstocks at variable pyrolysis temperatures. J Anal Appl Pyrol 133:136–146
5. Crawford JH (1996) Composting of agricultural wastes-a review. In: The science of composting. Springer, Cham
6. Weiland P (2006) Biomass digestion in agriculture: a successful pathway for the energy production and waste treatment in Germany. Eng Life Sci 6(3):302–309
7. Suman Bajracharya SB, Mohita Sharma MS, Gunda Mohanakrishna GM, Benneton XD, Strik DP, Sarma PM, Deepak Pant DP (2016) An overview on emerging bioelectrochemical systems (BESs): Technology for sustainable electricity, waste remediation, resource recovery, chemical production and beyond. Renew Energy. 98:153–170
8. Li S, Chen G (2018) Factors affecting the effectiveness of bioelectrochemical system applications: data synthesis and meta-analysis. Batteries 4(3):34
9. Monier V, Mudgal S, Escalon V, O'Connor C, Gibon T, Anderson G, Montoux H, Reisinger H, Dolley P, Ogilvie S, Morton G (2010) Preparatory study on food waste across EU 27: final report. http://ec.europa.eu/environment/eussd/pdf/bio_foodwaste_report.pdf (data dostępu: październik 2015)
10. Food Wastage Footprint (2014) Food wastage footprint full-cost accounting. Food & Agriculture Organization of the UN (FAO), Rome
11. Huang CH, Liu SM, Hsu NY (2020) Understanding global food surplus and food waste to tackle economic and environmental sustainability. Sustainability 12(7):2892
12. Hiç C, Pradhan P, Rybski D, Kropp JP (2016) Food surplus and its climate burdens. Environ Sci Technol 50(8):4269–4277
13. Lipinski B, Hanson C, Waite R, Searchinger T, Lomax J (2013) Reducing food loss and waste. World Resource Institute, Washington
14. Mexda JA, Mukhaiyar R (2021) Evaluasi Pembelajaran Daring pada Perkuliahan di Laboratorium Dasar dan Pengukuran UNP. Ranah Res 4(1):1

15. Ramanaiah SV, Cordas CM, Matias SC, Reddy MV, Leitao JH, Fonseca LP (2021) Bioelectricity generation using long-term operated biocathode: RFLP based microbial diversity analysis. Biotechnol Rep 32:e00693
16. Logan BE, Rossi R, Ragab AA, Saikaly PE (2019) Electroactive microorganisms in bioelectrochemical systems. Nat Rev Microbiol 17(5):307–319
17. Du Z, Li H, Gu T (2007) A state of the art review on microbial fuel cells: a promising technology for wastewater treatment and bioenergy. Biotechnol Adv 25(5):464–482
18. Millanar-Marfa JM, Borea L, Castrogiovanni F, Hasan SW, Choo KH, Korshin GV, de Luna MD, Ballesteros FC Jr, Belgiorno V, Naddeo V (2022) Self-forming dynamic membranes for wastewater treatment. Separ Purif Rev 51(2):195–211
19. Ren L, Ahn Y, Logan BE (2014) A two-stage microbial fuel cell and anaerobic fluidized bed membrane bioreactor (MFC-AFMBR) system for effective domestic wastewater treatment. Environ Sci Technol 48(7):4199–4206
20. Santoro C, Arbizzani C, Erable B, Ieropoulos I (2017) Microbial fuel cells: from fundamentals to applications. A review. J Power Sour 356:225–244
21. Rachinski S, Carubelli A, Mangoni AP, Mangrich AS (2010) Microbial fuel cells used in the production of electricity from organic waste: a perspective of future. Química Nova 33:1773–1778
22. Li S, Chen G (2018 Jan) Effects of evolving quality of landfill leachate on microbial fuel cell performance. Waste Manag Res 36(1):59–67
23. Logan BE, Wallack MJ, Kim KY, He W, Feng Y, Saikaly PE (2015) Assessment of microbial fuel cell configurations and power densities. Environ Sci Technol Lett 2(8):206–214
24. Tota-Maharaj K, Paul P (2015) Performance of pilot-scale microbial fuel cells treating wastewater with associated bioenergy production in the Caribbean context. Int J Energy Environ Eng 6:213–220
25. Park Y, Park S, Kim JR, Kim HS, Kim BG, Yu J, Lee T (2017) Effect of gradual transition of substrate on performance of flat-panel air-cathode microbial fuel cells to treat domestic wastewater. Bioresour Technol 226:158–163
26. Pant D, Singh A, Van Bogaert G, Olsen SI, Nigam PS, Diels L, Vanbroekhoven K (2012) Bioelectrochemical systems (BES) for sustainable energy production and product recovery from organic wastes and industrial wastewaters. RSC Adv 2(4):1248–1263
27. Min B, Kim J, Oh S, Regan JM, Logan BE (2005) Electricity generation from swine wastewater using microbial fuel cells. Water Res 39(20):4961–4968
28. Lu L, Ren ZJ (2016) Microbial electrolysis cells for waste biorefinery: a state of the art review. Bioresour Technol 315:254–264
29. Show KY, Lee DJ, Tay JH, Lin CY, Chang JS (2012) Biohydrogen production: current perspectives and the way forward. Int J Hydrog Energy 37(20):15616–15631
30. Dinesh GK, Chauhan R, Chakma S (2018) Influence and strategies for enhanced biohydrogen production from food waste. Renew Sust Energ Rev 92:807–822
31. Sadhukhan J, Lloyd JR, Scott K, Premier GC, Eileen HY, Curtis T, Head IM (2016) A critical review of integration analysis of microbial electrosynthesis (MES) systems with waste biorefineries for the production of biofuel and chemical from reuse of CO2. Renew Sust Energ Rev 56:116–132
32. May HD, Evans PJ, LaBelle EV (2016) The bioelectrosynthesis of acetate. Curr Opin Biotechnol 42:225–233
33. Das S, Ghangrekar MM (2017) Value added product recovery and carbon dioxide sequestration from biogas using microbial electrosynthesis
34. Kumar G, Saratale RG, Kadier A, Sivagurunathan P, Zhen G, Kim SH, Saratale GD (2017) A review on bio-electrochemical systems (BESs) for the syngas and value added biochemicals production. Chemosphere 177:84–92
35. Rabaey K, Rozendal RA (2010) Microbial electrosynthesis—revisiting the electrical route for microbial production. Nat Rev Microbiol 8(10):706–716

36. Li S, Chen G, Anandhi A (2018) Applications of emerging bioelectrochemical technologies in agricultural systems: a current review. Energies 11(11):2951
37. Luo H, Xu P, Ren Z (2012) Long-term performance and characterization of microbial desalination cells in treating domestic wastewater. Bioresour Technol 120:187–193
38. Brastad KS, He Z (2013) Water softening using microbial desalination cell technology. Desalination 309:32–37
39. Strik DP, Timmers RA, Helder M, Steinbusch KJ, Hamelers HV, Buisman CJ (2011) Microbial solar cells: applying photosynthetic and electrochemically active organisms. Trends Biotechnol 29(1):41–49
40. Neto SA, Forti JC, De Andrade AR (2010) An overview of enzymatic biofuel cells. Electrocatalysis 1:87–94
41. Wang X, Feng Y, Wang H, Qu Y, Yu Y, Ren N, Li N, Wang E, Lee H, Logan BE (2009) Bioaugmentation for electricity generation from corn Stover biomass using microbial fuel cells. Environ Sci Technol 43(15):6088–6093
42. Zhang Y, Min B, Huang L, Angelidaki I (2009) Generation of electricity and analysis of microbial communities in wheat straw biomass-powered microbial fuel cells. Appl Environ Microbiol 75(11):3389–3395
43. Lu L, Ren N, Xing D, Logan BE (2009) Hydrogen production with effluent from an ethanol–H2-coproducing fermentation reactor using a single-chamber microbial electrolysis cell. Biosens Bioelectron 24(10):3055–3060
44. Behera M, Jana PS, More TT, Ghangrekar MM (2010) Rice mill wastewater treatment in microbial fuel cells fabricated using proton exchange membrane and earthen pot at different pH. Bioelectrochemistry 79(2):228–233
45. Boghani H, Kim JR, Dinsdale RM, Guwy AJ, Premier GC (2017) Reducing the burden of food processing washdown wastewaters using microbial fuel cells. Biochem Eng J 117:210–217
46. Wu CH, Liu SH, Chu HL, Li YC, Lin CW (2017) Feasibility study of electricity generation and organics removal for a molasses wastewater by a waterfall-type microbial fuel cell. J Taiwan Inst Chem Eng 78:150–156
47. Cai W, Liu W, Yang C, Wang L, Liang B, Thangavel S, Guo Z, Wang A (2016) Biocathodic methanogenic community in an integrated anaerobic digestion and microbial electrolysis system for enhancement of methane production from waste sludge. ACS Sustain Chem Eng 4(9):4913–4921
48. Lu L, Xing D, Xie T, Ren N, Logan BE (2010) Hydrogen production from proteins via electrohydrogenesis in microbial electrolysis cells. Biosens Bioelectron 25(12):2690–2695
49. Cai W, Liu W, Cui D, Wang A (2016) Hydrogen production from buffer-free anaerobic fermentation liquid of waste activated sludge using microbial electrolysis system. RSC Adv 6(45):38769–38773
50. Kadier A, Simayi Y, Kalil MS, Abdeshahian P, Hamid AA (2014) A review of the substrates used in microbial electrolysis cells (MECs) for producing sustainable and clean hydrogen gas. Renew Energy 71:466–472
51. Logan BE, Call D, Cheng S, Hamelers HV, Sleutels TH, Jeremiasse AW, Rozendal RA (2008) Microbial electrolysis cells for high yield hydrogen gas production from organic matter. Environ Sci Technol 42(23):8630–8640
52. Lu L, Xing D, Ren N, Logan BE (2012) Syntrophic interactions drive the hydrogen production from glucose at low temperature in microbial electrolysis cells. Bioresour Technol 124:68–76
53. Khan MA, Ngo HH, Guo W, Liu Y, Zhang X, Guo J, Chang SW, Nguyen DD, Wang J (2018) Biohydrogen production from anaerobic digestion and its potential as renewable energy. Renew Energy 129:754–768
54. Fornero JJ, Rosenbaum M, Angenent LT (2010) Electric power generation from municipal, food, and animal wastewaters using microbial fuel cells. Electroanalysis: an international journal devoted to fundamental and practical aspects of. Electroanalysis 22(7-8):832–843
55. Paucar NE, Sato C (2021) Microbial fuel cell for energy production, nutrient removal and recovery from wastewater: a review. PRO 9(8):1318

56. Bajracharya S, Sharma M, Mohanakrishna G, Benneton XD, Strik DP, Sarma PM, Pant D (2016) An overview on emerging bioelectrochemical systems (BESs): technology for sustainable electricity, waste remediation, resource recovery, chemical production and beyond. Renew Energy 98:153–170
57. Sevda S, Yuan H, He Z, Abu-Reesh IM (2015) Microbial desalination cells as a versatile technology: functions, optimization and prospective. Desalination 371:9–17
58. Hou C, Liu A (2017) An integrated device of enzymatic biofuel cells and supercapacitor for both efficient electric energy conversion and storage. Electrochim Acta 245:303–308
59. Mohan SV, Mohanakrishna G, Sarma PN (2010) Composite vegetable waste as renewable resource for bioelectricity generation through non-catalyzed open-air cathode microbial fuel cell. Bioresour Technol 101(3):970–976
60. Zhang Y, Angelidaki I (2015) Recovery of ammonia and sulfate from waste streams and bioenergy production via bipolar bioelectrodialysis. Water Res 85:177–184
61. Gildemyn S, Luther AK, Andersen SJ, Desloover J, Rabaey K (2015) Electrochemically and bioelectrochemically induced ammonium recovery. J Vis Exp 95:52405
62. Kelly PT, He Z (2014) Nutrients removal and recovery in bioelectrochemical systems: a review. Bioresour Technol 153:351–360
63. Zheng Y, Wang C, Zheng ZY, Che J, Xiao Y, Yang ZH, Zhao F (2014) Ameliorating acidic soil using bioelectrochemistry systems. RSC Adv 4(107):62544–62549
64. Lu L, Yazdi H, Jin S, Zuo Y, Fallgren PH, Ren ZJ (2014) Enhanced bioremediation of hydrocarbon-contaminated soil using pilot-scale bioelectrochemical systems. J Hazard Mater 274:8–15
65. Yuan HongYing YH, Li SiMeng LS, Liu JunLiang LJ, Song ChengYi SC, Chen Gang CG. Cry1Ab adsorption and transport in humic acid-coated geological formation of alumino-silica clays
66. Fane AG (2011) Membranes and the water cycle: challenges and opportunities. Appl Water Sci 1(1):3–9
67. Li S, Barreto V, Li R, Chen G, Hsieh YP (2018) Nitrogen retention of biochar derived from different feedstocks at variable pyrolysis temperatures. J Anal Appl Pyrolysis 133:136–146

MIX
Papier aus verantwortungsvollen Quellen
Paper from responsible sources
FSC® C105338

If you have any concerns about our products,
you can contact us on
ProductSafety@springernature.com

In case Publisher is established outside the EU,
the EU authorized representative is:
Springer Nature Customer Service Center GmbH
Europaplatz 3, 69115 Heidelberg, Germany

Printed by Libri Plureos GmbH
in Hamburg, Germany